Electrical Properties of Polymers

CONTRIBUTORS

STEPHEN H. CARR

ROBERT F. COZZENS

D. KEITH DAVIES

P. FISCHER

VINGIE Y. MERRITT

JEROME H. PERLSTEIN

DONALD A. SEANOR

G. M. SESSLER

Electrical Properties of Polymers

Edited by DONALD A. SEANOR
Reprographics Business Group
Xerox Corporation
Joseph C. Wilson Center for Technology
Rochester, New York

1982

 ACADEMIC PRESS

A Subsidiary of Harcourt Brace Jovanovich, Publishers

New York   London
Paris   San Diego   San Francisco   São Paulo   Sydney   Tokyo   Toronto

ACADEMIC PRESS, INC.
111 Fifth Avenue, New York, New York 10003

United Kingdom Edition published by
ACADEMIC PRESS, INC. (LONDON) LTD.
24/28 Oval Road, London NW1 7DX

Library of Congress Cataloging in Publication Data
Main entry under title:

Electrical properties of polymers.

   Includes index
   1. Polymers and polymerization--Electric properties.
I. Seanor, Donald A.
QD381.9.E38E35 1982   547.8'40457   82-13882
ISBN 0-12-633680-6

PRINTED IN THE UNITED STATES OF AMERICA

82 83 84 85   9 8 7 6 5 4 3 2 1

Contents

Chapter 4   Photovoltaic Phenomena in Organic Solids
Vingie Y. Merritt

Chapter 5   Thermally Stimulated Discharge Current Analysis of Polymers
Stephen H. Carr

Chapter 6   Polymeric Electrets
G. M. Sessler

Chapter 7   Contact Electrification of Polymers and Its Elimination
D. Keith Davies

Chapter 8   Dielectric Breakdown Phenomena in Polymers
P. Fischer

Contents

List of Contributors

Numbers in parentheses indicate the pages on which the authors' contributions begin.

STEPHEN H. CARR (215),   Department of Materials Science and Engineering, Northwestern University, Evanston, Illinois 60201

ROBERT F. COZZENS (93),   Department of Chemistry, George Mason University, Fairfax, Virginia 22030

D. KEITH DAVIES (285),   ERA Technology Ltd., Leatherhead, Surrey KT22 7SA, England

P. FISCHER (319),   Siemens Aktiengesellschaft, Research and Development Center, D-8520 Erlangen, West Germany

VINGIE Y. MERRITT (127),   IBM Corporation, Corporate Headquarters Division, IBM Journal of Research and Development, White Plains, New York 10601

JEROME H. PERLSTEIN (59),   Research Laboratories, Eastman Kodak Company, Rochester, New York 14650

DONALD A. SEANOR (1),   Reprographics Business Group, Xerox Corporation, Joseph C. Wilson Center for Technology, Rochester, New York 14603

G. M. SESSLER, (241),   Technical University of Darmstadt, D-6100 Darmstadt, West Germany

Preface

Over the years, the more traditional involvement of polymers in electrical applications has been as electrical insulation. Consequently, major emphasis has been placed on the study of electrical properties in situations in which polymers act as inert or passive electrical elements. Only in recent times, with the advent of highly conducting polymeric materials and materials that exhibit piezo- and pyroelectric properties, has their potential to perform as active elements been realized. To this range of properties has been added the capability of acting as photoactive devices, as polymeric photoconductors have become increasingly used in the electrophotographic industry. Add to this a developing need for large-area photoactive materials to capitalize on solar energy, as well as the unique properties associated with disordered materials, and it can be seen that the potential of polymeric materials as active electrical elements has been barely tapped.

While much of the work on polymers discussed in the book is often of a rudimentary nature, carried out on perhaps not the best-characterized materials, the scope for innovation is tremendous. In this truly interdisciplinary task, the abilities of the synthetic chemist, the molecular spectroscopist, and the solid-state physicist, as well as the polymer materials scientist, will be called upon and taxed to the utmost limit. Key to capitalizing on the potential of polymers as active electrical elements will be a detailed understanding of the basic phenomena, combined with the capability of synthesizing complex molecules in a form with controlled morphology. The whole gamut of sophisticated synthetic, experimental, and theoretical techniques will be required to understand the basic phenomena and to elucidate the structure–property relationships.

This is a task well begun. While significant progress has already been made, much still remains to be done. We hope that this book will provide a significant point of reference for future endeavors.

Chapter 1

Electrical Conduction in Polymers

Donald A. Seanor
REPROGRAPHICS BUSINESS GROUP
XEROX CORPORATION
JOSEPH C. WILSON CENTER FOR TECHNOLOGY
ROCHESTER, NEW YORK

I. INTRODUCTION

The range of electrical conductivity observed in materials covers a range of 25 orders of magnitude. This is one of the largest variations in any materials property. While one tends to regard polymers primarily as insulators, in recent years the discovery of highly conducting graphite intercalation compounds (Vogel, 1977), graphite superconductivity, metallic polysulfur nitrogen (Kronick *et al.,* 1962), and doped polyacetylenes that can be made to exhibit metallic conductivity as well as *n*-type or *p*-type

1

Donald A. Seanor

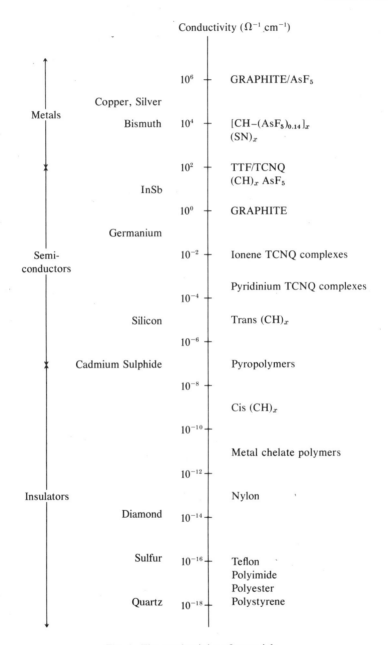

Fig. 1. The conductivity of materials.

semiconductivity (Shirakawa *et al.*, 1978), means that the range of conductivity of carbon-based polymers can be regarded as covering a similar span (Fig. 1).

The purpose of this chapter is to provide a simple theoretical framework within which to discuss electrical conduction and to present a limited number of examples of how polymer structure impacts conductivity. Before embarking on this journey, it should be emphasized that polymers are not the simple, covalent crystals of conventional solid-state physics. Conventional solid-state physics deals with the properties of well-defined, regular arrays of atoms. This certainly does not describe the usual situation in polymers.

Polymeric materials are unique because of the range of structural forms that can be synthesized and the way in which changes can be made in the structure in a local or general way. They can exist as amorphous materials, as crystalline materials, or as mixtures of crystalline and amorphous materials. Even highly crystalline polymers contain considerable amorphous material. They are molecular materials. That is, each polymer chain is its own individual entity, and the interaction with other polymer chains is usually weak. Polymer chains can take up different conformations and can be oriented mechanically. Within the individual polymer chains, the chemical units need not have a unique spatial arrangement. In contrast to many materials, polymers do not have a unique molecular weight but are described in terms of weighted average molecular weights representing the average and the distribution about the mean.

Thus, in comparison to well-ordered, covalent, or ionically bonded inorganic materials, polymers are weakly bonded, disordered materials. The functional groups from which they are made up need not have unique spatial relationships to each other.

These differences have profound effects on most of the properties of polymers. They are heat insulators and are usually soft and easily distorted. They are electrically insulating and poor conductors of charge.

We can think of the electrical properties in terms of a progression from organized, strongly interacting materials (such as silicon and even carbon in its diamond allotrobe), through weakly ordered molecular materials (such as anthracene), to weakly bonded long-chain polymers.

With strongly interacting atoms, such as covalently bonded silicon in its tetragonal form, a band structure describing the allowed electron energy levels can be set up. The energy levels are broad because the interactions are strong and the atoms have a regular spacing (Fig. 2a).

In a molecular crystal such as anthracene, the interactions within the molecule are strong. In fact, one could regard the orbital hybridization and formation of bonds as an attempt to write a band structure for an

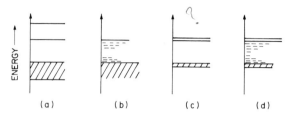

Fig. 2. The transition from ordered to disordered materials: (a) covalent ordered, (b) covalent disordered, (c) molecular ordered, and (d) molecular disordered.

aromatic molecule. The intermolecular interactions are weak, the intermolecular distances are large, and the energy bands of the crystal are narrow (Fig. 2c). This has yet another effect. Since the intermolecular forces are weak, the coupling between molecules is small, and the whole tends to behave as an array of individual molecules. The perturbations of the molecular spectra are small, and the absorption spectrum resembles that of the individual molecules, although some interaction is possible, and energy transfer from molecule to molecule can be quite efficient.

Then, when one considers disordered covalent materials, many localized states are introduced into the conduction band and the mode of charge transport becomes dominated by the localized states (Mott, 1967, 1969) (Fig. 2b).

Finally (Fig. 2d), the disordered molecular material or polymer may not have a band structure; there may be an array of molecular states and molecular ion states, as well as many localized dipole states associated with the disorder. Thus, many transport properties are unique to these disordered molecular materials. Such properties are complicated by the existence of molecular ion states and regions of differing polarity or polarizability. The free charges may prefer to exist as molecular ions, they may be trapped in polar regions, or they may be trapped as a result of polarization of the surrounding medium.

There are three levels of structure to be considered (Seanor, 1976). The first level is the basic chemical composition. The nature of the monomer unit(s) will determine the behavior of the polymer, particularly where photoconduction is involved. The second level of structure involves the spatial arrangement of the basic polymer units within the individual polymer molecules (the microstructure). This is determined to a great extent by the polymerization reaction and the precise reaction conditions. The third level of structure is the spatial arrangement of the polymer chains in the solid state (the macrostructure). In addition to the microstructure, the conditions (temperature, solvent, and cooling or evaporation rate) under which the solid is produced are the primary factors that affect the morphology.

We should note, in addition, that where insulators are concerned the origin of the charge carriers is by no means clear. A polymer with a molecular weight of 10^6 has $\sim 12 \times 10^{18}$ end groups per cubic centimeter. An insulator with a conductivity of $10^{-17}\,\Omega^{-1}\,cm^{-1}$ and a hypothetical mobility of $10^{-5}\,cm^2/V$ sec requires only $\sim 10^8$ mobile charge carriers per cubic centimeter. Most polymers, particularly condensation polymers, can easily contain such a level of ionic impurity. The extent to which extrinsic charge carriers (impurities, electrons, or holes injected from electrodes) control the conductivity of the more insulating polymers is seldom unambiguous. Thus, many discussions of electrical conduction in insulating polymers are open to criticism on the grounds of suspect purity and spurious electrode effects, unless it is clear where the charge carriers originate. In particular, the criterion of mass transport for ionic conduction has seldom been definitively used. In fact, even with an ability to detect analytically one impurity atom in 10^9, the detection of mass transport would be a difficult problem to solve. It is for this reason that much of the more definitive work in the field has involved studies on injected or photogenerated charge carriers. Given this limitation, this chapter will cover only a limited fraction of the large volume of work on polymer conductivity* and discuss generally applicable principles.

The specific electrical conductivity of a solid, $\sigma\,\Omega^{-1}\,cm^{-1}$, is defined as the current, in amps, flowing through a centimeter cube of the material under unit electrical potential; i.e.,

$$\sigma = iL/AV,$$

where the sample length is L (cm), its area is A (cm²), and the potential is V (V).

The specific conductivity σ is related to two basic parameters, the charge carrier density n (cm⁻³) and the charge carrier mobility μ cm²/V sec ($=$cm sec⁻¹/V cm⁻¹); i.e.,

$$\sigma = \sum_i q_i n_i \mu_i,$$

where q_i is the charge on the ith species. With polymeric materials, each parameter n_i or μ_i may be ambient-sensitive, may be potential-sensitive, and may be influenced by the precise conditions of fabrication.

* For example, since 1960, the "Digest of Literature on Dielectrics and Insulation," published by the National Academy of Sciences, has contained anywhere from 100 to 500 papers per year dealing with aspects of the electrical properties of polymers.

"Ambient-sensitive" means that, in addition to being temperature-sensitive, both the number and the mobility may be sensitive to the precise experimental conditions. This may mean sensitivity to lattice spacing, sample preparation, and ambient atmosphere (such as moisture or the electron-accepting and -donating properties of the surrounding gas). The mobility μ is a vector and is therefore direction-sensitive. The number of charge carriers is a pure number; i.e., we should write

$$n_i = n_i(V, T, A), \qquad \mu_i = \mu_i(V, T, A, z),$$

where A indicates ambient and z indicates direction. It should be noted that the system is defined only for good, single-crystal samples. The influence of crystalline–amorphous phase boundaries, and interfaces in general, may lead to large deviations from ideal behavior.

Thus, any phenomena that affect parameters such as lattice spacing, adsorption equilibrium, potential distribution, dipole orientation, and molecular species can impact electrical conductivity. Typically, processes such as first- and second-order phase transitions, chemical degradation, dipole alignment and molecular motion, charge-carrier trapping and detrapping, impurities, and electrodes have been shown to affect electrical measurements in polymers.

It is also necessary to define the parameters that should be controlled in order to specify the system uniquely. This also helps in understanding the behavior under nonequilibrium conditions. In some situations, the phase rule may help define the system and aid in interpretation. The phase rule relates the number of independent variables (the number of degrees of freedom) F, the number of components C, and the number of phases P by the equation

$$F = C - P + 2.$$

For example, at a phase transition for a single-component material, $P = 2$, $C = 1$, and $F = 1$. Hence the temperature completely defines the system. On the other hand, for a polymer in equilibrium with water vapor, $P = 2$, $C = 2$, and $F = 2$. Thus, the stoichiometry is defined only upon fixing both temperature and vapor partial pressure. A polymer with both crystalline and amorphous phases requires a definition of crystalline content and temperature. Such dependencies must be recognized in order to make meaningful experiments.

Studies on conductivity should be aimed at understanding the origin of the charge-carrying species, their number, and the way in which they move through the bulk of the material. Ultimately, these parameters should be related to the chemical composition, the microstructure, and the morphology of the particular material. Only when such relationships

are known and understood will it be possible to predict electrical properties in a rational manner and to set out to synthesize well-defined materials with tailor-made properties. Typical of this type of problem is the
need to understand, for example, the influence of cis–trans isomerism in
polyacetylene and doped polyacetylene materials, as well as their morphology (Roland *et al.*, 1980).

II. EXPERIMENTAL TECHNIQUES

Until recently, most studies on polymers have involved insulators
rather than conductors. The emphasis has been on dc techniques; ac
bridge techniques are not discussed here. The experimental methods described for high-resistance insulators using high-impedance electrometers
are usually more appropriate than the bridge techniques used with more
conductive materials. There are many papers dealing with the electrical
properties of such insulators as ceramics and polymers. Detailed references can be found in reviews of these topics, to which the reader in-
terested in greater detail is referred (Blumenthal and Seitz, 1974; Forster,
1969; Gutman and Lyons, 1967; Norman, 1970; Van Turnhout, 1974).

Phenomenologically, the conductivity is obtained by measuring the current flowing through a piece of the material and using the sample dimensions to calculate σ from the equation

$$\sigma = (d/AV)i,$$

where d (cm) is the sample thickness, A is its area (cm², and V is the
potential across the material. Hence the dependence of either the number
of carriers or the carrier mobility upon the potential will be shown by a
potential dependence of σ as defined above.

A typical experimental system (Fig. 3) consists of the following components:

(1) a sample holder consisting of sample, electrodes, leads, and a
means of securely locating the sample,

(2) a means of controlling the ambient atmosphere,

(3) a heater capable of giving a variable but linear rate of temperature
increase over a broad range of temperatures,

(4) a current detector capable of measuring currents from as small as
10^{-13} A to as high as, perhaps, 10^{-4} A—usually a high-impedance electrometer,

(5) a source of controlled, low-ripple dc potential,

(6) a means of recording the current and temperature, e.g., an XY
recorder or a computer interface,

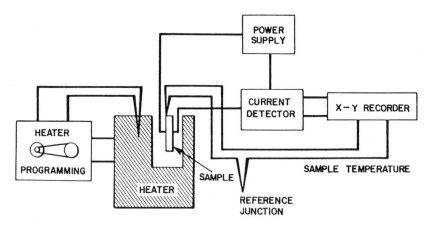

Fig. 3. Block diagram of electrical conductivity system.

(7) an additional means of preconditioning the sample—such as a source of irradiation in the case of photoconductivity or transport studies, the energy of which depends upon the nature of the material.

A. Sample Preparation and Electrodes

Sample preparation and the electrode system can play a critical role in determining the electrical parameters. It is necessary to ensure that the observed behavior is characteristic of the material, not of the way in which the sample is made and electroded. Cross-checking experiments using different electrodes and methods of sample preparation are required.

The sample may be in the form of a fiber, film, single crystal, block, or compacted disk of powdered material. Each presents its own peculiar problems.

In the case of fibers, electrical contact presents a serious problem, particularly if bundles of yarn are involved as opposed to monofilaments (Seanor, 1969).

Powders present a number of problems. Among these can be listed effects of particle size, contacts, interfacial effects, mixing, and compacting. The samples are often prepared using KBr pellet presses or other types of dies. The compacting pressure may affect behavior. The range of effective compacting pressure will vary from material to material. Under some conditions of relative humidity, the sample may stick to the die face. The use of mold releases is not recommended, as the surface of the sample may be changed. Thin Teflon®* sheets cut to size often help to eliminate sticking of the sample to the die face. It is not recommended that the sample be compacted between thin metal foil electrodes, as voids may easily be created at the sample–foil interface.

Composite materials can show a marked dependence upon processing parameters. For example, Norman (1970), in discussing the behavior of conductive rubbers, devotes three chapters to the influence of mixing, type of carbon black, volume loading, and curing on the electrical properties of filled rubbers.

In many respects, thin films, single crystals, or slabs cut from larger blocks present the least complicated systems. Even here, factors such as residual solvent and morphology may be important. Certainly in the case of polymers the presence of residual solvent, changes in the crystalline/ amorphous ratio, the sample cooling rate (by its influence on the glass transition temperature or crystallite size), and the ambient atmosphere all affect electrical behavior. The equivalent electrical circuit of an inhomogeneous sample can be represented as a resistance–capacitance network to which the factors discussed may each contribute as shown in Fig. 4. Depending upon the sample and precautions taken, the equivalent electrical circuit can be simplified and analyzed. For example, some compacted powder samples can be simplified to single-circuit elements in which the surface component dominates. Unfortunately, simplification is not always possible, particularly if interfacial polarization occurs. In addition, care must be taken to minimize and to allow for capacitance effects in the measuring system, leads, and sample holder.

Formation of the electrode is of paramount importance. In laying down the conductive material, the surface should not be chemically changed and there should be no gaps between the electrode and sample. Vacuum deposition of a metal such as silver, aluminum, or gold is commonly employed. The use of conductive paints is acceptable, provided that the solvent does not affect the substrate to be studied. Foils have also been employed but may create interfacial voids. For experiments involving light, one electrode must be semitransparent. Frequently, Nesa glass has been used for the semitransparent electrode, as have been thin layers of gold or grids of metal (evaporated or embedded). Photoinjection experiments use layers of an established photoconducting material, e.g., selenium. Table I lists the more commonly used electrode materials.

The electrode should be ohmic, that is, it should not perturb the potential distribution across the sample by injecting charge carriers or by creat-

---

* Teflon is a registered trademark of E. I. du Pont de Nemours, Inc.

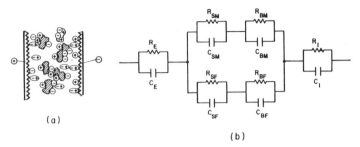

Fig. 4. Representation of an inhomogeneous sample. Filler particles, +, −; free charges, +, −; dipoles. (a) An array of irregular filler particles (F) in a continuous matrix (M). These may be crystalline regions in a semicrystalline polymer. (b) The equivalent circuit: E represents the electrode and measuring circuit contributions to capacitance and resistance, B the bulk contribution, and I the interface contribution.

ing potential barriers at the surface. It is also advisable to use guard-ring electrodes in order to minimize surface leakage currents.

B. Electrical Measurements

The design of the sample holder can present quite serious problems. Because of the wide range of criteria required for each specific situation, individual workers usually design their own. There are certain common

TABLE I *Electrode Materials*

Hg—Liquid metal, not recommended; easily oxidized, toxic vapors
Al—Vacuum-deposited; may oxidize
Ag—Vacuum-deposited, silver paint, silver epoxy cement; may react with some substrates; may diffuse into sample
Au—Vacuum-deposited, paint, paste; inert and not oxidized
Cu—Vacuum-deposited; reactive in many gases
Ni—Vacuum-deposited from solution; reactive at high temperature
Pd—Vacuum-deposited; inert except in certain atmospheres where chemisorption may occur; forms Pd–H_2 alloys which can be used as proton-injecting electrodes
Pt—Vacuum-deposited with difficulty, paint, paste; may be catalytically active
Ir—For very high temperatures
C or graphite—Applied from colloidal suspension; requires nonoxidizing atmosphere
Nesa—A semitransparent conductive tin oxide coating often applied to quartz or glass for contact electrodes; for photoconduction experiments
Hg, In, Sn, Gn—Low-melting metals used alone or as alloys; easily oxidized; vapors may be toxic

criteria: They all must support and locate the sample and the electrodes, there must be a means for easy thermal and ambient control, and sources of stray electrical capacitance must be minimized. The nature of the specific sample frequently dictates the type of sample holder to be used. Typical cells are shown in Figs. 5 and 6.

Figure 5 shows a typical glass cell used to study powders (Gray, 1969). The powder is lightly compressed by the spring-loaded, upper platinum electrode in a sintered-glass tube with a corresponding electrode at the bottom. Both electrodes are provided with shielded electrical and thermocouple leads. A platinum or gold-plated platinum screen surrounds the sintered-glass tube and acts as a guard electrode. The holder can be connected to a vacuum system for control of the ambient, while the sintered-glass reaction tube allows easy access for reactant gas.

Figure 6 shows a metal photoconductivity cell (Reucroft *et al.*, 1970) which can easily be used for transport studies if the appropriate electronic circuitry and light source are provided. The high-intensity monochromater can be easily calibrated and allowance made for window and electrode absorption.

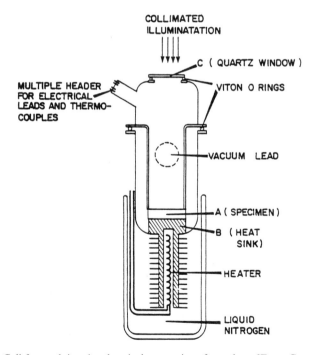

Fig. 5. Cell for studying the electrical properties of powders. [From Gray (1969).]

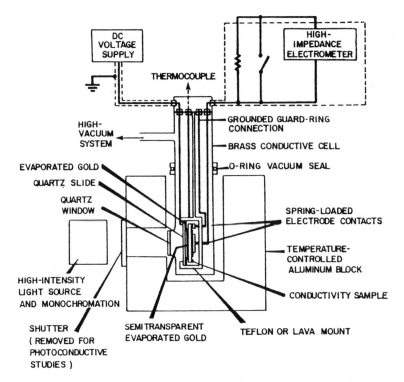

Fig. 6. Photoconductivity cell. [From Reucroft *et al.* (1970).]

C. The Time-of-Flight Technique

One of the more powerful experimental techniques available for studying charge transport is the so-called time-of-flight measurement pioneered by Spear (1957) and Kepler (1960). By this technique, it is possible to study the transit of charge carriers across a sample directly, as well as to obtain the quantum efficiency of intrinsic photoconductors.

The time-of-flight measurement system is shown schematically in Fig. 7. The experiment is performed by injecting a thin sheet of charge carriers into the material and following the current flow in the external circuit as the carriers migrate under the influence of an applied dc potential. For intrinsic photoconductors, the charge carriers are generated by a short (10^{-5} sec or less) light pulse in the absorption band of the photoconductor. Charge carriers can be injected into nonphotoconductors using a bilayer structure of a known photoconductor and the material to be studied. If the material under study is transparent to a wavelength at which the photo-

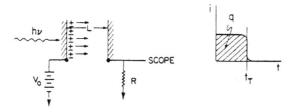

Fig. 7. Schematical representation of the time-of-flight technique. The voltage across the sample is constant V_0, and the drift of a carrier sheet across the sample is time-resolved.

conductor is active, the photoconductor–material interface can be illuminated directly. Otherwise, the rear face of the photoconductor must be illuminated and allowance made for transport through the photoconductor and any complicating factors at the photoconductor–material interface. Electrons can be injected from an electron-beam pulse; hole injection requires the generation of hole–electron pairs. Provided that the pulse is of short duration compared to the transit time, that the absorption depth is small compared to the sample dimension, and that the charge injected is insufficient to perturb the applied electric field, the transit time can be measured and an "effective" charge-carrier mobility calculated. Either of two circuit conditions is required. If the time constant of the external circuit is less than the transit time and the electrical relaxation time of the sample exceeds the transit time, the current in the external circuit reflects the charge-carrier migration across the sample. If the circuit time constant exceeds the transit time, the measurement reflects the current integrated over time, i.e., $\int_0^t i\, dt$. Typically, the sample relaxation time is greater than 1 sec, the circuit response time can be designed to be less than 10^{-4} sec, and the transit time of many polymers is in the 10^{-4}–10^{-1} sec range for reasonable voltages and sample thicknesses (1–100 μm). In the simplest case, the transit time τ is given by

$$\tau = d/V_d,$$

where V_d is the drift velocity.

It is assumed that the injected charge q_i is less than the capacitive charge stored in the sample (CV_0). Then, to a first approximation, the electric field E is not perturbed by the injected charge. It is uniform across the sample and given by $E = V/d$. The injected charge moves with a drift velocity $V_d = \mu_d E$, where μ_d is the drift mobility. The displacement current is $i = nqV_d/d$, where n is the number of injected charge carriers. When the injected sheet of charge carriers arrives at the back electrode, the current in the external circuit should drop abruptly to zero. Hence, from a plot of i versus t, the transit time t_T can be directly determined and the drift mobility, $\mu_d = d^2/Vt_T$, is obtained. Also, the number of charge carriers generated by the light pulse is calculated from the current–time integral and the number of photons in the light pulse.

In general, polymeric solids do not exhibit the ideal rectangular pulse anticipated. Deviations from ideality reveal novel and important information regarding the statistical transport of charge in polymeric systems (Mort and Pfister, 1979). A further variant of the time-of-flight measurement is the so-called xerographic discharge technique. In this method, a film of the material is cast onto a conducting substrate (or has an electrode evaporated onto it). The free surface is then charged to a potential V_0 using a corona device, and the change in surface potential is monitored as a function of time. Photoconduction experiments again use a pulse of light short compared to the transit time. It is implicitly assumed that the change in surface potential is small compared to V_0, in which case there is a little pertubation of the electric field. The quantum efficiency ϕ for charge generation $t = 0$ is obtained from the $t = 0$ discharge rate $(dV/dt)_{t=0} = \phi Nq/C$, where N is the number of photons absorbed and C is the sample capacitance. The drift mobility can be obtained from the xerographic discharge measurement, provided the system is operating in a space-charge-limited (SCL) mode, i.e., the number of injected charge carriers approximates CV_0. This condition is most easily obtained using a sensitizing injecting layer, such as selenium, to generate the charge carriers (Mort *et al.*, 1972). Deep trapping and field nonuniformities create perturbations in the shape of time-of-flight current–time curves and so complicate interpretation of the dV/dt curve (Chen, 1972; Chen *et al.*, 1973).

Measurements of conductivity using corona charging have also been made (Fig. 8). Two cautions relating to the nature of the conducting

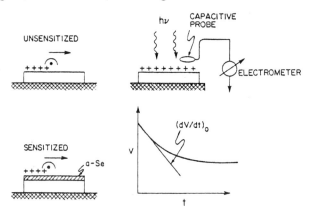

Fig. 8. Schematic representation of the xerographic discharge technique. The free, unelectroded surface of the sample is charged with a corotron to V_0, and the decay of the surface potential upon exposure to light is observed.

species and to chemical modification of the polymer surface are in order. With time-of-flight experiments in which a hole or electron is photoelectrically injected from a sensitizing layer, there is no doubt as to the nature of the charge-carrying species. However, in corona-charging experiments, it is not clear whether or not the corona ions $[(H_2O)_n H_3O^+$ or $CO_2^-]$ are themselves the charge-carrying entity or whether the chemical reaction required for injection takes place. In concept, injection involves a chemical reaction between the substrate, P, and the charging species; e.g., $(H_2O)_n H_3O^+ + P \rightarrow H + (n + 1)H_2O + P^+$, where P^+ represents an injected hole associated with a substrate molecule or atom. In most cases, the energetics for this exchange will be relatively unfavorable. Photons, however, can assist the reaction by providing sufficient energy to create an electron–hole pair, $P^{\mp}$, which can readily react:

$$(H_2O)_n H_3O^+ + P^{\mp} \rightarrow (n + 1)H_2O + H + P^+.$$

It should also be remembered that air corona discharges are chemical factories and provide chemically aggressive ambients; e.g., ozone, nitrogen oxides, nitric acid, and ammonium nitrate are all found to various degrees in corona discharges. Chemical attack on the surface of polymers by coronas is well documented in the literature. Surface modification via corona discharge to improve adhesion, to improve wettability, or to provide sites for chemical grafting is well known. Typical species created at the surface are ketonic, peroxy, and carboxylic acid groups. If the injecting properties of the surface are altered by the corona treatment, then the conduction process under study could well be dictated by the surface treatment.

III. ELECTRICAL CONDUCTIVITY

The type of electrical conductivity measurement reported in the literature usually involves a simple measurement of current as a function of time, temperature, ambient atmosphere, and potential. Attempts are then made to relate the conductivity to physical processes thought to be occurring in the polymer. Until recent years, there have been few attempts to study, for example, charge-carrier mobility as an independent variable.

Frequently it is found that electrical conductivity varies exponentially with temperature, is a function of time, and may vary with electrical field; i.e.,

$$\sigma = \sigma_0 \exp - E_\sigma/kT = f(\text{time}) = Af(E).$$

Changes in E_σ, the activation energy of conduction, are often observed in the neighborhood of glass-transition temperatures. Since the conductivity is made up of terms relating to both the number and the mobility of the charge carriers, any prediction regarding the conduction process that does not recognize these dependencies is meaningless. As more mobility measurements have been carried out, it has become recognized that the motion of the charge carriers is an activated process. Thus, the simple assumption that polymers can be described in terms similar to those used for crystalline, covalent semiconductors has been seriously questioned. Much has been learned from the study of disordered inorganic materials and by the extension of experimental techniques to polymers.

A. The Charge Carriers

Phenomenologically, the electrical conductivity of polymers increases exponentially with increasing temperature. This has been interpreted in terms of classical semiconductor theory to indicate that intrinsic charge-carrier creation occurs as described by conventional solid-state physics. The implication is that the charge carriers and the generation step are intrinsic to the polymer. More detailed studies have led to profound questioning of this explanation, particularly in insulating materials. Frequently nonlinear current–voltage characteristics are observed, electrode effects are noted, irreversible changes with time are sometimes seen, and behavior is often history-dependent.

The nonohmic behavior has often been attributed to ionic conductivity; seldom have mass-transport measurements been carried out to determine if mass transport, the best criterion for ionic conduction, takes place. There are a number of mechanisms in addition to ionic conduction by which nonlinear current–voltage characteristics can arise. These will be discussed in detail. First we will consider elements of solid-state theory, then alternative sources of charge carriers, and finally ionic conduction.

1. Intrinsic Charge-Carrier Generation

One way of classifying materials is on the basis of their electrical properties (Fig. 1). Materials are classified as metals, semiconductors, or insulators, depending upon the way in which their electrical resistivity changes with temperature. With metals, the resistivity increases with temperature; with semiconductors and insulators, the resistivity decreases exponentially with temperature. The difference between semiconductors and insulators is one of degree rather than kind. This classification has been explained in terms of a model called the band model which is discussed in textbooks on solid-state physics (see the Bibliography).

Briefly, the motion of an electron, detached from its parent atom or molecule but free to move in a periodically varying potential field such as that existing between atoms on a covalent lattice, can be described in terms of a modulated, traveling wave function. Such a wave function has real solutions for certain ranges of energy, termed energy bands. The ranges of energy for which there are no real roots are called forbidden energies. In a real solid, we are concerned with the highest filled band, called the *valence band,* the lowest empty band, called the *conduction band,* and the difference in energy between them, called the *forbidden energy gap.* The conducting properties are controlled by the width of the forbidden energy gap and the range of allowed energies in the conduction and valence bands. The width of the band at a given interatomic distance depends upon the interatomic interaction or, in the case of molecules, the intermolecular interaction. In a strongly interacting system, the allowed bands are wide; in a weakly interacting system, such as a molecular crystal, the bands are narrow. Within the band, the energy levels remain discrete but are closely spaced.

Another important parameter resulting from the band approach is the *effective mass* of the charge carrier, which is related to the width of the energy band. The effective mass is important because it relates the response of the charge carrier to the accelerating effect of an external electric field and is also related to the width of the band. For example, in a narrow band, the effective mass is high and the charge carrier responds slowly to an applied electric field. Consequently, the mobility is low. This is of particular significance in molecular materials where the intermolecular overlap integral is small and the bands are narrow. In polymers, it has been calculated that the bandwidth for a single linear polymethylene chain should be large (McCubbin and Manne, 1968). However, disorder and the relatively large distances between polymer chains suggest that the limiting step will be transfer of charge carriers between adjacent polymer segments or between specific sites which may be on the same or on a different polymer chain.

If the energy band is partially filled with electrons, the material is metallic and there is no forbidden energy gap. For a filled band below a small forbidden energy gap, the material is a semiconductor. If the energy gap is large, it is an insulator. Sodium, with one 3s electron per atom, is a metal, since the lowest energy band is only half filled and the electrons are free to move. On the other hand, a covalent material, such as silicon, contains two electrons in each bond and all the available orbitals are completely filled. An ionic material has no easily available electrons, since each shell is completely filled. Similarly, polymers based on carbon, oxygen, and nitrogen contain two electrons for each available state. This is the case regardless of the orbital hybridization. Such materials with all available energy levels completely filled are insulators. As we will see in the next section, the removal or addition of electrons from filled energy bands will lead to a situation in which there are more available states than electrons; a material with more available states then electrons to fill them will conduct charge. Therefore, for materials with completely filled states, no electrical conduction should occur until an electron is excited from a filled orbital into the lowest empty orbital. Consequently, such materials should be insulators. However, in reality, covalent and ionic materials do conduct, but at a low level. This low level of conductivity arises because in any material there will always be a small fraction of electrons with an energy in excess of the forbidden energy gap. Statistically, the number of charge carriers is controlled by the Boltzmann distribution law. If the forbidden energy gap is E_g, then the number of electrons with energy above E_g depends upon the temperature T. If we excite n electrons from the ground state to the conducting state (conduction band), we leave behind an equivalent number of vacancies p in the ground state (valence band), which also contribute to the conductivity; i.e.,

$$n = p \quad \text{and} \quad n^2 = n_0 \exp[-(E_g/kT)].$$

This is the origin of intrinsic semiconduction.

Another useful concept is that of the Fermi level. At 0 K, all the electrons fall into the lowest allowed energy level. As the temperature is increased, the probability of higher-energy electronic states being occupied is given by the Fermi–Dirac distribution function $f = 1/[1 + \exp(E - E_F)kT]$, where E_F is the Fermi energy. In a metal, it represents a real state; it is the energy level that corresponds to the highest filled state at 0 K and can be obtained directly by measuring the energy required to remove the electron from the metal to infinity. This parameter is called the work function ϕ. It is also equivalent to the chemical potential of the electrons and, as such, is a useful thermodynamic concept to bear in mind. In a semiconductor, the Fermi level lies within the forbidden energy gap and cannot be measured directly; but it is equal to half the sum of the ionization potential I and electron affinity E_a, i.e.,

$$E_F = (I + E_a)/2.$$

There is no clear distinction between insulators and semiconductors. The main difference lies in the width of the forbidden energy gap, hence in the number of charge carriers. It is calculated that, for a material with 10^{22} atoms/cm^{-3} and a forbidden energy gap of 14 kcal (0.6 eV), there are 10^{12} charge carriers cm^{-3} at 25°C. This is not an unreasonable number, which combined with a mobility of 10^2 cm^2/V sec, corresponds to a conductivity of 10^{-5} Ω^{-1} cm^{-1}. An energy gap of 42 kcal (1.8 eV) would correspond to 10^{-8} carriers cm^{-3} and a conductivity of 10^{-25} Ω^{-1} cm^{-1}, which is ridiculously low. Consequently, we must look beyond intrinsic generation for the origin of the charge carriers in insulators.

In summary, band theory suggests that conductors have more energy states available then there are electrons to fill these states. Semiconductors and insulators have the same number of electrons as energy states (allowing for the spin quantum number S of 2). Intrinsic conduction in semiconductors arises by excitation of an electron across the forbidden energy gap (either by thermal, photolytic, or other means), a process that creates an equal number of excited electrons and vacancies (positive holes) in the highest filled states. For ordered arrays of atoms, the terminology is that of band theory—the conduction band being the next band at higher unoccupied states and the valence band (ground state). For molecular materials, the equivalent terms are the highest filled molecular orbital (HFMO) and the lowest empty molecular orbital (LEMO). The intermolecular orbital overlap is small. Consequently, the energy bands are narrow; charge-carrier mobility is low and, important from a photoconduction viewpoint, the spectroscopic properties of a molecular crystal resemble those of an array of isolated molecules rather than those of a strongly interacting array.

If charge transport involves the charge carrier spending the greater part of its time on specific molecules, then on a time scale equivalent to the vibrational relaxation times (10^{-12} sec), the localized states involved are, in fact, radical anions and radical cations. Charge transfer may then be thought of as a localized oxidation or reduction reaction, e.g.,

$$M + M^+ \xrightarrow[\text{field}]{} M^+ + M \quad \text{for hole conduction,}$$

$$M^- + M \xrightarrow[\text{field}]{} M + M^- \quad \text{for electron conduction.}$$

The energetics of the oxidation–reduction couple due to the applied field can then play critical roles in determining the electrical properties.

2. Impurity Conduction

So far in our discussion of semiconductors and insulators, it has been implicitly assumed that, for each electron that is excited to the conduction band from the ground state, a vacancy is created in the valence band. This vacancy is called a positive hole. Conduction can occur by means of electrons moving in the conduction band or by means of electrons in the valence band moving to vacancies in the ground state. The latter case is perhaps a little harder to envision, but the net result is the movement of a positive hole in the direction of the field.

Similarly, in molecular systems, the creation of an electron and a positive hole in conducting states can be envisioned. If an electron donor or an electron acceptor is introduced into the host matrix, the electrical properties can be controlled in a way similar to that in which doping of inorganic crystals by altervalent ions is used to create *n*- or *p*-type conduction.

Another way in which conduction is increased is to produce materials that are mixtures of easily ionized electron donors and electron acceptors. These materials, in which the charge is shared between the donor and acceptor, are called charge-transfer complexes. A number of these compounds have been shown to be conductive. The conductivity depends upon the ionic character of the complex. Fully ionic compounds are more conductive in the ground state than weakly ionic compounds (Williams, 1976).

For example, the extremely high conductivity of graphite–arsenic pentafluoride materials is thought to involve a sequence of reactions:

$$2AsF_5 \rightarrow AsF_6^-(ads) + AsF_4^+(ads),$$
$$Cg + AsF_6^-(ads) + AsF_4^+ \rightarrow Cg^+AsF_6^- + AsF_4,$$
$$Cg^+ + AsF_6^- \rightarrow Cg^+AsF_6^-,$$
$$2AsF_4 \rightarrow AsF_3 + AsF_5.$$

The outcome is that three AsF_5 molecules create one AsF_3 molecule (detected) and two conducting states (Forsman, 1977; Bartlett *et al.*, 1978).

Similarly, doping of polyacetylene by sodium or arsenic pentafluoride creates *n*- or *p*-type conductivity. The *p–n* junction behavior shown in Fig. 9 was obtained by mechanically pressing *p*-type (AsF_5-doped) material against *n*-type (Na-doped) material (Chaing *et al.*, 1978). The current–voltage characteristics resemble those of typical inorganic *p–n* junction devices.

The key to the behavior of these donor–acceptor materials lies in the ionization potential and electron affinity. There are well-known methods for deciding which elements and structures will be useful for lowering ionization potentials and increasing the electron affinity of the donor and acceptor (Turner, 1966; Briegleb, 1964). At one time it was also suggested that stable free radicals would provide a means of doping molecular materials and polymers (Eley and Parfitt, 1955). Certainly rapid motion of the

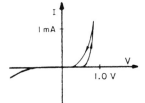

Fig. 9. *I–V* curve for a doped polyacetylene *p–n* junction. *n*-Type, Na doped; *p*-type, AsF_5-doped. [From Chiang *et al.* (1978).]

unpaired electron along a carbon chain is suggested by the ESR hyperfine structure of many free radicals. However, little intermolecular transfer is observed, since the energy required to detach the unpaired electron from its parent free radical is comparable to the energy required to ionize a molecule (Eley and Willis, 1961).

B. Localized States

In discussing electrical conduction in the context of band theory, the need for regular arrays of similar atoms is clear. It is also clear that, as the interaction between atoms decreases, the width of the allowed energy band decreases and the effective mass of the charge carrier (the parameter relating to the response of the charge carrier to an external field) increases. A molecular material, such as anthracene, is calculated to be about the limit to which band theory applies (LeBlanc, 1961; Friedman, 1964). In the case of a polymer, the condition of regular spacing of like atoms applies only to the case of the simplest carbon–hydrogen polymers such as polymethylene $(-CH_2-)_n$, polyenes, polyacetylenes, etc., and only along the chain axis.

Breakdowns in local structure, e.g., cis–trans forms in polyacetylene or dipoles in the chain, will all impact the system energetics (Duke and Fabish, 1976; Fabish, 1979). Thus it is not to be expected that even the most highly crystalline nondipolar polymers will have high mobilities and that the mobilities will be isotropic (Baughman and Chance, 1978). This indeed appears to be the situation in polyacetylenes in which mobilities are ~ 1 cm^2/V sec in the chain direction and the anisotropy is $> 10^2$ (μ_{chain}/μ_h) (Reimer and Bassler, 1975, 1976; Siddiqui, 1980). Thus, for such materials, the existence of narrow bands similar to those of molecular crystals would be expected. However, other than the large crystals described by Baughman (1977), single-crystal polymers greater than tens of microns in diameter have not been made. In general, even highly crystalline polymers do not show greater than 90% bulk crystallinity and even the crystalline regions are likely to be disordered. The typical schematic representation of a semicrystalline polymer (Fig. 10) contains crystalline lamallae, the remainder being amorphous material. Carrier trapping can occur at phase boundaries. Chain folding can lead to spherulite formation in which the growth of chain-folded crystallites proceeds outward from the point of initiation. In addition, many polymers contain polar groups; each dipole can act as an electron or hole trap. Fox and Turner (1966) have calculated that a dipole of 1.65 D is capable of binding an electron (with the additional effect of dielectric relaxation around the trapped charge carrier).

Fig. 10. Schematic representation of a semicrystalline polymer. [From Florey (1953).]

The bulk structure of polymers is responsible for many of the physical properties of the material. It is of course dependent upon factors such as the chemistry of the monomer and the stereochemistry of the polymer. In addition, it can also be controlled by the precise processing conditions. Solvents can control the morphology of copolymers; cooling rates can be used to alter the extent of crystallization and crystallite size. Extrusion rolling and stretching can be used to alter molecular and chain orientation. The addition of plasticizers and solvents will affect molecular motions in the polymer, as will the presence of hydrogen bonding, chain branching, and cross-linking.

Many polymers consist of both crystalline and noncrystalline (amorphous) regions. Such materials are referred to as semicrystalline polymers. Extremes of this morphology are possible—wholly crystalline and wholly amorphous polymers. However, even the more crystalline polymers contain considerable amounts of disordered materials. For example, polyethylene crystallizes to form spherulites that show considerable disorder at the boundaries. Single crystals of polyethylene can be obtained that show considerable order. However, their size is small (a few microns). The problem of growing large, single crystals is associated mainly with the thermodynamics—interaction energies are low and entropic effects are high. Typically, the features of polymer single crystals are (Sharples, 1972)

(a)   small size (microns in length and breadth, 100 Å in thickness),
(b)   constant thickness for a given set of conditions, increasing with temperature of crystallization,
(c)   growth by extension of length and breadth,
(d)   molecular chains ordered parallel to the thickness,
(e)   chain folding, since the chain length exceeds the thickness,

(f)   high crystallinity, although less than 100%, with considerable surface disorder.

While the existence of polymer single crystals is of real theoretical significance, the practical consequences are negligible. More typically, polymers are semicrystalline (Fig. 10). It is convenient to think of semicrystalline polymers as a continuous matrix of an amorphous polymer in which the properties are modified by the crystalline regions that act as sites for reinforcing the amorphous matrix. As far as electrical properties are concerned, the effect of the crystallinity is to lower the conductivity. The precise mechanism by which this occurs depends on the precise conduction mechanism; if conduction is ionic, ion mobility through the crystalline regions will be low; if electronic, it will perhaps be faster, but the crystalline–amorphous interface may act as a trapping region—a phenomenon akin to Maxwell–Wagner interfacial polarization.

Wholly amorphous polymers can exist in two states depending upon the temperature. At low temperatures, they are hard, glassy materials. At a temperature referred to as the glass-transition temperature T_g, they undergo a transition to a rubberlike state. Unlike the crystalline–melt transition, the glass–rubber transition is a second-order transition and there are no discontinuities in properties such as specific heat, volume, and mechanical properties. As far as electrical properties are concerned, the relevance of glassy and rubbery states is still open to speculation. Evidence is accumulating that changes in conductivity can be related to the onset of certain molecular motions. For example, the thermal discharge of electrets can be related to the onset of molecular motions in poly(vinylidine fluoride), polyethylene, and other polymers. Current thoughts on this topic are summarized in Chapters 5 and 6.

The transition between the glassy and amorphous states is accompanied by the "freeing-up" of gross molecular motions of the polymer chain. Other types of transitions have also been identified with the onset of other molecular motions (McCrum *et al.*, 1967). Five types of motion appear possible in amorphous materials. These are, in order of increasing temperature (Fig. 11),

(a)   side-chain motions, e.g., rotation of

$$-CH_3,\quad -\!\!\bigcirc,\quad \text{or}\quad \bigcirc\!\!-\!\!\overset{\overset{\displaystyle |}{N}}{}\!\!-\!\!\bigcirc ,$$

(b)   motion of two four-carbon moieties in the main chain (the Schatzki crankshaft effect),

Polymer	Dipole geometry	Structure	
Polymethyl acrylate	One component in flexible side chain, one component rigid perpendicular	$-[CH_2-CH-]_n-$ with side group C double-bonded to O and single-bonded to $O-CH_3$	
Poly(vinyl chloride)	Rigid perpendicular	$-[CH_2-CH-]_n-$ with side group Cl	
Polyethylene oxide	Rigid perpendicular	$\left[CH_2 \;^{O}\!\!\diagdown\, CH_2 \right]_n$	
Polyesters	Rigid perpendicular and parallel	$-[R-C(\!\!=\!\!O)-O-R'-]_n-$	
Poly (p-chloro-phenyl acetylene)	Rigid perpendicular and parallel	$-[CH=C-]_n-$ with phenyl ring substituted with Cl	

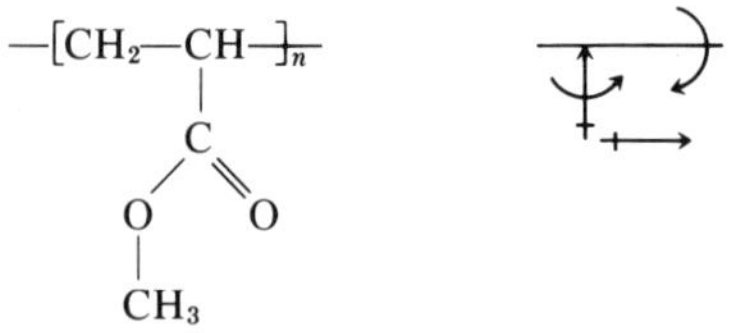

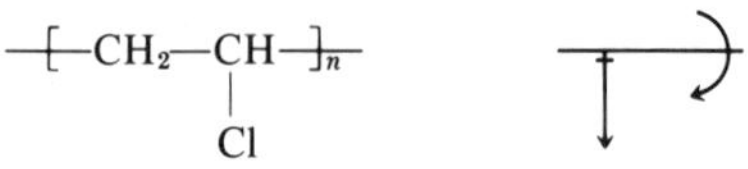

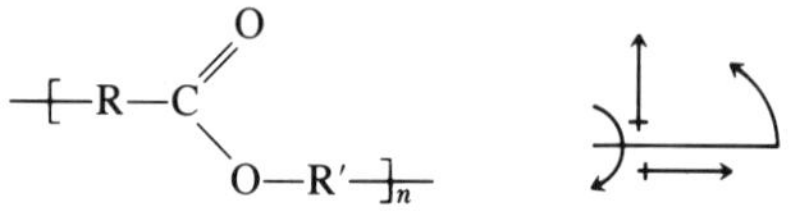

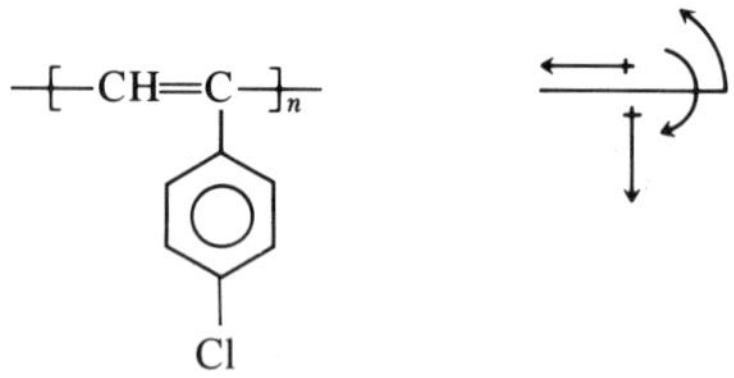

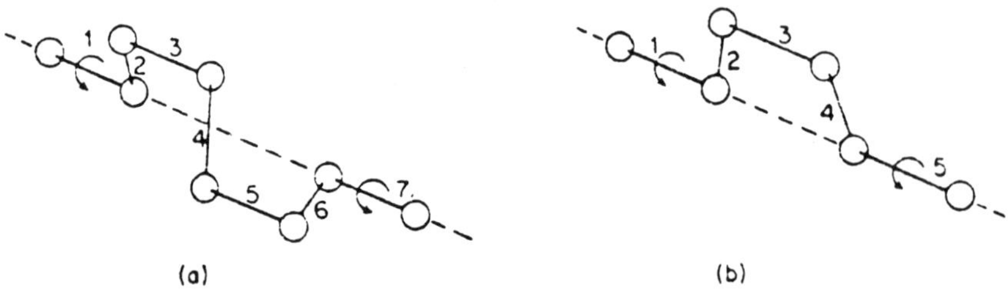

Fig. 11. Molecular motions in polymers. The crankshaft mechanisms of (a) Shatzki and (b) Boyer. [From McCrum *et al.* (1967).]

(c)   motion of moieties containing heteroatoms in the polymer chain,

(d)   motion of segments containing 50–100 backbone atoms (corresponding to T_g,

(e)   motion of the entire chain as a unit.

In crystalline polymers, additional possibilities exist:

(a)   crystal melting,

(b)   change in crystalline structure,

(c)   motion of side chains in the crystallite,

(d)   crystalline–amorphous interactions including interfacial friction,

(e)   intracrystallite interactions.

Thus localized energy states, i.e., states not forming extended bandlike states, can be envisioned associated with at least the following on a molecular level:

(1)   surface states induced by strain or chemical reactions,

(2)   surface dipole states,

(3)   bulk dipole states,

(4)   bulk molecular ion states,

(5)   impurities (different chemical groups, polar groups, ionic groups),

(6)   chain ends,

(7)   chain branches,

(8)   chain folds,

(9)   changes in tacticity or stereochemistry,

(10)   crystalline–amorphous boundaries,

(11)   broken bonds,

(12)   polaron states (trapped charge and region of surrounding polarized dielectric), and

(13)   local density fluctuations.

Escape from these localized states may be purely thermal; it may depend on the local environment and specific molecular motions. The local field may assist in detrapping. Thus the energy diagram appropriate for a polymer might be represented by Fig. 12. This figure indicates a uniform crystallite surrounded by amorphous regions. This crystallite contains some defects but, by and large, has a regular array of atoms. There are interfacial states at the crystalline–amorphous boundary. Within the amorphous region, the localized states are of neither uniform depth nor distribution (Seanor, 1967).

The localized states may act in trapping carriers from the extended states of the crystalline region. In the language of semiconductor theory, a neutral acceptor state or an ionized donor state will trap electrons and a

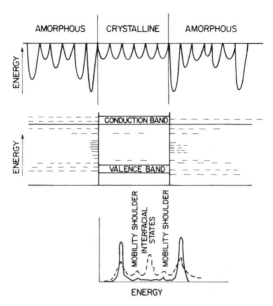

Fig. 12. The energy diagram of a polymer with crystalline and amorphous regions. Crystalline region density of states; ---, amorphous region density of states. It is not known whether the mobility shoulder is real or not.

neutral donor state or an ionized acceptor state will trap holes. A trapped charge can also act as a recombination center. Lattice polarization around a charge in a localized state can lead to an increase time in depth trap and tend to make transport more difficult (polarons). In the amorphous region, the density fluctuations will create localized states and possibly a mobility band as described by Mott (1967, 1969) and by Mott and Davies (1971).

In the case of dipole-containing polymers, it is not clear that, even in the crystalline regions, energy bands can be derived in concept. In addition, Duke and Fabish (1976) argue that side groups can act as traps creating ion-radical states which, because of the differences in local environment, could have energy-level variations of as much as 1 eV. Consequently, the hopping probability is much reduced. An important outcome of their hypothesis is that localized states deep within the energy gap are an inherent property of the polymer and are related directly to its chemical structure. It is not necessary to invoke impurities or disorder to explain traps, although lack of long-range order will contribute to the range of energy associated with a particular chemical group. Because the energy of closely adjacent sites is unlikely to be the same, both the distance and energy terms in the equation describing hopping probability,

$$p_n \propto \exp(-\Delta E)\exp(-\gamma\rho),$$

will be small. Here ΔE represents the energy difference between the initial and final states, γ describes the fall-off in wave function with distance, and ρ is the average separation of the sites.

Another way of stating the Duke–Fabish concepts is to suggest that disorder leads to a range of energetically different environments for each chemically different group in the polymer. Once a charge carrier is trapped, the polaron states created are also spread over a broad range of energy (Duke, 1978).

In addition to bulk states, there can also be surface states and interfacial states. It must also be appreciated that the crystalline and amorphous regions are distributed in three dimensions. Thus a two-dimensional representation, such as Fig. 12, is oversimplified.

C. Injection Processes

So far we have only considered thermal generation of the charge carriers within the material itself. Other methods of charge-carrier generation involve emission from electrodes or generation under the influence of light (photoconduction). Direct emission from electrodes is particularly important at high fields, and only recently has the extent to which this type of process affects conduction been realized.

The process of charge-carrier injection under high fields at low temperatures has given rise to its own discipline—low-temperature inelastic electron tunneling spectroscopy (LTIETS)—in which current fluctuations as a function of voltage are correlated with interfacial and bulk infrared group vibrational frequencies (Hansma, 1977; Keil *et al.*, 1976; Wolf, 1978). Figure 13 shows a comparison of the IET spectrum of cytosine adsorbed on aluminum (Simonsen and Coleman, 1973) with the infrared and Raman spectra. The plot is in terms of d^2I/dV^2.

At higher temperatures, the influence of electrode processes on the apparent electrical properties has received much attention. It is assumed that, once the charge carrier is inside the solid, it does not know its origin. It recognizes only the influence of temperature, the local electric field (which may be affected by the way in which the charge carrier enters the solid), and the molecular nature of the material. This means that it is also possible to use controlled injection to study charge-carrier behavior within the solid.

Electrons in a metal are free to move throughout the bulk of the solid. However, when they reach the surface, they are subjected to a constraint imposed by the noncontinuity of the solid. In order to move beyond the surface, they require excess energy. The work done to remove an electron to infinity is called the work function ϕ_M. If, instead of a vacuum being adjacent to the metal, a dielectric is placed there, the potential barrier will

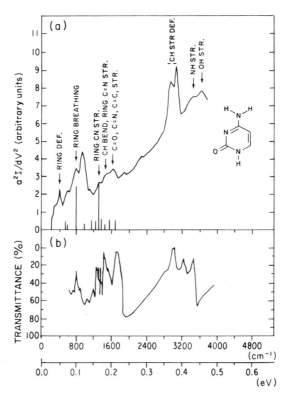

Fig. 13. IET spectrum for cytosine at 4.2 K. The Raman (a) and Nujol mull (pure IR) (b) infrared spectra are used for comparison. [From Simonsen and Coleman (1973).]

be modified, and both electrons and holes can be injected, depending upon the energy levels in the dielectric.

Such processes can be represented by the equations

$$M^+ + P \rightarrow M + P^+ \quad \text{at the anode,}$$
$$M^- + P \rightarrow M + P^- \quad \text{at the cathode,}$$

where M represents the electrode and P the polymer.

In the absence of surface states, the energetics of such processes can be represented by the conditions that for hole injection

$$\phi_P - (\phi_M + E_{act}) \leq 0,$$

and for electron injection

$$\phi_M - (E_A + E_{act}) \leq 0,$$

where ϕ_P is the work function of the polymer, E_A is its electron affinity, and E_{act} is the energy required to cross the potential barrier at the electrode polymer interface. However, as described in Chapter 7, the surface of the polymer is complex and surface states extend some distance into the polymer. On making contact between the electrode and the polymer, electron transfer will take place until thermodynamic equilibrium is attained. The contact charge will cause some lowering of the barrier to injection.

There are three ways in which the energy required to escape from the metal may be supplied:

(1)   by thermal methods in a process known as thermionic emission,

(2)   by the application of high electric fields in field emission, and

(3)   by photon absorption at sufficiently short wavelengths in the photoemission process.

Mechanisms (1) and (3) may be modified by the Schottky effect which arises from the field-dependent lowering of the potential barrier to injection.

A number of experimental observations on polymeric systems have been explained in terms of emission processes. Miyoshi and Chino (1967) have documented Schottky emission into polyethylene. Lily and McDowell (1968) have studied the Schottky effect in Mylar®* polyester films and have suggested that injection into surface states takes place followed by a field-dependent detrapping process.

Some of the data obtained on polyethylene single crystals show the relationship between the resistivity and sample thickness for different values of V in the equation for Schottky emission:

$$i = AT^2 \exp\{-[(\phi/kT) - (q^{3/2}V^{1/2}/\varepsilon^{1/2}d^{1/2}kT)]\},$$

where V represents the potential across a film of thickness d, ε is the permittivity, and ϕ is the potential barrier at the electrode–polymer interface.

The first terms of the equation represent Richardson's equation, $i = AT^2 \exp[-(\phi/kT)]$, for thermionic emission, in which $A = 120$ A cm^{-2} deg^{-2} and ϕ represents the work function of the electrode. In the presence of an applied field, the potential barrier is lowered by an amount $\Delta\phi = q^{3/2}V^{1/2}/\varepsilon^{1/2}d^{1/2}$. Consequently, an exponential dependence on $V^{1/2}$ is observed.

To allow for surface states, trapped charge, and local polarization effects, Taylor and Lewis (1971) have argued in favor of a more generalized

---

* Mylar is a registered trade name of E. I. du Pont de Nemours, Inc.

barrier-lowering term of the form

$$E(x) = \phi qK/(ax)^n.$$

The equivalent form of the equation for barrier lowering becomes

$$i = i_0 \exp\{-[(\phi - q\beta E^{n/n+1})/kT]\},$$

where $\beta = (n + 1)(K/ax)^{n/n+1}$. Data for polyethylene and polyester yield $n = 0.45$ and 0.15, respectively, and values for the energy barriers compatible with the polymer work functions of Davies (1969). (See Chapter 7.)

The exponential dependence of current on $E^{1/2}$ is not characteristic of just Schottky barrier lowering. If the charge carrier becomes trapped in a coulombic potential well, then the field-induced lowering of the barrier to detrapping has the same functional dependence. The difference arises because in the coulombic well the force between the charge carrier and the charged site varies as q^2/x^2, whereas the force between the injected charge and its image charge in the electrode varies as $q^2/(2x)^2$. Consequently, the formalism of the Poole–Frenkel trap barrier lowering exhibits the same $E^{1/2}$ dependence. If the local dielectric constant is known, it is possible to distinguish between the two effects. Alternatively, electrodes of different work function should allow injection currents to be distinguished from detrapping phenomena (although surface states may interfere with this simple picture).

At high fields, the width of the potential barrier decreases. As a result, the probability of finding an electron on the other side of the potential barrier by quantum mechanical tunneling increases. The tunneling process is independent of temperature, and the tunneling current is characterized by the equation

$$i = (q^2V/h^2d)(2m^*\phi)^{1/2} \exp[-4\pi d(2m^*\phi)^{l2}/h]$$

(if $T \rightarrow 0$ and $qV \ll \phi$), where m^* is the effective mass of the charge carrier and h is Planck's constant. The experimental results at low temperatures do not agree with the theoretically expected tunneling current curve. Miyoshi and Chino (1967) concluded that Schottky emission was responsible for most of the current in films 500–1200 Å thick. In thinner films or at low voltages, the tunneling current was greater than the Schottky current. Somewhat similar conclusions were reached by Gregor and Kaplan (1968) in their study on thin films produced by electron bombardment of epichlorohydrin vapor; i.e., charge-carrier tunneling predominated at low temperatures and low voltages, whereas Schottky barrier lowering was observed at higher temperatures and voltages. Schottky emission is also thought to account for the conductivity of thin (100–

$$n \; (CH_2\!-\!CH\!-\!CH_2Cl) \xrightarrow[\text{bombardment}]{\text{electron}} \; +O\!-\!CH_2\!-\!CH)_n$$

epichlorohydrin   polymer

500 Å) films of polydivinylbenzene synthesized by passing divinylbenzene through a glow discharge (Gregor, 1968).

The energy needed to overcome the surface potential barrier may also be provided by light of the appropriate wavelength.

Williams and Dresner (1967) studied the photoemission of holes into anthracene from metals of different work functions. The photoemission threshold was calculated by plotting the square root of the quantum yield versus the photon energy. In this case, by summing the threshold energy and the work function, they calculated the valence band of anthracene to be 5.80 ± 0.34 eV below the vacuum level. This agrees well with the value obtained by direct photoemission from anthracene into a vacuum (Vilesov, 1964).

In a similar set of experiments, Lakatos and Mort (1968) observed the emission of holes into poly-N-vinylcarbazole. The photoemission threshold was obtained for the series of metals shown in Table II.

poly-N-vinylcarbazole

The plots of the square root of the quantum yield versus the wavelengths for different potentials show that the threshold for photoemission of holes shifts to lower frequencies as the applied field increases. If the value of E_H for an applied potential of 5 V is taken as zero, plots of ΔE_H versus $E^{1/2}$ are linear up to fields of 2×10^4 V/cm and pass through the origin. At higher fields, the curve levels off. These data indicate that Schottky barrier lowering takes place at the electrode–polymer interface.

The energetics of the emission process show the ionization potential I_p to be $\phi_M + E_H$, where E_H, the threshold energy for photoemission, equals $h\nu_0$ and ν_0 is the frequency at which photoemission is first observed (for low applied fields). The near constancy of $(\phi_M + E_H)$ for three metals is shown in Table II in which the average value of 6.1 eV is in excellent agreement with the ionization potential, $I_p = I_g - P$, given by Lyons

TABLE II   *The Vacuum Work Function ϕ_M of Various Metals, the Threshold Energy E_H for Photoemission of Holes from These Metals into Poly-N-Vinylcarbazole, and $\phi_M + \phi_H$[a]*

Metal	ϕ_M (eV)	E_H (eV)	$\phi_M + E_H$ (eV)
Au	5.22[b]	1.28	6.50
Cu	4.45[c]	1.48	5.93
Al	4.20[d]	1.53	5.73

[a] From Lakatos and Mort (1968).
[b] E. E. Huber, *Appl. Phys. Lett.* **8**, 169 (1966).
[c] J. C. Riviere, *Proc. Phys. Soc. London Ser. B* **70**, 676 (1957).
[d] A. Hermann and S. Sapher, "The Oxide Coated Cathode." Chapman and Hall, London, 1951.

(1963), where I_g is the gas-phase ionization potential of 7.6 eV (Lardon *et al.*, 1967) and P, the polarization energy, equals 1.5 eV (Sharp, 1967).

Based on Lyons' (1963) equation for the forbidden energy gap, $E_g = I_g - 2P$, the conducting state for electrons should lie 4.6 eV above the ground state of solid poly-*N*-vinylcarbazole. Therefore, electron injection should be observed at a threshold frequency ν_e given by

$$h\nu_e = \phi_M - P.$$

In fact, electron injection was observed very close to 3.3 eV using gold electrodes. The suggested energy levels in poly-*N*-vinylcarbazole are shown in Fig. 14.

D. Steady-State Conduction

Having introduced the concepts of charge-carrier traps and barriers to injection, we can begin to discuss steady-state conduction. Assume that the number of charge carriers and the local electric field are functions of location in the sample and are represented by the terms $n(x)$ and $E(x)$, where x is the distance from a reference electrode.

Then the steady-state current density J cannot be a function of position; i.e.,

$$J = qn(x)\mu E(x) \neq f(x).$$

This means only that the charge flux is constant across the sample. How-

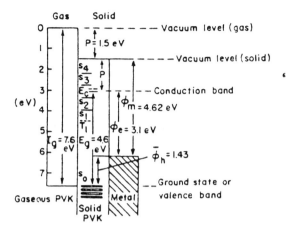

Fig. 14. Energy-band diagram for the Cu–PVK interface. The ionization potential (i.e., the threshold for photoemission into a vacuum) for PVK is $I_c = I_g - P$, where I_g is the ionization potential of the PVK molecule and P is the polarization energy of PVK by a point charge. T and s_n represent triplet and singlet energy states: ϕ_m is the metal work function. [From Mort and Lakatos (1970).]

ever, the number of charge carriers and their drift velocity is sensitive to the electrical environment.

The relationship between the field gradient and the number of charge carriers is defined by Poisson's equation:

$$dE(x)/dx = qn(x)/\varepsilon,$$

where ε is the dielectric constant.

The integral of the local field across the sample is the applied potential V_0; i.e.,

$$\int_0^x E(x)\, dx = V.$$

These three equations, in combination with any set of boundary conditions, define the current–voltage relationships.

In the simplest case, $n(x)$ is independent of position and constant across the sample. In this case, $E(x)$ is not a function of position, $dE/dx = 0$, and

$$J = qn\mu E.$$

If the drift velocity were controlled by potential-assisted detrapping (the Poole–Frenkel effect), it would be reflected in the $I–V$ characteristics. Similarly, when n is controlled by thermally assisted detrapping or by injection, the dependence is shown in the $I–V$ curve.

An ohmic electrode is an electrode that has no potential barrier to injection and is capable of providing (in theory) an infinite supply of charge carriers. This means that, immediately adjacent to the electrode, the electrical field is zero and there is an infinite supply of charge carriers; i.e., the boundary conditions are

$$n(0) = \infty \quad \text{and} \quad E(0) = 0.$$

Substituting $n(x)$ in Poisson's equation into the equation for current, we have

$$J = \varepsilon \mu E(x)[dE(x)/dx].$$

Integrating, $Jx = \varepsilon \mu E^2(x)/2$ and

$$V = \int_0^d (2J/\varepsilon\mu)^{1/2} x^{1/2} \, dx,$$

which yields

$$J = 9\varepsilon \mu V^2/8d^3.$$

This is known as Child's law for a trap-free insulator (Mott and Gurney, 1940). The deviations from Ohm's law arise because the solid is unable to transport all the injected charge. The accumulated charge (the space charge) builds up adjacent to the ohmic electrode. Consequently, the field distribution changes. Now $n(x)$ and $E(x)$ are given by

$$n(x) = 3\varepsilon V/4qd^{3/2}x^{1/2}, \qquad E(x) = 3Vx^{1/2}/2d^{3/2}.$$

There has been much discussion in the literature on space-charge-limited currents (SCLC) (Lampert, 1964; Lampert and Mark, 1970; Kryszewski and Zymanski, 1970). The analysis provides a powerful means by which to interpret nonlinear current–voltage characteristics and particularly to quantify the numbers and energies of traps.

If the applied potential is sufficiently low, conduction is ohmic. If, at some higher potential, the solid is unable to transport all the charge, the current becomes nonlinear in voltage; at a potential V_x, the mobility and the voltage can be calculated from the crossover current J_x; i.e.,

$$J_x = qn\mu V_x/d = 9\varepsilon\mu V_x^2/8d^3,$$
$$V_x = 8qd^2n/9\varepsilon,$$
$$n = 9V_x\varepsilon/8qd^2,$$
$$\mu = 8J_xd^3/9V_x^2\varepsilon.$$

Alternatively, μ can be obtained from the slope of the J–V^2 curve, $dJ/d(V^2)$:

$$\mu = [8d^3 dJ/d(V^2)]/9\varepsilon.$$

If there are traps in the solid, the space-charge-limited current may be decreased by several orders of magnitude. Rose (1955) argued that neither the space-charge density nor the field distribution should be altered by trapping, but that the equation relating current to voltage should be modified by a trap-limiting parameter θ relating the proportion of trapped charge to free charge. J is now written

$$J = 9\varepsilon\mu V^2\theta/8d^3.$$

The simplest trapping model is that of a single set of traps lying at an energy E_T below the conducting state. At low injection levels, most of the traps are empty and θ is given by

$$\theta = n_{\text{eff}}\exp[-(E_T/kT)]/\{N + n_{\text{eff}}\exp[-(E_T/kT)]\},$$

where n_{eff} is the density of states in the conduction band and N is the density of traps. Assuming $N \gg n_{\text{eff}}\exp[-(E_T/kT)]$, we have

$$\theta = n_{\text{eff}}\exp[-(E_T/kT)/N].$$

Substituting for θ yields

$$J = 9n_{\text{eff}}\exp[-(E_T/kT)]\varepsilon\mu V^2/8d^3N.$$

This equation is referred to as Child's law for an insulator with shallow traps.

The equations for obtaining the density of charge carriers and mobility from the $I-V$ curves differ from the trap-free equations only by the factor θ.

As the traps fill, the current should approach the trap-filled limit of the trap-limited Child's law region of the current–voltage curve as shown in Fig. 15. At this point, there is a tremendous increase in the current as the last traps are filled. The current rapidly approaches the trap-free Child's law current. It often appears that dielectric breakdown occurs when the traps become filled. However, the $I-V$ curves should be reproducible.

The voltage at which transition to the trap-filled limit current occurs can be used to calculate the trap density by noting that the average charge-carrier density $\bar{n}$ is

$$\bar{n} = \int_0^d n(x)\,dx\Big/\int_0^d dx = 3\varepsilon V/2qd^2.$$

Hence, the trap density N at the trap-filled limit is given by

$$N = 3\varepsilon V_{\text{TFL}}/2qd^2.$$

More complex situations can be envisioned in which there are discrete arrays of traps with differing energy (Fig. 16) (Siddiqui, 1980). Situa-

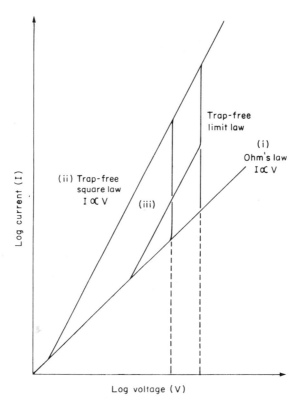

Fig. 15. Schematic *I–V* curves for space-charge-limited currents. (i) Ohm's law region; (ii) trap-free Child's law; (iii) Child's law with trapping. Single carrier, single set of traps.

tions where there is a distribution of traps energies and other more complicated trap models have also been discussed. (Lampert, 1964; Tredgold, 1966; Lampert and Mark, 1970).

A detailed study of charge-limited currents can yield considerable information about the conduction process, and the technique is very versatile. For example, it is possible, by the choice of appropriate electrodes, to inject electrons, holes, or both electrons and holes simultaneously. The latter technique has been used to study electron–hole recombination in anthracene using positive and negative anthracene radical ions as the injecting electrodes (Helfrich and Schneider, 1965).

The reason for the interest in space-charge-limited currents in polymers lies in the wealth of information which can be obtained from the current–voltage curves. The relationship of the number and depth of the traps to the pretreatment of the polymer, its molecular structure, and morphology are of prime importance in understanding the electrical behavior of polymers. Changes in crystallinity, orientation, or even tacticity can each affect the electrical properties in addition to effects relating to the molecular structure. Study of space-charge-limited currents represents one way of tackling these difficult problems.

For example, Fig. 17 shows a series of current–voltage curves obtained by Setter (1967) for polyethylene at different temperatures. Time effects were observed, and the current–voltage curves show well-defined changes from ohmic to square law behavior. At high voltages, the current levels off rather than showing a change to trap-filled current characteristics. The leveling off of current suggests that saturation of the electrodes occurs. Such behavior is very similar to that reported by Helfrich (1967) for thick anthracene crystals when using aqueous I_2/KI electrodes. When the curves of Fig. 17 are analyzed based on space-charge-limited theory,

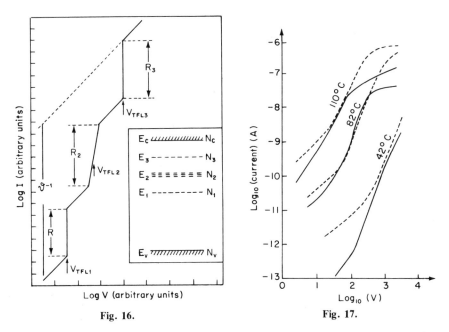

Fig. 16.   Fig. 17.

Fig. 16. Schematic I–V plot for a wide-bandgap semiinsulator with an arbitrary number (three) of discrete sets of traps, as these are filled completely in sequence. The inset shows energy levels for the conduction (N_c, E_c) and valence (N_v, E_v) bands as well as the three trap levels. The broken curve shows the trap-free square law behavior.

Fig. 17. Current–voltage curves for 50-μm polyethylene film with time and temperature as parameters. ---, 10 sec after applying potential; ——, 20 min after application of potential. [From Setter (1967).]

an effective mobility with an activation energy of 1.02–1.12 eV can be deduced. This suggests that a trapping level about 1.1 eV below the conducting state controls the conductivity. This could be caused by traces of impurities or oxidation products or be a function of the disorder in the system. Similarly, the current–voltage characteristics of poly(vinyl acetate) have been analyzed in terms of space-charge-limited currents (Chutia and Barva, 1980).

IV. CHARGE-CARRIER MOBILITY

A. The Influence of Traps on Charge-Carrier Mobility

Following the first reported direct study on charge-carrier mobility in poly-N-vinylcarbazole (Regensberger, 1968), studies of charge transport in polymers appeared with increasing frequency. Despite the complexity and breadth of the materials now studied, a number of consistent themes emerge. The charge-carrier mobility is low (10^{-2}–10^{-10} cm^2/V sec); it is thermally activated, is potential-dependent, and may not scale with sample thickness.

These observations strongly suggest that charge transport occurs by a hopping mechanism operating between localized sites. The potential dependence may arise as a consequence of lowering the potential barrier to detrapping. The traps may be a consequence of disorder, or disorder itself may contribute to the observed behavior.

There have been few attempts rigorously to relate trap sites to structure, morphology, or specific chemical treatments. However, valuable insights into the role of trapping by impurities have been obtained in a series of papers dealing with molecularly doped polymers (Mort, 1972; Mort *et al.*, 1976; Pfister, 1977; Pfister and Griffiths, 1978; Limburg *et al.*, 1978; Mort and Pfister, 1979). In such studies, the distance between dopant molecules can be (on the average) controlled by the concentration. The energetics of the host and dopant molecules can be selected so as to amplify the important energy-level considerations. In such studies, the anticipated concentration dependences are observed and

$$\mu \propto \rho^2 \exp(-\gamma\rho) \exp(-E/kT),$$

where ρ is the average separation of the dopant molecules, is observed. Here ρ is the average molecular separation given by $N^{-1/3}$, N being the concentration of dopant, and γ is the parameter describing the exponential decay of the molecular wave function with distance. The energy term E is related to the difference in energy between the initial and final states of the charge-carrier–site pair. According to the experimental results of Pfister (1977) the term μ/ρ^2 varies directly as exp $-\rho$.

Ideally a system with traps consists of a set of point traps contained within a continuous invariant matrix. In a real molecular matrix, this is not a justified assumption; the dopant molecules have finite size and are not spherical. A polymer is not a continuum at the molecular level. Typically the dopant molecules are 5–10 Å in diameter and are not spherically symmetrical. Depending upon the temperature, the polymeric matrix will undergo local density fluctuations. Thus changes in molecular orientation and intermolecular orbital overlap integral are likely. The overlap term between two dopant molecules can vary by orders of magnitude as a function of orientation and temperature. Similar considerations apply to a charge trapped at a group pendant to a polymer chain, trapped at a dipole on the polymer chain, or even trapped at a chain fold. The distance between and the energetics of sites in a polymeric matrix will be a fluctuating function of time. A temporal fluctuation in the hopping distance or energy difference between sites on the same or different polymer chains is easy to envision and would lead to a strong correlation among charge transport, detrapping, and the molecular relaxation properties of the polymer (Seanor, 1965). If such a mechanism existed, the waiting time between hops would be dominated by molecular motions, since the charge carrier would await the condition of maximum hopping probability before moving. The rate-limiting step would be controlled by local fluctuations and would be reflected in the observed activation energy of mobility. Certainly the activation energy of conduction is high, and frequently changes in the activation energy of conduction are correlated with the onset of a new molecular relaxation process. Self-polarization of the medium (polaron formation) and disorder can also add to the activation energy of conduction. Duke and Fabish (1976) have argued that trapping sites associated with chemical groups attached to the polymer backbone will not, because of different spatial configurations, have a unique energy. Rather, in a polymeric matrix, the energy of traps associated with side groups, such as the benzene rings in polystyrene, may be distributed over a range of as much as ± 1 eV. Such a model would lead to a prediction that localization would be more pronounced at lower temperatures. This appears to be the case for the Lexan^{TM}–triphenylmethane system (Pfister, 1977) and for polystyrene (Watson, 1978). In the case of electrons injected into polystyrene, charge transport is dominated by deep traps below T_g. Above T_g, deep traps no longer dominate the transport process and the transport exhibits features characteristic of disordered systems (*vide infra*).

Before discussing transport in disordered systems, it is appropriate to review the transport behavior of a molecularly doped polymer and to use this as a model for a polymer containing intrinsic traps. The system considered is Lexan polycarbonate doped with N-isopropylcarbazole (NIPC), triphenylamine (TPA), or combinations of both (Mort and Pfister, 1979). Studies were carried out by the time-of-flight technique using a selenium injection layer as a controlled source of charge carriers. It is assumed that transport involves the movement of an electron from the highest filled molecular orbital of a neighboring cation radical ion (an electron-deficient molecule).

First, it is established that in both NIPA–Lexan and TPA–Lexan the concentration dependence is as expected. Log μ/ρ^2 varies as (the number of dopant molecules)$^{-1/3}$ (Mort et al., 1976, Pfister, 1977). Since transport in such doped materials is by design unipolar, it is predicted that only molecules with an ionization potential less than that of the dopant molecule will act as trapping sites. Thus in Fig. 18, the strong dependence of hole drift velocity (solid circles) on TPA concentration at zero NIPC concentration reflects hopping between TPA molecules. At a fixed NIPC concentration, the drift velocity decreases with increasing TPA concentration (open circles in Fig. 18). Since the ionization potential of TPA is less

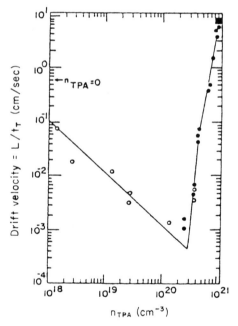

Fig. 18. The TPA concentration dependence of hole velocity in Lexan (●) and NIPC–Lexan where the NIPC concentration was fixed at 1×10^{21} molecules cm^{-3} (*) and 2×10^{20} molecules cm^{-3}. ○, $n_{\text{NIPC}} \cong 1 \times 10^{21}$ cm^{-3}; ●, $n_{\text{NIPC}} = 0$; ■, $n_{\text{NIPC}} = 2 \times 10^{20}$ cm^{-3}. [From Pfister et al. (1976).]

than that of NIPC, the velocity at TPA concentrations less than the NIPC concentration should be inhibited. At a fixed concentration of TPA, the mobility should be unaffected by the presence of the NIPC molecules. With no NIPC, the drift velocity increases by about four orders of magnitude as the distance between the TPA molecules decreases, indicating that transport between TPA molecules dominates. In NIPC-doped Lexan, transport between NIPC molecules dominates. As TPA is added to the NIPC-doped polymer, the drift velocity decreases, indicating that the TPA molecules act as hole traps. Since the overlap between TPA molecules is small, the charge carriers require thermal excitation to the NIPC level in order to be transported between NIPC molecules. As the TPA concentration increases, transport between TPA molecules becomes increasingly important until, at sufficiently high levels, transport between TPA pairs dominates. Under these conditions, transport becomes independent of the NIPC concentration.

These experiments clearly show the influence of molecular doping and trapping on the charge-transport process. Provided the experiments are not complicated by crystallization and interactions between the dopant molecule(s) and the matrix (or each other), such systems provide excellent models on which to build an understanding of the polymers themselves.

B. The Influence of Disorder on Charge-Carrier Transport

Conventional solid-state physics deals primarily with ordered (crystalline) materials. The disordered (amorphous) state received little attention until the late 1960s (Mott, 1967, 1969; Mott and Davies, 1971), when the discovery of switching in chalcogenide glasses gave impetus to study of the amorphous state. Central to theoretical treatments of the amorphous state is the concept of localized states within the band structure of the crystal as its order degenerates. To a great degree, it is the localized states within the band structure and, more importantly, within the forbidden energy gap, that determine the conductivity of the solid and that dominate the transport processes. The realization that most polymers have considerable disorder has led to the use of disordered solid-state concepts to help elucidate their electrical properties.

As charge carriers move through an amorphous solid, they experience a spatial and, possibly, a temporal fluctuation in local density, orientation, and energy of side groups and dipoles associated with the polymer. Thus, in a time-of-flight experiment, the resulting distribution of hopping times and waiting times causes the charge-carrier packet to disperse as it traverses the solid—some carriers arrive earlier than the average, some later. This results in the shape of the *i–t* traces shown in Fig. 19. Of particular note is the long tail. Analysis suggests that a gaussian distribution of hopping times would lead to a cutoff of the current pulse much earlier than is actually observed (Pfister and Scher, 1977, 1978). The position of the center of the gaussian distribution should vary linearly with

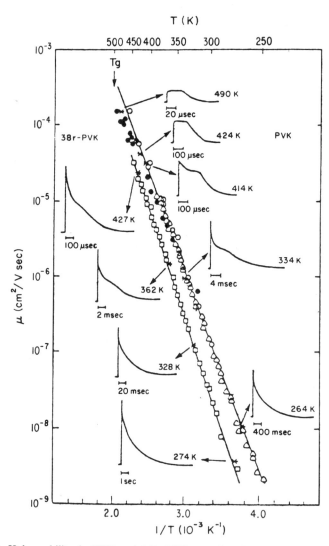

Fig. 19. Hole mobility in PVK and 3-Br-PVK as a function of temperature. Several representative hold current signals are shown. $E = 20$ V/μm and $L = 2.5$ μm. The loss of the transit CUSP is related to the system disorder. [From Pfister and Griffiths (1978).]

time, whereas the width should vary only as $t^{-1/2}$. In addition, a spreading of the pulse should not be seen for $t < t_{transit}$. Once the leading edge of the injected charge packet reaches the rear electrode, the transient current should drop in a manner reflecting the charge-carrier spatial distribution; i.e., with a gaussian distribution, in normalized time units $(t/t_{transit})$, the width of the tail should vary as $t^{-1/2}$. In most disordered solids, the current pulse is neither rectangular (a unique drift velocity) nor characteristic of the gaussian distribution. Dependent upon the specific solid and temperature, the current pulse may vary from a well-defined transit with nongaussian decay to a rather featureless monotonic decay curve. For example, the poly-N-vinylcarbazole (PVK) and 3-bromo-poly-N-vinylcarbazole (3-Br-PVK) traces shown in Fig. 19 indicate a well-defined transit for PVK above room temperature but no well-defined trace for 3-Br-PVK.

Scher and Montroll (1975) have provided a general theoretical framework within which to analyze this dispersive transient current. Furthermore, specific relationships between the pulse shape, its field dependence, and the sample thickness dependence of the transient pulse can be made when normalized in terms of $t/t_{transit}$ and $i/i_{transit}$. Their basic argument suggests that relatively small fluctuations in terms of hopping distance and site energy can introduce large variations in transport because of the exponential amplification via the $\exp[-(\gamma p)]\,\exp(-\Delta/kT)$ term of the transport equation.

The Scher–Montroll treatment uses a continuous-time, random-walk formulation in which the hopping time distribution function $\Psi(t)$ formulation dictates the outcome of the I–t curve. For example, if $\Psi(t)$ is given by $\exp -t/\tau$, conventional gaussian behavior results for $t > \tau$. However, if a power function, $\Psi(t) \propto t^n$, is used, non-gaussian transport results. Scher and Montroll use the formalism $\Psi(t) \propto t^{-(1+\alpha)}$ which results in the relationships

$$
\begin{aligned}
i(t) &\propto t^{-(1-\alpha)} \qquad (t < t_{transit}) \\
&\propto t^{-(1+\alpha)} \qquad (t > t_{transit}),
\end{aligned}
$$

where

$$
t_{transit} \sim [L/\rho(E)]^{1/\alpha}\,\exp(\Delta_0/kT),
$$

where L represents the sample thickness, $\rho(E)$ is the mean displacement per hop, Δ_0 is an energy term, and α is a constant which appears to be characteristic of the degree of disorder. The more disordered the system, the smaller is α; the more dispersive the transport, and the stronger the dependence on thickness and field. Note the predictions of the model:

1. The transit time will vary superlinearly with sample thickness. This has been difficult to show in polymers but has been convincingly demonstrated for hole transport in α-As_2Se_3 (Pfister and Scher, 1977, 1978).

2. Translation of the normalized current ($i/i_{transit}$), normalized time ($t/t_{transit}$) into a single master curve with a dependence of $-(1 - \alpha)$ for $t < t_{transit}$ and a $-(1 + \alpha)$ dependence for $t > t_{transit}$ as shown in Fig. 20.

3. The apparent voltage dependence of the transit time.

Given a knowledge of the microscopic transport, the mean displacement per hop can be calculated. The phenomenological relationship

$$\mu(E) \sim L^{-1/\alpha} \sinh(q\rho E^{1/\alpha}/2kT) \exp(-\Delta_0/kT)$$

has been established (where ρ is the hopping distance). This can be compared to the relationship characteristic of potential-assisted hopping between sites of energy $-U$:

$$\mu = 2a\nu \sinh(qaE/2kT) \exp[-(U/kT)],$$

where a is the jump distance and ν is the escape frequency (Pfister, 1977).

The transport in molecularly doped Lexan polycarbonate has successfully been explained in terms of dispersive transport (Pfister, 1977). Inconsistent differences between transport in poly-N-vinylcarbazole and 3-bromo-poly-N-vinylcarbazole have been explained in terms of morphological differences (Pfister and Griffiths, 1978). PVK is more crystal-

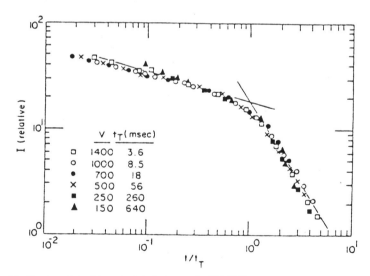

Fig. 20. Master plot of transient hole traces in NIPC–Lexan at room temperature. The traces were normalized to t_T and shifted along the log I axis for best superposition. 0.3 weight ratio; $L = 19.2$ μm; 296 K. [From Mort *et al.* (1976).]

line than 3-Br-PVK and shows a transition from dispersive to gaussian behavior at higher temperatures. 3-Br-PVK shows no signs of gaussian transport. Polystyrene below its glass transition has transport characterized by a single, deep trap, whereas in the rubbery state above the glass transition the dispersive mode appears to better describe the charge-carrier transport (Watson, 1978). Carrier mobility in electron acceptor doped polyethylene and polystyrene has been shown to increase with doping (Yoshino *et al.*, 1980). This is in contrast to the situation in poly-*N*-vinylcarbazole doped with strong electron acceptors. Consequently, it appears likely (although by no means proven) that subtle changes in polymer morphology can have a pronounced influence on the charge-transport characteristics. More work along these lines is required.

V. IONIC CONDUCTION

Ionic conduction has frequently been invoked to account for the low but finite conductivity of many of the more insulating polymers. In some cases, deviations from Ohm's law have also been explained in terms of ionic conductivity. However, as previously discussed, deviations from Ohm's law can arise from many causes, only one of which is ionic conductivity, and the numbers involved make unambiguous assignment of a conduction mechanism difficult.

On the other hand, there are polymeric systems in which ionic conductivity can be and has been unambiguously observed. These cases relate primarily to polymers that contain ions (ionomers, polyelectrolytes), contain groups capable of ionizing, or to which ionic materials have been added. In all such materials, the presence of water plays an important role, as shown in Fig. 21. Water may act as a source of ions, as a high-dielectric-constant impurity, as a plasticizer, or as a local structure modifier. In such cases, the definitive criterion for ionic conduction, that of mass transfer, can unambiguously be demonstrated.

For example, conduction in nylon 66 has been shown to involve proton transfer above 120°C (Seanor, 1968). Three temperature regimes were noted as shown in Fig. 22: (a) below 90°C where conductivity decreased with time and no gas evolution was detected, (b) a region between 90 and 120°C where conductivity was constant over long periods and nominal gas evolution was detected, and (c) a region above 120°C where the gas evolved was approximately half that calculated for complete protonic conduction and where the conductivity decreased with time. It was concluded that in the first two regions conductivity was electronic, but that in the last one conduction involved protons and electrons. Studies on dc conductivity of nylon 66 (in air and under reduced pressure) showed the importance of morphology in the conduction process, e.g., by determining

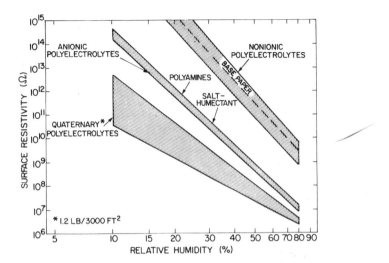

Fig. 21. The surface resistivity of groups of polyelectrolytes as a function of relative humidity. The particular anions and cations also play important roles. [From Hoover and Carr (1968).]

the conductivity of doubly oriented nylon as a function of direction (Seanor, 1967).

The proton-conduction mechanism proposed involved the self-dissociation of hydrogen-bonded amide groups and subsequent bond rearrangement.

Self-ionization:

$$\begin{bmatrix} O=C & N-H\cdots O=C & N-H \\ & & \end{bmatrix} \rightleftharpoons \begin{bmatrix} O=C & N^\ominus \end{bmatrix}\begin{bmatrix} H-O-C & {}^\oplus N-H \end{bmatrix}$$

$$\rightleftharpoons \begin{bmatrix} {}^\oplus_{O}-C & N + O=C & {}^\oplus NH_2 \end{bmatrix}$$

Equivalent steps occurring throughout hydrogen-bonded networks can be envisaged for the subsequent charge transport.

Proton transfer:

$$\begin{bmatrix} O=C & {}^\oplus NH_2\cdots O-C & N-H \end{bmatrix} \rightleftharpoons \begin{bmatrix} O=C & NH \quad HO-C & {}^\oplus NH \end{bmatrix}$$

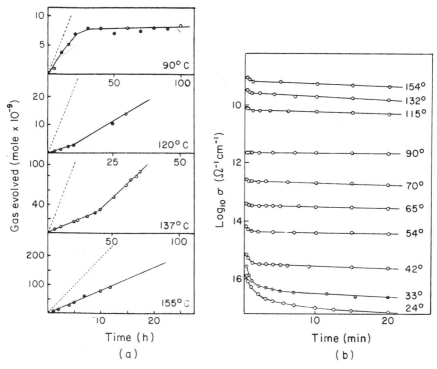

Fig. 22. Evidence for ionic and electronic conduction in nylon 66. (a) Gas evolution versus time. (b) Current versus time. [From Seanor (1968).]

Transfer of proton and electron:

$$\left[\begin{array}{c} N \\ \diagdown \\ C-O^{\ominus} \cdots\cdots H-N \\ \diagdown \\ C=O \end{array} \right] \rightleftharpoons \left[\begin{array}{c} N \\ \diagdown \\ C-OH \quad N^{\ominus} \\ \diagdown \\ C=O \end{array} \right] \rightleftharpoons \left[\begin{array}{c} NH \\ \diagdown \\ C=O \quad N \\ \diagdown \\ C-O^{\ominus} \end{array} \right]$$

A similar mechanism has been invoked for conduction in poly(olefin oxides) (Binks and Sharples, 1968).

$$\begin{array}{ccc} & & H-CH \\ | & | & | \\ H-CH-H & H-C-H & \oplus O-H \\ \diagdown & \diagdown & | \\ O & O & \longrightarrow H-C-H & + & \diagup C-H \\ \diagup & \diagup & \diagup & \diagdown \\ H-C-H & H-C-H & H-C-H & H-C-H \\ | & | & \diagup & \diagdown \end{array}$$

If there is sufficient segmental motion to allow the groups to take up energetically favorable spatial positions, charge transfer can take place, e.g., poly(methylene oxide) is highly crystalline, molecular motion is difficult, and the material is an insulator. As the number of adjacent methylene groups increases, the molecule becomes more flexible, the melting temperature drops, and the conductivity increases.

In general, there is good agreement between segmented motion and the conduction process (see Carr, Chapter 5). There is also limited evidence that the conductivity of many polymers is equivalent when compared at temperatures normalized to the glass-transition temperature. While not conclusive evidence for ionic conductivity, this suggests the involvement of segmental motion in the conduction process (Warfield and Hartman, 1980).

In ionomers (polymers in which the ionic group is part of the chain) and polymers doped with ionic impurities, ion migration can be studied directly. Because of the high viscosity, back-diffusion is limited and the boundary between ionic and nonionic regions can be directly observed.

Crowley *et al.* (1976) studied the migration of silver ions in ion-exchanged sulfonated polystyrene, as shown in Fig. 23. The conductivity, Ag^+ mobility, and Ag^+ concentration are shown in Fig. 23 as a function of the water concentration. Above 2% by weight water, the ion mobility increases, suggesting that plasticization of the polymer is important. Similarly, Seanor (unpublished) studied the migration of the colored cation of rhodamine F3B chloride in lightly linked polyurethane, concluding from the combination of migration kinetics and conductivity that the colored cation was responsible for the bulk (80%) of the charge transport.

In a related series of papers, Wallace (1971, 1974) has discussed the conductivity of sulfonated polystyrene in detail. Ion content, ion mobility, and water content are clearly related. Again, it appears that water aids dissociation, as well as increasing ion mobility at the higher levels. Further support for this view is added by the relationship of the activation energy of conduction and the volume of the hydrated cation (Crowley *et al.*, 1976). (See Fig. 24.)

The general relationship between conductivity and moisture content is of the form

$$\ln \sigma = \ln \sigma_{dry} - \alpha w_{ads}.$$

This has been a widely observed relationship, rationalized in part by relating the dielectric constant to the water uptake by an equation of the form

$$\alpha w_{ads} = A(1/\varepsilon_{dry} - 1/\varepsilon_{wet}),$$

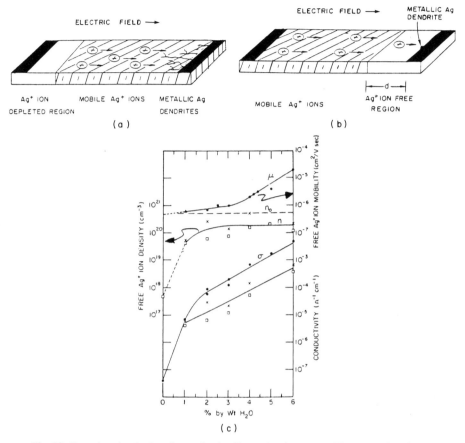

Fig. 23. Ion migration in the silver salt of sulfonated polystyrene. The upper drawings, (a) and (b), indicate the technique. (c) The graphs show the free-ion mobility, the carrier density, and the conductivity as a function of % H_2O by weight for Ag^+ counterion sample at 300 K. ●, high-voltage (>2.0 V) dc measurements; ×, low-voltage (<2.0 V) dc measurements; □, ac measurements; ▲, mobility measurements requiring the low-water-content technique. [From Crowley *et al.* (1976).]

where α and A are constants, w_{ads} is the weight of the water per unit mass of polymer, and ε is the dielectric constant (Rosenberg, 1962a).

On the other hand, Eley and Leslie (1964) derived an equation,

$$\log \sigma = \text{const} + \log(w_{ads})^{1/2} + Bw_{ads}^{x},$$

that gives almost as good a fit to the experimental data. This equation is derived from a model in which water contributes charge carriers.

Lewis and Toomer (1981) have used the time-of-flight technique to

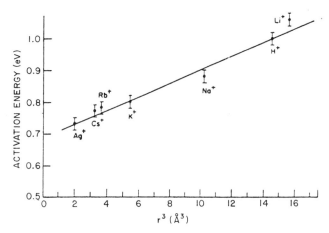

Fig. 24. Activation energy for conduction in dry ion-exchanged sulfonated polystyrene membranes as a function of the cube of the hydrated ion radius. [From Crowley *et al.* (1976).]

study charge transport in hydrated bovine serum albumin. Using both negatively and positively charged ions, they obtained clear evidence for dispersive charge transport. The macroscopic mobility appeared to *increase* with increasing hydration following the relationship

$$\mu = \mu_{\mathrm{d}} \exp(\beta m),$$

where μ_{d} is the mobility in the dry state, m (wt %) is the level of hydration, and β is a constant. Their studies indicate that the disorder parameter α decreases with increasing hydration. At high levels of hydration the experimentally determined mobility agrees well with theoretical calculations (Kertesz *et al.*, 1978; Sohai, 1974). Charge injection, i.e., the transport of electrons or positive holes, is suggested to occur. In view of the high macroscopic mobility the transport of the deposited ions is considered unlikely. Thus it seems that one effect of water adsorption is to aid the transport of charge as well as to modify the local structure.

The effect of water on the conduction process is by no means clear, particularly if electronic conduction is involved. The local dielectric constant will increase. The effect of moisture on the mechanical properties of most polymers indicates that plasticization occurs, and one could speculate that the charge-carrier hopping distance or path length would increase. Water can also act as an electron donor and source of charge carriers. The complexity of the problem is illustrated by the apparent change from electronic to ionic conduction in hydrated proteins as the water content increases (Rosenberg, 1962b, 1964).

Ionic conduction in polymers does not appear to follow simple theoretical analysis. Waldren's rule relating charge mobility to viscosity is not obeyed. Nor are the simple concepts of electrolyte conduction, which relate conductance to ion concentration [see, e.g., Szwarc (1968)], obeyed by ionomers or salt-doped polymers.

Waldren's rule relates the viscous drag on an ion to the accelerating force of the electric field. At equilibrium, the viscous force f_v acting on an ion of radius R is given by Stokes's law:

$$f_v = 6\pi\eta Rv,$$

where η is the viscosity and v is the effective velocity ($=\mu E$). The electrical force f_e is

$$f_e = zqE,$$

where z is the number of charges on the ion. At equilibrium, $f_e = f_v$ and

$$zqE = 6\pi\eta R\mu E.$$

If we put

$$\sigma = qn\mu,$$

then

$$\sigma\eta = z(q^2n/6)\pi R.$$

The drift mobility μ is related to the diffusion constant D by the Einstein equation

$$\mu/D = q/kT.$$

Substituting for $\mu = \sigma/nq$, we have the Nernst–Einstein equation

$$\sigma/D = nq^2/kT.$$

The Einstein equation is useful in obtaining an estimate of the mobilities involved. For many polymers, D is typically on the order of 10^{-8} cm² sec⁻¹. For $kT \approx 600$ cal ($T = 27°$ C), $\mu \sim 4 \times 10^{-7}$ cm²/V sec, which is very close to observed values of the mobility. For nonohmic ionic conduction, the mobility depends upon the applied field; i.e., $\mu = \mu_0 f(E)$, which gives $qD/kT = \mu_0 f(E)$. Combined with the equation $\mu = 2av$ $\sinh(qaE/2kT) \exp-(V/kT)/E$, representative of field-assisted ion migration,

$$f(e) = 2 \sinh(aqE/2kT)kT/qaE.$$

This equation has been utilized by Barker and Thomas (1964a, b) to estimate the distances between potential wells in alkali-halide-doped cellulose acetate. Jump distances of 80–200 Å were calculated that were the same order of magnitude as those obtained by Amborski (1962) for poly(ethylene terephthalate) and by Foss and Dannhauser (1963) for polypropylene. The latter authors also used calculations based on deviations from Ohm's law to obtain the jump distance. Deviations from Ohm's law ohmic by field-assisted ionic dissociation have also been discussed by Barker and Thomas (1964a, b), who concluded that this effect was important only at very high fields ($>10^6$ V cm^{-1}).

The available evidence does not agree well with the predictions of Waldren's rule. In general, the $\sigma\eta$ product is not constant over a large range of σ and η. If we write $\sigma = q\mu_0 \exp[-(E\mu/kT)]n_0 \exp(-En/kt)$ and $\eta = \eta_0 \exp(E_\eta/kTm)$, then $\sigma\eta = \sigma_0\eta_0 \exp[-(E_\mu + E_\eta - E_\eta)/kT]$. Plots of ln $\sigma\eta$ versus $1/T$ should be linear. The data of Kallweit (1966) for plasticized poly(vinyl chloride) do not show the anticipated linear relationships (Fig. 25). A plot of ln $\sigma\eta$ versus $1/T$ exhibits a change in sign of the overall activation energy, as well as a region independent of temperature. In part, this may be understood on the basis of the molecular processes occurring in the polymeric matrix and the way in which the viscosity is measured. Strictly speaking, the viscosity required in Stokes's law equation is that of the local environment in which the charge carrier finds itself. In the case of liquids, a macroscopic measurement is adequate because the molecular dimensions of the medium are approximately the same as those of the ion, and it is adequate to assume that the matrix behaves as a hydrodynamic continuum. In the case of polymers in the solid state, this is not a good assumption. There are holes in the matrix equivalent in size to the dimensions of the ions. Consequently, local density fluctuations can dominate ionic motion. Macroscopic viscosity measurements will not reflect local effects. It might, however, be possible to use ion motion as a probe in examining local structure.

Considerations of the effect of ions on local structure impact polymer mechanical properties, which in turn are reflected in the electrical properties (Barker and Thomas, 1964a). For example, the addition of ions to cellulose acetate not only increased conductivity (Barker and Thomas, 1964b) but also increased the glass-transition temperature of the polymer and the activation energy of conduction. The size of the ion and its polarizability were also shown to be important. Barker (1976) discusses the local structure hypothesis in detail.

The relationship between ionic concentration and conductivity tends to follow a $c^{1/2}$ relationship indicative of weak electrolyte behavior (Barker and Thomas, 1969b). However, this is not always the case. Price and Dannhauser (1967), for example, suggested that ion interactions played an important role in the conduction processes of the polystyrene–phenyl-

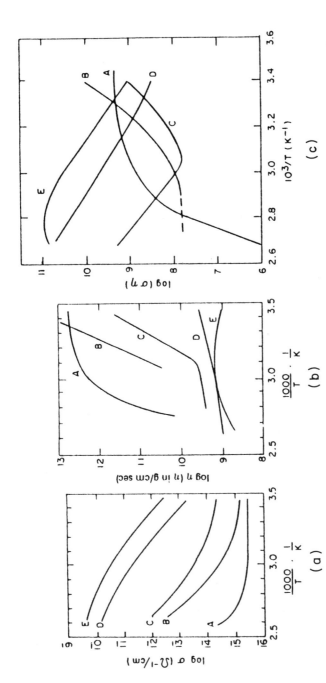

Fig. 25. Conductivity σ, and viscosity η of plasticized poly(vinyl chloride). (a) Log σ versus $1/T$; (b) log η versus $1/T$; (c) log $\sigma\eta$ versus $1/T$. Percent plasticizer by weight: $A = 0$, $B = 8$, $C = 16$, $D = 26$, $E = 27$ (Kallweit, 1966).

methane–tetra-*n*-butylammonium thiocyanate system and that field-assisted dissociation may occur (Onsager, 1934).

Analysis based on electrolyte theory is possible but is flawed by the following considerations:

1. Theories are based on continuum concepts. At the molecular level, polymers do not provide the required continuous matrix either spatially or dielectrically, as shown in Fig. 26.

2. Adequate analysis requires some understanding of effective ion radius. In hydrated systems, it may be adequate to use the radius of the hydrated ion.

3. A knowledge of the influence of the ion or ion pair on local structure is required.

4. A knowledge of the local dielectric environment is also required, since the dielectric constant enters into the equations describing salt dissociation. This is particularly relevant to polymers containing polar and nonpolar regions, as well as to amorphous–crystalline materials. Barker (1976) suggests that the ion would try to minimize its Gibbs's free energy by migration to regions of high dielectric constant (the crystalline regions rather than the amorphous). However, because of the higher energy required to create a hole in the crystallite into which the ion can migrate, it will likely remain at the interface.

5. In the case of polyelectrolytes, space-charge effects must be considered. In such materials, one ion is free to move, but the counterion is fixed in space. As the mobile ions migrate, a space charge must build up at the electrode of similar sign and the field distribution will become distorted.

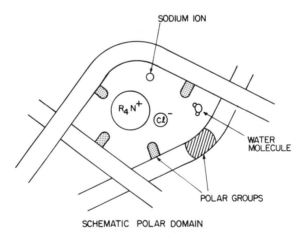

Fig. 26. Schematic ionic domain approximately to scale.

Overall ionic conduction has been studied in some cases, but detailed discussions at the molecular level are still required.

VI. SUMMARY

It has been the intent of this chapter to emphasize the generalities of polymer conductivity rather than the specifics of each and every material. Despite the complexity of polymeric materials, it appears that it is possible to treat electrical conduction in terms of polymer molecular structure and morphology. No longer is it necessary to invoke impurities to account for each material. The concepts of trapping and the involvement of pendant side groups, dipoles, and chain folds, as well as establishment of the role of disorder, can be used equally well to account for the electrical behavior of the more insulating materials.

The recent discovery of the highly conducting doped compounds of polyacetylene, poly(phenylene sulfide), and graphite adds an exciting new dimension to the properties of carbon-based materials. This aspect has barely been touched upon in this chapter. The field has advanced so rapidly that any review would quickly be outdated [see Street and Clarke (1981) and references therein, *Journal of Synthetic Metals* (1980, 1981)]. The future holds exceptional promise for these low-density, highly conductive materials.

REFERENCES

Amborski, L. E. (1962). *J. Polym. Sci.* **62**, 331.
Barker, R. E. (1976). *Pure Appl. Chem.* **46**, 157.
Barker, R. E., and Thomas, C. R. (1964a). *J. Appl. Phys.* **34**, 87.
Barker, R. E., and Thomas, C. R. (1964b). *J. Appl. Phys.* **35**, 3403.
Bartlett, N., Biagioni, R. N., McQuillan, B. W., Robertson, A. S., and Thompson, A. C. (1978). *Chem. Commun.* 200.
Baughman, R. H. (1977). *In* "Contemporary Topics in Polymer Science" (E. M. Pearce and J. R. Schaefgren, eds.), Vol. 2, p. 205. Plenum Press, New York.
Baughman, R. H., and Chance, R. R. (1978). *Ann. N.Y. Acad. Sci.* **313**, 705.
Binks, A. E., and Shapples, A. (1968). *J. Polym. Sci.* **A2** (6), 407.
Blumenthal, R. N., and Seitz, M. A. (1974). "Electrical Conductivity in Ceramics and Glass" (N. M. Tillen, ed.). Dekker, New York.
Briegleb, G. (1964). *Angew. Chem. Int. Ed. Engl.* **3**, 617.
Chen, I. (1972). *J. Appl. Phys.* **43**, 1137.
Chen, I., Emerald, R. L., and Mort, J. (1973). *J. Appl. Phys.* **44**, 3490.
Chiang, C. K., Gau, S. C., Fincher, C. R., Part, Y. W., MacDiarmid, A. G., and Heeger, A. J. (1978). *Appl. Phys. Lett.* **33**, 18.
Chutia, J., and Barva, K. (1980). *J. Phys. D. Appl. Phys.* **13**, L9.
Crowley, J. R., Wallace, R. A., and Bube, R. H. (1976). *J. Polym. Sci. Polym. Phys. Ed.* **14**, 1769.

Davies, D. K. (1969). *Br. J. Appl. Phys.* **2**, 1533.

Duke, C. B. (1978). *Surf. Sci.* **70**, 674.

Duke, C. B., and Fabish, T. J. (1976). *Phys. Rev. Lett.* **37**, 1075.

Eley, D. D., and Leslie, D. (1964). *Adv. Chem. Phys.* **7**, 238.

Eley, D. D., and Parfitt, G. D. (1961). *Trans. Faraday Soc.* **57**, 2280.

Eley, D. D., and Willis, M. R., (1961). *Symp. Elec. Conduct. Organ. Solids* (H. Kallman and M. Silver, ed.), p. 257. Wiley (Interscience), New York.

Fabish, T. J. (1979). *Crit. Rev. Solid State Mater. Sci.* **8**, 383.

Florey, P. J. (1953). "Principles of Polymer Chemistry." Cornell Univ. Press, Ithaca, New York.

Forsman, W. C. (1977). *Extended Abstr.: Biennial Conf. Carbon, 13th* p. 153.

Forster, E. O. (1969). "Physical Methods of Macromolecular Chemistry" (B. Carroll, ed.), Vol. 1, Chapter 3. Dekker, New York.

Foss, R. A., and Dannhauser, W. (1963). *J. Appl. Polym. Sci.* **7**, 1015.

Fox, K., and Turner, J. E. (1966). *J. Chem. Phys.* **45**, 1142.

Friedman, L. (1964). *Phys. Rev.* **133A**, 1168.

Gray, T. J. (1969). *Exp. Methods Catal. Res.* **1**, 293.

Gregor, L. V. (1968). *Thin Solid Films* **2**, 235.

Gregor, L. V., and Kaplan, L. V. (1968). *Thin Solid Films* **2**, 95.

Gutman, F., and Lyons, L. E. (1967). "Organic Semiconductors." Wiley, New York.

Hansma, P. K. (1977). *Phys. Rev.* **C30**, 145.

Helfrich, W. (1967). "Physics and Chemistry of the Organic Solid State" (D. Fox, M. M. Labes, and A. Weissberger, eds.), Vol. 3, Chapter 1. Wiley (Interscience), New York.

Helfrich, W., and Schneider, W. G. (1965). *Phys. Rev. Lett.* **14**, 229.

Hoover, M. F., and Carr, H. E. (1968). *Tappi* **51**, 552.

Kallweit, J. H. (1966). *J. Polym. Sci. Part A* **1**, 337.

Keil, R. G., Graham, T. P. H., and Roenker, K. P. (1976). *Appl. Spectrosc.* **30**, 1.

Kepler, R. G. (1968). *Phys. Rev.* **119**, 1226.

Kertesz, M., Koller, J., and Azman, A. (1978). *Phys. Rev. B* **18**, 5649.

Kronic, P. L., Kaye, H., Chapman, E., Maintha, S. B., and Labes, M. M. (1962). *J. Chem. Phys.* **36**, 2235.

Kryszewski, M., and Zymanski, A. (1970). *J. Polym. Sci.* **14**, 245.

Lakatos, A., and Mort, J. (1968). *Phys. Rev. Lett.* **21**, 1444.

Lampert, M. (1964). *Rep. Prog. Phys.* **27**, 329.

Lampert, M., and Mark, P. (1970). "Current Injection in Solids." Academic Press, New York.

Lardon, M., Lell-Döller, E., and Weigl, J. W. (1967). *Mol. Cryst.* **2**, 241.

LeBlanc, O. H. (1961). *J. Chem. Phys.* **35**, 1275.

Lewis, T. J., and Toomer, R. (1981). *J. Chem. Soc., Faraday Trans.* **1** (77), 2087.

Lily, A. C., Jr., and McDowell, J. R. (1968). *J. Appl. Phys.* **39**, 141.

Limburg, W. *et al.* (1978). *Organ. Coat. Plast. Chem.* **38**, 534.

Lyons, L. E. (1963). "Physics and Chemistry of the Organic Solid State" (D. Fox, M. M. Lares, and A. Weissberger, eds.), Vol. 1, Chapter 13. Wiley (Interscience), New York.

McCrum, N. G., Read, B. E., and Williams, A. (1967). "Anelastic and Dielectric Effects in Polymeric Solids." Wiley, New York.

McCubbin, W. L., and Manne, R. (1968). *Chem. Phys. Lett.* **2**, 230.

Miyoshi, U., and Chino, K. (1967). *Jpn. J. Appl. Phys.* **6**, 191.

Mort, J. (1972). *Phys. Rev. B* **5**, 3329.

Mort, J., and Lakatos, A. (1970). *J. Non-Cryst. Solids* **4**, 117.

Mort, J., and Pfister, G. (1979). *Polym./Plast. Technol. Eng.* **12**, 89.

Mort, J., Chen, I., Emerald, R. L., and Sharpe, J. (1972). *J. Appl. Phys.* **43**, 2285.
Mort, J., Pfister, G., and Grammatica, S. (1976). *Solid State Commun.* **18**, 693.
Mott, N. F. (1967). *Adv. Phys.* **16**, 49.
Mott, N. F. (1969). *Contemp. Phys.* **10**, 125.
Mott, N. F., and Davies, E. A. (1971). "Electronic Processes in Non-Crystalline Materials." Oxford Univ. Press (Clarendon), London and New York.
Mott, N. F., and Gurney, R. W. (1940). "Electronic Processes in Ionic Crystals." Oxford Univ. Press (Clarendon), London and New York.
> Norman, R. H. (1970). "Conductive Rubber and Plastics," Chapters 3–5. Elsevier, Amsterdam.
Onsager, L. (1934). *J. Chem. Phys.* **2**, 599.
Pfister, G. (1977). *Phys. Rev.* **816**, 3676.
Pfister, G., and Griffiths, C. H. (1978). *Phys. Rev. Lett.* **40**, 659.
Pfister, G., and Scher, H. (1977). *Phys. Rev. B* **15**, 2026.
Pfister, G., and Scher, H. (1978). *Adv. Phys.* **27**, 747.
Pfister, G., Mort, J., and Grammatica, S. (1976). *Phys. Rev. Lett.* **37**, 1360.
Price, J. W., and Dannhauser, W. (1967). *J. Phys. Chem.* **71**, 3530.
Regensburger, P. J. (1968). *Photochem. Photobiol.* **8**, 429.
Reimer, B., and Bässler, H. (1975). *Phys. Status Soliol.* **32a**, 435.
Reimer, B., and Bässler, H. (1976). *Chem. Phys. Lett.* **43**, 81.
Reucroft, P. J., Scott, H., and Serafin, F. L. (1970). *J. Polym. Sci. Part C* **30**, 261.
> Roland, M., Bernier, P., Lefrant, S., and Aloissi, M. (1980). Polymer **21**, 111.
Rose, A. (1955). *Phys. Rev.* **97**, 1538.
Rosenberg, B. (1962a). *J. Chem. Phys.* **36**, 816.
Rosenberg, B. (1962b). *Biopolym. Symp.* **1**, 453.
Rosenberg, B. (1964). *Nature (London)* **193**, 364.
Scher, H., and Montroll, E. W. (1975). *Phys. Rev. B* **12**, 2455.
Seanor, D. A. (1965). *Adv. Polym. Sci.* **4**, 317.
Seanor, D. A. (1967). *Proc. 1967 NAS/NRC Conf. Insulat. Dielec. Phenomean* p. 12.
Seanor, D. A. (1968). *J. Polym. Sci. Part A* **2**, 463.
Seanor, D. A. (1969). "Thermal Methods of Polymer Analysis" (P. J. Slade and L. Jenkins, eds.), Vol. 2, p. 293. Dekker, New York.
Seanor, D. A. (1976). "Photoconductivity in Polymers" (A. V. Patsis and D. A. Seanor, eds.), Chapter 3. Technomic Press, Westport, Connecticut.
Setter, G. (1967). *Kolloid Z.* **215**, 112.
Sharp, J. W. (1967). *J. Phys. Chem.* **71**, 2587.
Sharples, A. (1972). "Polymer Science" (A. D. Jenkins, ed.), Vol. 1, Chapter 4. North-Holland Publ., Amsterdam.
Shirakawa, H., Louis, E. J., MacDiarmid, A. G., Chaing, C. K., and Heeger, A. J. (1978). *Chem. Commun.* 578.
Siddiqui, A. S. (1980). *J. Phys. C Solid State Phys.* **13**, L 1079.
Simonsen, M. G., and Coleman, R. V. (1973). *Phys. Rev. B* **8**, 5875.
> Street, G. B., and Clarke, T. C. (1981). *IBM J. Res. Dev.* **25**, 51.
Spear, W. E. (1969). *J. Non-Cryst. Solids* **1**, 197.
Suhai, S. (1974). *Theor. Chim. Acta.* **34**, 157.
Szwarc, M. (1968). "Carbanions, Living Polymers and Electron Transfer Processes." Wiley (Interscience), New York.
Taylor, D. M., and Lewis, T. J. (1971). *J. Phys. D Appl. Phys.* **3**, 1346.
Tredgold, R. H. (1966). "Space Charge Conduction in Solids." Elsevier, Amsterdam.
Turner, D. W. (1966). *Adv. Phys. Org. Chem.* **4**, 31.

Van Turnhout, J. (1974). "Thermally Stimulated Discharge of Polymeric Electrets." Elsevier, Amsterdam.
Vilesov, F. I. (1964). *Sov. Phys.—Usp.* **6**, 888.
Vogel, F. L. (1977). *J. Mater. Sci.* **12**, 982.
Wallace, R. A. (1971). *J. Appl. Polym. Sci.* **42**, 3121.
Wallace, R. A. (1974). *J. Appl. Polym. Sci.* **18**, 2855.
Warfield, R. W., and Hartman, B. (1980). *Polymer* **21**, 31.
Watson, P. K. (1978). "Polymer Surfaces" (D. T. Clark and W. J. Feast, eds.), p. 91. Wiley, New York.
Williams, D. J. (1976). "Photoconductivity in Polymers" (A. V. Patsis and D. A. Seanor, eds.), p. 278. Technomic Press, Westport, Connecticut.
Williams, R., and Dresner, J. (1967). *J. Chem. Phys.* **46**, 2133.
Wolf, E. L. (1978). *Rep. Prog. Phys.* **41**, 1440.
Yoshino, K., Harada, S., Kyokane, J., Iwakawa, S., and Inuishi, Y. (1980). *J. Appl. Phys.* **51**, 2714.

BIBLIOGRAPHY

Barker, R. E. (1976). *Pure Appl. Chem.* **46**, 157.
Baughman, R. H. (1977). The solid state synthesis of photoconductivity, metallic and superconducting polymer crystals, *Contemp. Topics Polym. Sci.* **2**, 205.
Duke, C. B. (1978). Polymers and molecular solids: new frontiers in surface science, *Surf. Sci.* **70**, 674.
Duke, C. B. (1979). Localization in molecular solids, *Proc. 5th Int. Conf. Chem. Organ. Solid State, Mol. Cryst. Liq. Cryst.* **50**, 63.
Fabish, T. J. (1979). The electronic structure of polymers, *Critical Rev. Solid State Mater. Sci.* **8**, 383.
Goodings, E. P. (1976). Conductivity and superconductivity in polymers, *Chem. Soc. Rev.* **5**, 95.
Gutman, F., and Lyons, L. E. (1967). "Organic Semiconductors." Wiley, New York.
Kryszewski, M., and Szymanski, A. (1970). Space charge limited currents in polymers, *J. Polym. Sci.* **14**, 245.
Labes, M. M., Love, P., and Nichols, L. F. (1979). Polysulfur nitride-A. Metallic superconducting polymer, *Chem. Rev.* **79**, 1.
Lewis, T. J. (1977). Charge transport in polymers, *1976 Annu. Rep. Conf. Elec. Insulat. Dielec. Phenomena* p. 533. National Academy of Sciences, Washington, D.C.
Miller, J. S. (ed.) (1978). "Synthesis and Properties of Low-Dimensional Materials," *N.Y. Acad. Sci.* **313**.
Mort, J. (1980). Polymers as electronic materials, *Adv. Phys.* **29**, 367.
Mort, J., and Pfister, G. (1979). Photoelectronic properties of disordered molecular solids, *Polym./Plast. Technol. Eng.* **12**, 89.
Patsis, A. V., and Seanor, D. A. (ed.) (1976). "Photoconductivity in Polymers An Interdisciplinary Approach." Technomic Press, Westport, Connecticut.
Perlstein, J. J. (1977). Organic metals, *Angew. Chem. Int. Ed.* **16**, 519.
Seanor, D. A. (1972). "Polymer Science" (A. D. Jenkins, ed.), Chapter 17, p. 1187. North-Holland Publ., Amsterdam.
Stolka, M., and Pai, D. (1978). Polymers with photoconducting properties, *Adv. Polym. Sci.* **29**, 1.
Vogel, F. L. (ed.) (1977). Graphite intercalation compounds, *J. Mater. Sci.* **31**.

Chapter 2

Structure and Charge Generation in Low-Dimensional Organic Molecular Self-Assemblies

Jerome H. Perlstein
RESEARCH LABORATORIES
EASTMAN KODAK COMPANY
ROCHESTER, NEW YORK

I. INTRODUCTION

Nearly 20 yr have passed since the discovery of tetracyanoquino-dimethane (TCNQ) by du Pont workers (Melby *et al.*, 1962), and almost 25 yr since the discovery of the first organic conductors by Matsunaga and co-workers in Japan (Akamatu *et al.*, 1956). Within this period there has been an explosion in the study of charge-transport processes in organic materials, most notably in dark conduction but also in photoconduction. Much of this research stems from two apparently unrelated sources: (a) the attempts by physicists to establish a theoretical picture of the nature of charge transport in crystalline organic molecular solids (Berlinsky, 1976; Toombs, 1978) and (b) the technological aspects asso-

59

ciated with the development of new organic photoconductive devices, particularly in electrophotography (Schaffert, 1975), and organic photo-voltaic devices (Ghosh and Fent, 1978; Morel *et al.*, 1978; Merritt and Hovel, 1976).

It is interesting to note the rather strong divergence in the studies on dark-conducting materials versus photoconductors. The former have been almost exclusively studied by an examination of the chemistry and physics of donor–acceptor interaction (Perlstein, 1977; Torrance, 1979), with a few studies only recently undertaken on polymeric dark conduction (Chiang *et al.*, 1978a; Perlstein *et al.*, 1978; Street *et al.*, 1981), whereas the latter have been studied in great detail in amorphous polymeric systems (Patsis and Seanor, 1976) and the few thorough studies on single crystals have been limited to aromatic polyacene systems, particularly anthracene (Castro, 1971).

These two areas of research, however, are now beginning to converge in ways not entirely surprising to those who have been exposed to both areas of research simultaneously. On the one hand, crystalline polymeric materials with small amounts of disorder have been shown to have ground states that are either superconducting as in $(SN)_x$ (Kamimura, 1976) or metallic as in doped polyacetylenes (Chiang *et al.*, 1978b). On the other hand, amorphous polymers that incorporate small amounts of crystalline order have been shown to have orders of magnitude enhancement of their photoconductive charge-generation efficiencies (Dulmage *et al.*, 1978). It will thus not come as any great surprise if future technological and scientific advances in charge generation and transport of organic solids come from the study of the chemistry and physics of these processes in polymeric systems.

Future advances, for example, in organic photovoltaic devices as discussed in Chapter 4, will require an understanding of the dark charge-transport process as well as of the photoconductive charge-generation process. There do not appear to be any theoretical reasons why organic photovoltaic devices cannot have power-conversion efficiencies approaching those of inorganic materials, with the architectural advantages associated with organic molecular design.

Associated with the complexity of organic molecular structure is a richness of properties not always present in simple inorganic materials. They can have the properties of a metal (Ferraris *et al.*, 1973) and the magnetic properties of an insulator (Perlstein *et al.*, 1972; Scott *et al.*, 1978) at the same time. They can have optical properties that can be tailor-made for absorption at any wavelength in the infrared, visible, or ultraviolet (Griffiths, 1976). If cast as polymers, they can have properties of metals as well as many of the important physical characteristics of polymeric films

(Street *et al.*, 1981; Jeszka *et al.*, 1981). Organic materials can be stretched, oriented, and molded; they can be made *p*-type or *n*-type; they can be used as lasers; they can, in fact, do many of the things that their inorganic cousins can do and sometimes do them better.

In this chapter I wish to consider one small area of this gold mine, namely, the unusual physical properties in the generation and transport of charge when organic molecules are assembled in ordered arrays, especially as this occurs within polymeric matrices.

Many of the ideas expressed here are based on my thinking on charge-transport processes in organic materials over a period of many years. As such, some of these thoughts are speculative, but it is hoped that such speculations will lead to new experiments that will shed light on the nature of the structure and properties of these fascinating materials.

II. SELF-ASSEMBLIES OF ORGANIC MOLECULES: THE PHOTOCONDUCTIVE STATE

A. Introduction

A self-assembly of organic molecules into complex but regular organizational structures forms the basis for many of the unusual properties such molecules display. In biology they are typified by the supramolecular structures of fatty acids and various micelle-type organizations, e.g., phospholipids in various membrane structures (Tredgold, 1977). In polymer and biopolymer chemistry, they often form helical structures [e.g., DNA, poly(γ-benzyl glutamate)] (Liu *et al.*, 1967). In silver halide photography they form aggregates on the surface of silver halide grains (Emerson *et al.*, 1967), and in solid-state chemistry and physics they form the homosoric stacks that produce the unusual electrical and magnetic properties associated with quasi-one-dimensional conductors (Dahm *et al.*, 1975).

In this chapter, we wish to consider the self-assemblies of organic molecules that lead to a high quantum efficiency for photocharge generation, also known as aggregate photoconductors (Dulmage *et al.*, 1978).

B. Structural, Optical, and Electrochemical Properties of Thiapyrylium Assemblies

Figure 1 (color plate) shows an example of such a system (Dulmage *et al.*, 1978). It consists of a quasi-two-dimensional stacked array of the thiapyrylium dye molecule (I).

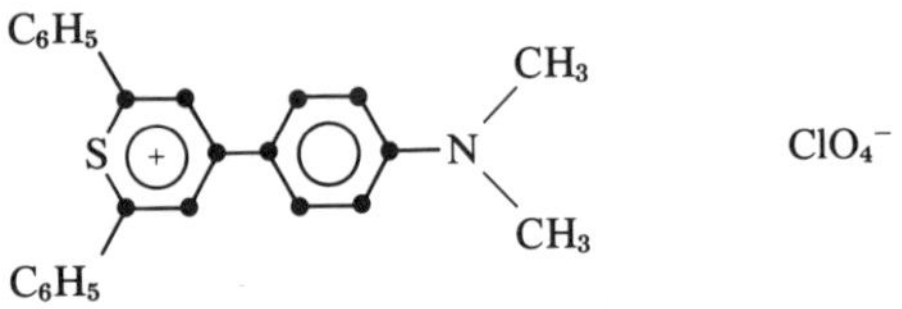

(I)

Each layer of dye molecules is separated from other layers along the *b* axis by layers of bisphenol A polycarbonate molecules (II). The dye inter-layer spacing is 15.8 Å.

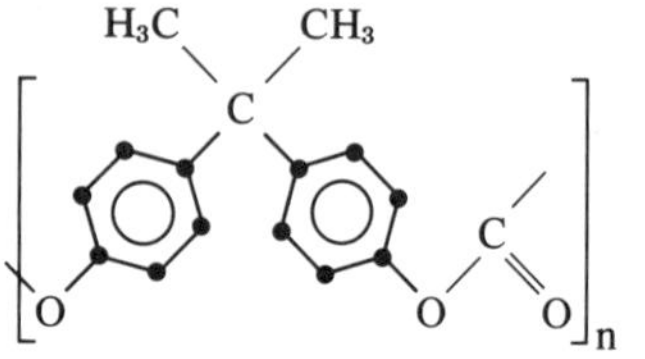

(II)

Figure 2 (color plate) shows the same crystal structure, but projected down the *b* axis where the arrangement of the dye molecules within a layer is more apparent. Within each layer, the molecules form a quasi-one-dimensional assembly along the *a* axis with an intermolecular spacing of 3.44 Å.

The x-ray crystal structure of this unusual complex between dye and polymer was determined from crystals of (I) complexed with the diphenyl carbonate of bisphenol A (III). The monoclinic crystals belong to space

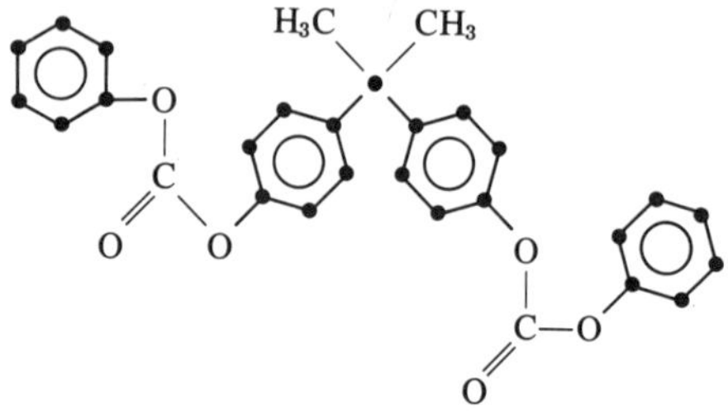

(III)

group $P2_1/a$, with $a = 11.01$ Å, $b = 31.64$ Å, $c = 10.75$ Å, and $\beta = 112.7°$.

The similarity between the optical properties of this crystalline complex and the optical properties of the complex formed with the polymer (II) suggests that at least the quasi-one-dimensional structures of both are identical. From the structure, it is to be expected that the intermolecular interactions responsible for the absorption and photoconduction will be strongest along the a axis and weakest along the b axis.

Chemically, the thiapyrylium dye (I) and many other structures like it, including cyanine dyes, have some unusual properties. From a redox point of view (I) is amphoteric. It consists of a donor portion, D, the dimethylaniline (IV), coupled to an acceptor portion, A, the thiapyrylium ring (V).

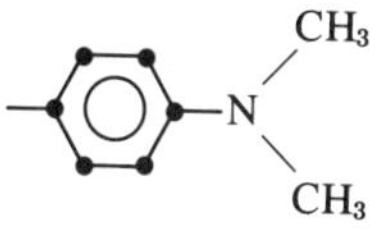

(IV)

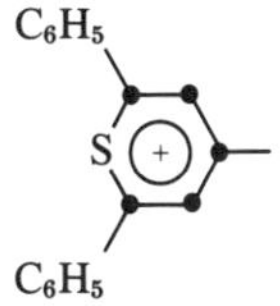

(V)

In fact, the peak oxidation potential of (I) is $+1.24$ V versus SCE in CH_3CN (Chang and Mina, private communication), whereas its peak-reduction potential in CH_3CN is only -0.52 V (Large and Beal, private communication). The difference between the oxidation and reduction potentials of $+1.76$ V corresponds to a photon energy of ca. 14,000 cm^{-1}, which is in the visible portion of the optical spectrum.

The dye can thus be represented as a strong coupling between a donor and an acceptor as in (VI) (Griffiths, 1976):

$$A^+ - D.$$
(VI)

Optical excitation of (I) can produce the excited singlet state by intramolecular charge transfer, as shown by the limiting structures in (VII).

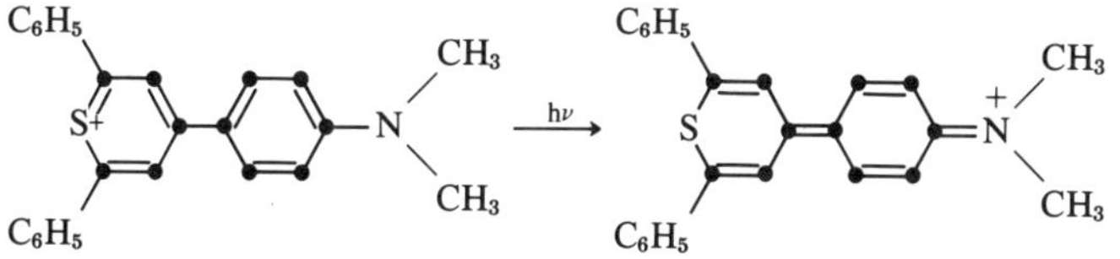

$$A^+ - D \qquad\qquad A - D^+$$
$$^1S_0 \qquad\qquad\qquad ^1S_1$$

(VII)

Generally such intramolecular charge transfer produces dye molecules with a very large molar extinction coefficient ε and large oscillator

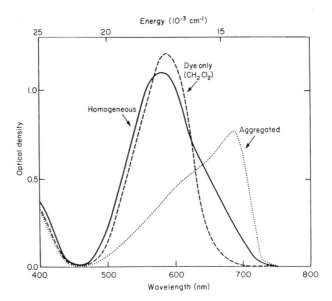

Fig. 3. Optical density versus wavelength and energy of thiapyrylium dye (I). ---, in dichloromethane; ———, in Lexan polycarbonate (II) unaggregated, ..., aggregated. Note the red shift in λ_{max} upon aggregation. [From Dulmage *et al.* (1978).]

strengths f (McGlynn and Smith, 1965). The dashed line in Fig. 3 shows the absorption spectra of (I) in CH_2Cl_2; $\lambda_{max} = 575$ nm, $\varepsilon = 56,100$, and $f = 0.64$.

From the structure, it might even be expected that the lowest triplet state will be close in energy to the lowest singlet.

To go from singlet (VIII) to triplet (IX), double bonds a and b are broken, but all this energy is recovered by the formation of a double bond in ring C

of (IX) and the subsequent delocalization energy of the Hückel $(4n + 2)$ circuit of this ring. The remaining energy of interaction is the two-electron exchange term

$$K_{12} = \int \Psi_A(1)\Psi_D(1)(e^2/r_{12})\Psi_A(2)\Psi_D(2)\, d\tau, \tag{1}$$

where Ψ_A and Ψ_D are the two molecular orbitals occupied by spin 1 and 2 on the acceptor and donor portions in Eq. (1), respectively. This term will be small, since examination of (IX) shows that the two spins occupy spatially different orbitals and thus $\Psi_A(1)\Psi_D(1)$ in Eq. (1), the overlap term, does not contain much charge density in the same regions of space for the two electrons.

Although direct experimental evidence for a small singlet–triplet splitting is lacking, it is known that no phosphorescence of (I) can be observed that has a lifetime of $>10^{-4}$ sec. The fluorescence occurs at 630 nm with a lifetime of 38 nsec (Thompson and Wilt, private communication).

Within each dye layer assembly, the molecules form a quasi-one-dimensional stacking wherein the acceptor portion (V) of each molecule sits over the donor portion (IV) of a neighboring molecule, as shown schematically in Fig. 4. This slipped-deck-of-cards stacking is very similar to what occurs in quasi-one-dimensional dark conductors (Stucky *et al.*, 1977). However, for the thiapyrylium assembly with light polarized parallel to the chain axis, there is no obvious low-energy charge-transfer band (Walker, private communication).

Figure 5 is a comparison of the optical absorption of a polycarbonate film containing (I) complexed with the polymer and the optical absorption (obtained by Kramers–Kronig transformation; Dulmage *et al.*, 1978) of (I)

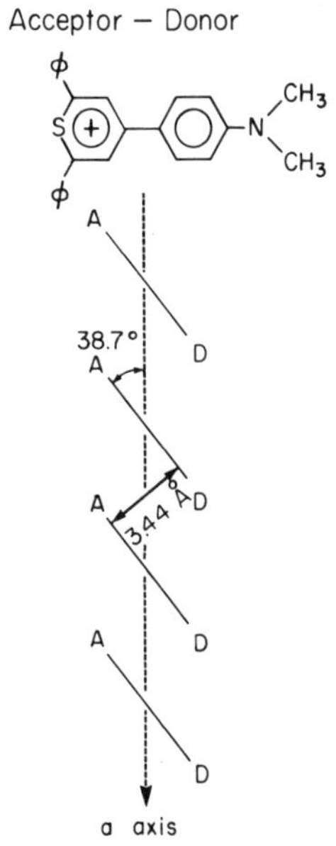

Fig. 4. Schematic of quasi-one-dimensional thiapyrylium assembly projected onto the *a* axis. The donor portion of each molecule is situated over the acceptor portion of a nearest neighbor.

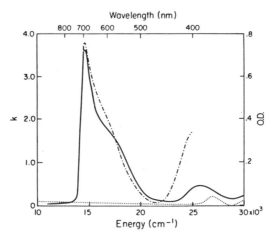

Fig. 5. Absorption coefficient versus wavelength and energy of thiapyrylium dye (I). $-\cdot-\cdot-$, Complexed with Lexan polycarbonate; ——, complexed with bisphenol A diphenyl carbonate with light polarized along the molecular axis; . . ., with light polarized perpendicular to the molecular axis. [From Dulmage *et al.* (1978).] No obvious charge-transfer band is seen in this polarization.

complexed with (III) as a single crystal. The two spectra agree quite nicely for the strong extinction at 685 nm, which in the crystal in Fig. 2 is almost along the long axis (i.e., the D–A axis; Walker, private communication) of the molecule. Perpendicular to this axis there is no charge-transfer absorption (dotted line in Fig. 5; Walker, private communication).

C. Preparation

The preparation of a polymer film containing the thiapyrylium assembly has been described in detail by Dulmage *et al.* (1978). The dye (typically 1–5% by weight) and polymer are dissolved in a solvent such as dichloromethane, and the solution is coated onto a metallized support. Rapid evaporation of the solvent produces a homogeneous film, shown in Fig. 6a (color plate), whose optical properties are similar to those of a solution of the dye (see Fig. 3), with λ_{max} at 575 nm. If this homogeneous film is then swelled by allowing it to come in contact with the vapors of a solvent or polymer-swelling agent, the dye molecules assemble into the layer structure of Figs. 1 and 2, with $\lambda_{max} = 685$ nm shifted 100 nm toward the red (see Fig. 3). Figures 6b and 6c show 2500× cross sections of this transformation. The dye assemblies have a rather diffuse, fibrous morphology. Figure 7 shows a transmission electron micrograph at 19,000× of the transformation (Marino, private communication). The fibrous struc-

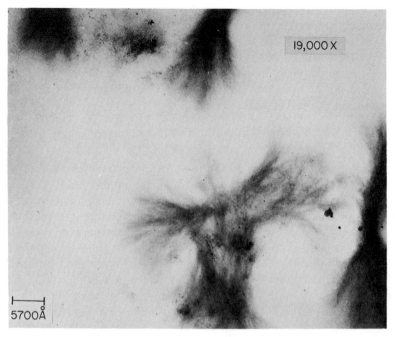

Fig. 7. Transmission electron micrograph at 19,000× of Lexan polycarbonate containing 3% thiapyrylium (I) after aggregation with chloroform vapor (Marino, private communication). Smallest filaments are ~70 Å.

ture is more evident, with the smallest fibers about 70 Å in diameter. Although Dulmage *et al.* (1978) describe this as a two-phase system consisting of a cocrystalline complex of dye and polymer as one phase embedded in an amorphous polymer matrix as the second phase, from the diffuseness of the photomicrographs and the absence of any clear crystal faces or edges it is not entirely certain that a true first-order phase boundary exists. Although it is difficult to assert anything about phase separation in the absence of any thermodynamic data, from the crystal structure in Figs. 1 and 2 and what is known about interchain coupling in low-dimensional dark conductors, I prefer to view the dye–polymer interaction as an interpenetrating network of dye layers and polymer molecules with the dye layers multiply connected throughout the polymer matrix. Viewed in this way, the dye layers may be independent; that is, there is no long-range coherence perpendicular to the layers, hence no three-dimensional long-range order. In this sense the dye–polymer interaction can be considered to form a two-dimensional crystalline complex. The questions of whether the assemblies are three-dimensionally or two-dimensionally ordered and the nature of their morphology have important consequences for the photoconduction. Generally, dispersions of dye crystals (which are three-dimensionally ordered structures) in polymer matrices tend to give films of lower quantum efficiency and with larger charge-carrier trapping cross sections either for one or both signs of the charge carrier, compared to the aggregate structure described here which has quantum efficiencies approaching 0.6.

D. Excitonic Properties of the Thiapyrylium Assembly

As reported by Dulmage *et al.* (1978), the long-wavelength absorption maximum for the thiapyrylium assembly occurs at 685 nm (1.8 eV), 100 nm (0.3 eV) toward the red of the monomer absorption in dichloromethane (see Fig. 3). This shift is *not* due to intermolecular charge-transfer interaction, as might be expected from the schematic of the stacking (the polarization of this band is not along the stacking axis but almost along the long axis of the molecule). Rather, the shift is best interpreted in terms of molecular exciton theory as follows (McRae and Kasha, 1964): For a stacked array of N molecules with no interaction between them, the first excited state would be N-fold degenerate as depicted in Fig. 8a. The major effect of turning on the coulomb interaction between molecules is to remove the excited-state degeneracy, producing an exciton band. The most important contribution to the coulomb potential for optical excitation is the interaction of the transition dipole on one molecule with those of its neighbors. With a point dipole approximation for the transition dipole M on each molecule separated by a distance r, the exciton bandwidth for the stacked array as $N \rightarrow \infty$ is given by

$$W = (4M^2/r^3)(3 \cos^2 \Phi - \sin \alpha), \qquad (2)$$

where Φ is the angle between the long axis of the molecule and the crystallographic a axis and α is the angle between the long axes of two nearest nonequivalent neighbors. (In the assembly of Fig. 1, only every other molecule is translationally equivalent.) From the structure, $\Phi = 38.7°$ and $\alpha = 56°$, so that

$$W = 4M^2/r^3. \qquad (3)$$

Thus the exciton bandwidth is directly proportional to the square of the experimental transition moment. Since the angle between the transition moments on neighboring molecules is not zero, two optical transitions to the top and bottom of the exciton band are allowed, but only the lower-energy transition at 685 nm is observed experimentally. The absence of the higher-energy transition is attributed to the long-axis molecular polari-

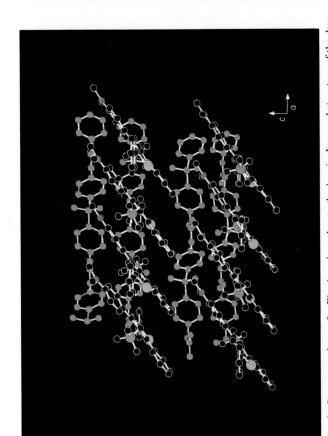

Fig. 2. Same complex as for Fig. 1 projected onto the *b* axis. Layered structure of the dye is evident with a quasi-one-dimensional stacked array of dye molecules along the *a* axis. Interplanar spacing along *a* is 3.44 Å.

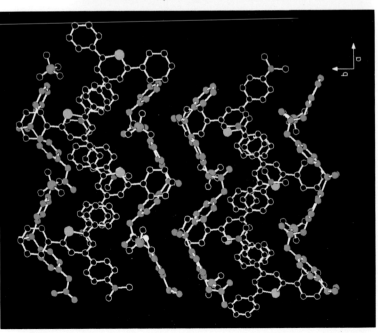

Fig. 1. *c*-axis projection of thiapyrylium dye (I)–bisphenol A diphenyl carbonate (III) complex. Yellow, sulfur; pink, nitrogen; blue, polycarbonate; green, chlorine of perchlorate. This projection shows the sandwiching of dye molecules between polycarbonate layers.

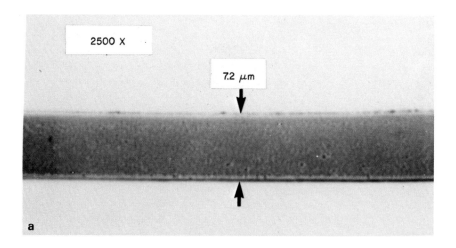

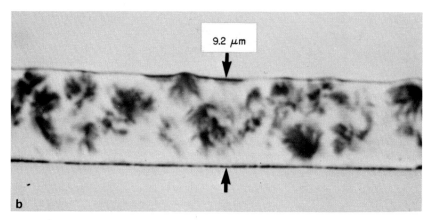

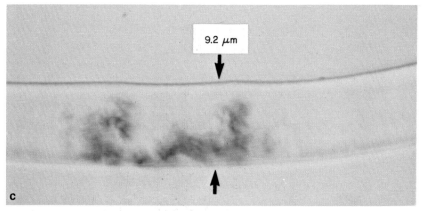

Fig. 6. (a) 2500× cross section of thiapyrylium dye (I) coated with Lexan polycarbonate (II) (note the presence of small aggregate nuclei). (b) After vapor treatment with dioxane with dye concentration ~3% by weight. (c) Same as in (b) but with dye concentration ~0.07%. Note the filamentary structure.

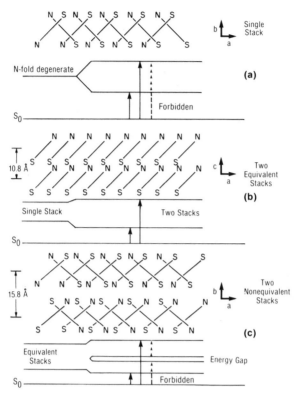

Fig. 8. (a) Exciton band formation of a stacked array of thiapyrylium molecules. For the arrangement of molecules shown two transitions are allowed, but the high-energy transition is expected to be very weak (Dulmage *et al.*, 1978). (b) Broadening of the exciton band due to the interaction of two equivalent stacks. (c) Exciton band splitting due to coulomb interaction between nonequivalent stacks.

zation shifted about 10° away from the N–S axis because one phenyl ring is twisted out of the molecular plane.

If one now includes coulomb interaction between translationally equivalent chains as depicted by the structure in Fig. 2, the overall effect is to broaden the exciton band and lower the energy of the first excited state (Fig. 8b) (Norland *et al.*, 1970). Finally, including the much weaker interactions with the nontranslationally equivalent chains (along b in Fig. 1) splits the exciton band and lowers the first excited state even more (Fig. 8c). The contribution of this interlayer interaction is small considering the large interlayer separation (15.8 Å). This point dipole approximation is good only for qualitative predictions of the exciton band structure. It breaks down quantitatively when the intermolecular spacing is small compared to the dipole length (i.e., the molecular length). For instance, using $M = 9.7 \times 10^{-18}$ esu calculated from the oscillator strength definition $f = (4.704 \times 10^{29})\nu M^2$, where ν is the transition frequency (in cm^{-1}), and with the intermolecular spacing $r = 3.44$ Å, $W = 5.8$ eV, which is considerably larger than the transition energy. A more appropriate theory would involve determination of the coulomb interaction for a transition charge density spread out over the entire molecule rather than localized at a point. As shown by Norland *et al.* (1970), the effect of distributing the charge is to decrease considerably the exciton bandwidth over that determined by Eq. (3).

Still unsettled is the question of whether there is an excited-state charge-transfer band associated with intermolecular D–A overlap, shown schematically in Fig. 4. Thiapyryliums form charge-transfer complexes with some amines (Van Allan *et al.*, 1977). The presence of charge transfer in the excited state (excimer formation) would make an autoionization mechanism of charge-carrier generation plausible (see below).

E.   Photoconduction

1.   Introduction

Unlike crystalline inorganic materials where charge generation occurs by interband transitions from the valence band to the conduction band upon light exposure, the single-photon charge-generation step in both crystalline and amorphous organic systems usually occurs indirectly by either surface annihilation of excitons or bulk autoionization of excitons (Inokuchi and Maruyama, 1976). For example, in crystalline anthracene, the most widely studied material to date, the wavelength dependence (Mulder, 1968) of the photoconductivity at low energies (<3.7 eV) follows the *optical density* of the crystal, as shown in Fig. 9. The generally accepted interpretation of this phenomenon is shown schematically in Fig. 10. Light absorption by the crystal produces a singlet exciton. The excitons migrate in all directions. Those that have an opportunity to migrate to the surface of the crystal eventually transform at the surface into a mobile hole and trapped electron or, conversely, into a mobile electron and trapped hole, depending on the nature of the surface and the polarity of the applied electric field. If the surface contains acceptor molecules, e.g., nitrobenzene, I_2, O_2, or anthraquinone (Gutmann and Lyons, 1967), the exciton will decompose into an electron which will be trapped at the acceptor site and a hole which with the proper field polarity will migrate through the anthracene crystal. Conversely, if the surface contains donor molecules, the exciton will decompose into a hole which is trapped at the

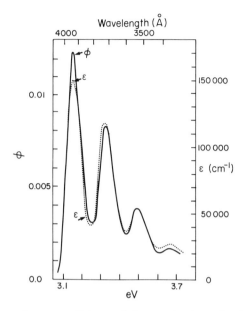

Fig. 9. Comparison of quantum efficiency of photoconduction and extinction coefficient as a function of energy for single-crystal anthracene doped with 10^{-3} mole of tetracene. Light is polarized along the b axis. [From Mulder (1968).]

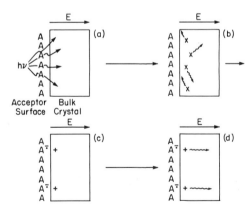

Fig. 10. Schematic of hole photogeneration process and transport. In (a) light is absorbed by the bulk crystal, producing excitons as in (b). Some of these excitons migrate to the surface, interacting with surface acceptors to yield electron–hole pairs as in (c). In (d) the electric field separates the trapped electron and the mobile hole which then migrates in the field through the bulk crystal. The same model applies for mobile electron photogeneration at a donor surface and reversed field polarity.

donor site and a mobile electron which can move through the anthracene crystal. These transformations can be represented chemically as in Eqs. (4)–(7):

$$\text{Excitation: anthracene} \xrightarrow{h\nu} \text{anthracene}^*, \tag{4}$$

Energy migration
and radical–ion-
pair formation:
$$\begin{cases} \text{anthracene}^* + A \rightarrow \text{anthracene}^{\dot{+}} + A^{\dot{-}} \text{ or} & (5) \\ \text{anthracene}^* + D \rightarrow \text{anthracene}^{\dot{-}} + D^{\dot{+}}, & (6) \end{cases}$$

$$\text{Transport: anthracene}^{\dot{\pm}} + \text{anthracene}^0 \rightarrow \text{anthracene}^0 \\ + \text{anthracene}^{\dot{\pm}}. \tag{7}$$

In Eq. (5) anthracene acts as a donor whose excited state interacts with an acceptor to form a radical–ion pair. The electron is trapped on A, and in the presence of a strong enough electric field the members of the electron–hole pair separate and the hole (radical cation) migrates through the neutral anthracene crystal. Conversely, for a donorlike surface, the hole is trapped and the electron (radical anion) migrates through the anthracene as in Eq. (7), either by a series of hops or by band delocalization.

Excitons that are created close to the crystal surface (i.e., for strongly absorbed light, large extinction coefficient) have a higher probability of producing the ion-pair state than excitons created far away from the surface (i.e., for weakly absorbed light, small extinction coefficient). Mulder (1968) has worked out a quantitative theory demonstrating the congruence of the optical density and the photoconductivity.

As a consequence of this interpretation, if donors or acceptors are rigorously excluded from the surface, an exciton reaching the surface will not form a radical–ion pair. For clean anthracene where an oxygen acceptor is rigorously excluded, no photocarrier generation occurs by surface dissociation of an exciton (Braun, 1968).

At high energies (>3.7 eV), bulk generation of charges occurs by field-assisted autoionization of excitons (Braun and Chance, 1974). In this mechanism, both electron and holes are mobile unless trapped by donor or acceptor molecules, respectively.

2. *The Photodischarge Experiment–Application to Thiapyrylium Assemblies in Polymer Matrices*

The most direct experiment (as well as the easiest to assemble and to obtain data from) for determination of the quantum efficiency of charge generation is the potential discharge experiment shown schematically in Fig. 11. The polycarbonate film (II) containing the thiapyrylium assembly (I), which has been preformed on a transparent metal electrode, is

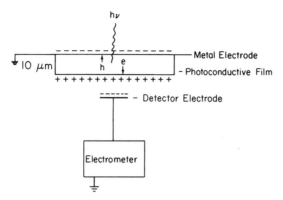

Fig. 11. Schematic of photodischarge experiment for direct determination of quantum efficiency. For strongly absorbed light, this experiment allows determination of the quantum efficiency of electrons and holes independently.

corona-charged positively or negatively in the dark. The film surface is then exposed to light through the metal electrodes or through a transparent detector electrode placed close to the film surface (typically a metallized quartz disk). The detector electrode is connected to an electrometer and "sees" the potential across the film by capacitative coupling. Upon light exposure, electrons and holes created in the film migrate to the surface and reduce the potential by combining with the surface charge.

Figure 12 shows a typical result for the photodischarge of the thiapyrylium assembly both for positive and negative surface charging and light exposure at 690 nm. The essential features are an initial dark decay and a nonlinear photodecay. Some numbers will help to put the experiment in perspective. When 1000 V is applied across a 10-μm (10^{-3}-cm)- thick film, it produces a field E of 10^6 V/cm. This is a very large field within an order of magnitude of the largest field that can be tolerated by most materials before breakdown occurs. The charge per unit area on the film surface Q/A, where Q is the total surface charge and A is the film surface area, is given by

$$Q/A = CV/A, \qquad (8)$$

where C is the film capacitance and V is the applied potential. If all the charge resides on the surface of the film the geometric capacitance is given by

$$C = \kappa\varepsilon_0 A/d, \qquad (9)$$

where κ is the film dielectric constant, ε_0 is the permittivity of free space, and d is the film thickness.

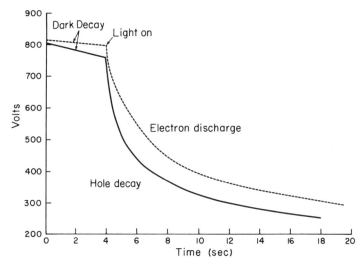

Fig. 12. Photodischarge curve of thiapyrylium (I) complexed with Lexan polycarbonate (II). ---, negatively charged surface; ——, positively charged surface exposure. Irradiance was 225 ergs/cm² sec.

With the reported $\kappa = 2.9$ and $\varepsilon_0 = 8.85 \times 10^{-12}$ C²/N m²,

$$Q/A = \kappa\varepsilon_0 E = (2.9)(8.85 \times 10^{-12})(10^8) = 2.6 \times 10^{-3} \quad \text{C/m}^2. \quad (10)$$

The electron charge is 1.6×10^{-19} C, so that the surface charge density at this field is 1.6×10^{12} charges/cm². If this entire charge were homogeneously distributed throughout the bulk film, the charge density would be 1.6×10^{15} charges/cm³, which is not only a very small fraction (ca. 10^{-6}) of typical organic molecular densities but also smaller than the impurity concentrations which are usually $>0.1\%$. Since purification procedures for most organic materials (including the materials discussed in this chapter) are not as well developed as those for inorganic semiconductors, the role of impurities in the charge-generation and transport processes in these assemblies remains to be resolved.

If the quantum efficiency for charge generation were unity and there were no field dependence of the quantum efficiency, then 1.6×10^{12} photons/cm² would be required to discharge the film to zero potential. At $\lambda_{max} = 685$ nm (Fig. 3) this corresponds to an exposure A of

$$A = \frac{nhc}{\lambda} = \frac{(1.6 \times 10^{12})(6.6 \times 10^{-27})(3 \times 10^{10})}{6850 \times 10^{-8}} = 4.6 \quad \text{ergs/cm}^2. \quad (11)$$

In reality, the quantum efficiency decreases as the field across the film decreases, as is observed from the nonlinearity of the voltage-versus-time data in Fig. 12. If the irradiance I of the light source is known in photons/m^2 sec, then the field dependence of the quantum efficiency ϕ can be determined by computing the slope dV/dt at each point along Fig. 12. From Eq. (8)

$$d(Q/A)/dt = (C/A)(dV/dt)$$

and from Eq. (9)

$$\phi = \frac{d(Q/A)}{dt} \bigg/ eI = \frac{\kappa \varepsilon_0}{eId} \frac{dV}{dt} . \tag{12}$$

The only assumptions used in deriving Eq. (12) are that (a) all photogenerated charges reach the surface (i.e., there is no range limitation; this is easily tested by measuring ϕ as a function of film thickness) and (b) the amount of charge moving through the film at any time is much less than the surface charge at that time [so-called emission-limited discharge, which is also easily tested by measuring ϕ as a function of light intensity (Mort and Chen, 1975)].

Figure 13 shows ϕ versus E on a log scale for both positive and negative film surface charging of the thiapyrylium assembly (Borsenberger *et al.*, 1978b). Exposure was at $\lambda_{max} = 690$ nm and was emission-limited, and ϕ

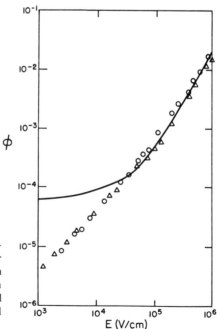

Fig. 13. Electron (O) and hole (△) quantum efficiency ϕ versus electric field for thiapyrylium (I) in polycarbonate (II) [From Borsenberger *et al.* (1978b).] Solid line is an attempt to fit the data to a three-dimensional Onsager model with $\phi_0 = 4 \times 10^{-2}$ and $r = 35$ Å.

was only 10^{-2} at 10^6 V/cm but showed no evidence of saturation even at this high field. At low fields, e.g., $\sim 10^3$ V/cm, ϕ fell to $\sim 4 \times 10^{-6}$ with no indication of a zero-field saturation. The solid line is an attempt to fit these data with Onsager's theory of geminate recombination (see below). The same results were obtained for film thicknesses up to 9.2 μm, indicating that that there was no range limitation (i.e., no deep trapping) of *either* electrons or holes for this thickness. Although a comprehensive study of the wavelength dependence of ϕ has not been done at low fields, at high fields (5×10^5 V/cm) ϕ is roughly independent of λ from 400 to 700 nm.

These results indicate that a mechanism whereby an exciton migrates to the macroscopic film surface is not operative, since the quantum efficiency is the same for weakly absorbed light ($\lambda = 450$ nm) and for strongly absorbed light ($\lambda = 690$ nm).

Two interpretations of this are possible: either (a) charge generation occurs by autoionization in the bulk of the thiapyrylium assembly or (b) charge generation occurs by exciton annihilation via donor or acceptor impurities at *internal* surfaces.

3. Effects of Doping on the Photoconduction of Thiapyrylium Assemblies

So far we have considered a chemically simple (but structurally and morphologically complex) photoconductive self-assembly of thiapyrylium dye molecules complexed with a polycarbonate whose high-field quantum efficiency ϕ for photogeneration is about 1%. We now want to consider what happens to ϕ when the thiapyrylium assembly is doped* with a quantity of donor molecules or acceptor molecules.

Typical donor dopants are some form of aromatic amine; many of these are known, and Nielsen (1976) has listed them. In the work of Dulmage *et al.* (1978) and Borsenberger *et al.* (1978b), the donor used as a dopant is an amino-substituted arylmethane (X). (Virtually any of the amines described by Nielsen could be used as well.)

(X)

---

* We use the word "doped" loosely, as the dopant is in much higher concentration than the dye.

Amines as donor organic dopants for thiapyrylium photoconductivity are useful because (a) they readily form radical cations which are stable for long times compared to charge-transit times across polymer films, and (b) the oxidation potentials of the amines are less than or equal to the oxidation potentials of the thiapyrylium dyes.

This donor radical cation property of amines is the chemical signature of a hole-only transporting material; this property has been amply demonstrated for amines (Borsenberger *et al.*, 1978a; Pfister, 1977).

Conversely, electron acceptors that readily form radical anions are electron-only transporting materials. Trinitrofluorenone (Schaffert, 1971) and boron diketonates (Halm, 1977) are examples of these, with trinitrofluorenone having a reduction potential close to that for the thiapyrylium dye (I). Table I shows a comparison of the redox properties of some donors and acceptors with pyrylium dyes.

If a polycarbonate film containing a few percent of the thiapyrylium dye (I) and about 40% amine (X) is transformed into the thiapyrylium self-assembly, a remarkably large increase in the quantum efficiency for electron and hole charge generation occurs (Dulmage *et al.*, 1978). The results for holes are shown in Fig. 14, where log ϕ versus log E is plotted for

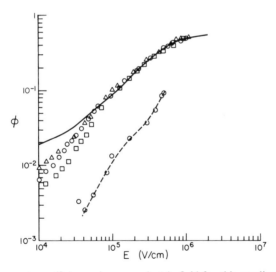

Fig. 14. Hole quantum efficiency ϕ versus electric field for thiapyrylium (I) complexed with Lexan polycarbonate containing ~40% amine by weight (X) (Borsenberger *et al.*, 1978b). Data are shown for three thicknesses. Solid line is fit to three-dimensional Onsager theory with $\phi_0 = 0.59$ and $r = 54$ Å. Dotted line is data for polyvinylcarbazole–trinitrofluorenone as a comparison. Film thickness: $\triangle$, 8.6 μm; $\bigcirc$, 10.5 μm; $\square$, 13.3 μm. [From Melz (1972).]

TABLE I *Oxidation and Reduction Potentials of Donors and Acceptors*

Donors	E_p (V)[a]	Ref.
	+1.25	Beal *et al.* Private communication.
	+1.00	Beal *et al.* Private communication.
(10)	+0.68	Beal *et al.* Private communication.
	+0.66	Albrecht and Gruenbaum. Private communication.

Acceptors	E_p (V)[b]	Ref.
	−0.45	Groner. Private communication.
	−0.84	Halm (1977).

TABLE I (*Continued*)

Amphoteric dyes	E_p (V)^b	E_p (V)^a	Ref.
	-0.70	$+1.30$	Chang and Mina. Private communication.
(1)	-0.52	$+1.24$	Chang and Mina. Private communication.
	-0.55	$+1.24$	Chang and Mina. Private communication.

^{a} Oxidation, SCE reference.
^{b} Reduction, SCE reference.

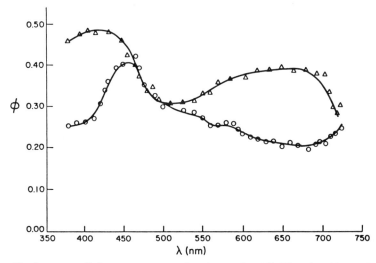

Fig. 15. Quantum efficiency versus wavelength at $\sim 5 \times 10^5$ V/cm for thiapyrylium (I) complexed with Lexan polycarbonate (II) containing 40% amine (X). $\triangle$, positive exposures; $\bigcirc$, negative exposures. [From Borsenberger *et al.* (1978b).]

exposure at $\lambda_{max} = 690$ nm (Borsenberger *et al.*, 1978b). Compared with the aggregate-only film (Fig. 13), ϕ increased to 0.6 at 10^6 V/cm, almost two orders of magnitude over the aggregate-only film. As in the aggregate-only film, ϕ at 5×10^5 V/cm is also roughly independent of wavelength (Fig. 15).

III. DISCUSSION

A. Three-Dimensional Onsager Model

Although no presently known model of photogeneration efficiency is consistent with all the data for the thiapyrylium assembly, it is generally recognized that an Onsager model for geminate recombination is a good starting point for evaluating ϕ (Onsager, 1934, 1938). In this model, photon excitation produces a bound electron–hole pair with excess kinetic energy. The electron and hole then diffuse apart, thermally equilibrating with their surroundings at a separation distance r. It is at this thermalized r that Onsager determined the probability of escape of the electron from the hole in the presence of further diffusion *and* an electric field. If ϕ_0 is the probability that a photon produces a thermalized ion pair separated by a distance r, then for a three-dimensional isotropic distribution of such pairs the probability of escape is given by (Geacintov and Pope, 1971; Pai and Enck, 1975)

$$
\Phi = \phi_0 \frac{kT}{eEr} \exp \frac{e^2}{4\pi\kappa\varepsilon_0 kT} \exp\left(\frac{-eEr}{kT}\right)
$$

$$
\times \sum_{m=0}^{\infty} \frac{1}{m!} \left(\frac{-e^2}{4\pi\kappa\varepsilon_0 kT}\right)^m \sum_{n=0}^{\infty} \sum_{l=m+n+1}^{\infty} \frac{1}{l!} \frac{eEr^l}{kT}. \tag{13}
$$

The first few terms of this complex expression in powers of E are given by (Pai and Enck, 1975)

$$
\Phi = \phi_0 \exp -\frac{\omega}{kT} \left\{ 1 + \frac{er}{2!kT}\left(\frac{\omega}{kT}\right) E \right.
$$

$$
- \frac{1}{3!}\left(\frac{er}{kT}\right)^2 \left(\frac{\omega}{kT}\right)^2 \left(\frac{kT}{\omega} - \frac{1}{2}\right) E^2
$$

$$
\left. + \frac{1}{4!}\left(\frac{er}{kT}\right)^3 \left(\frac{\omega}{kT}\right)^3 \left[\left(\frac{kT}{\omega}\right)^2 - \frac{kT}{\omega} + \frac{1}{6}\right] E^3 - \cdots \right\}, \tag{14}
$$

where $\omega = e^2/4\pi\kappa\varepsilon_0 r$ is the coulomb energy of the electron–hole pair. For large E, $\phi \to \phi_0$, whereas for small E,

$$\phi = \phi_0 \exp \frac{-\omega}{kT} \left[1 + \frac{er}{2!kT} \left(\frac{\omega}{kT} \right) E \right] \tag{15}$$

or

$$\phi \rightarrow \phi_0 \exp(-\omega/kT) \qquad \text{as} \quad E \rightarrow 0. \tag{16}$$

Borsenberger *et al.* (1978b) applied this model to the thiapyrylium self-assembly both with the added donor (X) (solid line Fig. 14) and without the donor (solid line Fig. 13). For the amine-doped film the fit is good at high fields, with $\phi_0 = 0.59$ and $r = 54$ Å, but deviates significantly at low fields, possibly owing to bulk charge trapping. For the undoped film, ϕ_0 and r were chosen to be 0.04 and 35 Å, respectively, but the high-field and low-field saturation regions are not observed experimentally. Several other salient features about the data should be pointed out. The fact that ϕ at high fields is independent of photon energy (within a factor of 2; Fig. 15) implies that ϕ_0 is independent of wavelength. Whether r is also independent of wavelength is not entirely clear, although it appears that, since ϕ versus λ was measured at fields well below the saturation field in Fig. 15, r is also independent of photon energy. What then is the meaning of the thermalization distance r? If the dye assembly is excited to the second singlet state, the same ϕ is obtained when the assembly is excited to the first singlet state, a result suggesting an intramolecular radiationless transition that is rapid compared to the charge-separation time. The parameter r then becomes meaningless. As pointed out by Scher (1976), the generation process itself should be viewed as a hopping process. This can qualitatively occur as follows. A dye molecule is excited to the first or second singlet state as

$$\text{dye} \xrightarrow{h\nu} \text{dye}^*.$$

Radiationless processes produce a bound electron–hole pair on the same molecule:

$$\text{dye} \rightarrow \text{dye}^{+\,\cdot\,-}.$$

For example, for dye (I) strong intramolecular coupling to vibrational modes (polaron formation) could produce an electron–hole pair (a biradical) on the same molecule as in (XI).

singlet state → polaron formation → bound electron / bound hole

(XI)

The bound electron–hole pair is configurationally identical to the triplet state of the dye as in (IX) (McGlynn *et al.*, 1969). Polaron binding energies of an electron to an aromatic molecule are estimated to be 0.1–0.3 eV (Lipari *et al.*, 1976, 1977). Table II shows the singlet–triplet splitting for a series of molecules similar to dye (I), indicating that the singlet–triplet splitting in (I) may be quite small. Singlet exciton band formation (Fig. 9) will surely enhance this effect.

When one thus includes polaron effects and exciton band formation, it appears that intramolecular biradical formation could compete favorably with any other decay processes. Thus ϕ_0 would be the probability of biradical formation:

$$\text{dye} \xrightarrow{h\nu} \text{dye*} \xrightarrow{\phi_0} \text{biradical}. \tag{17}$$

The next step, according to Scher, is for the trapped hole to hop or tunnel to a donor (i.e., for an electron from the donor to hop to the biradical) or in the presence of a suitable acceptor for the trapped electron to hop or tunnel to the acceptor, as shown schematically in Fig. 16 and in Eq. (18):

$$\text{biradical} + \text{donor} \rightarrow \text{free electron and hole}. \tag{18}$$

A major requirement here is that the oxidation potential of the donor be less than or equal to the oxidation potential of the dye, but not so much smaller that recombination of the hole formed on the donor with the trapped electron on the dye is appreciable. Similarly, for an acceptor the

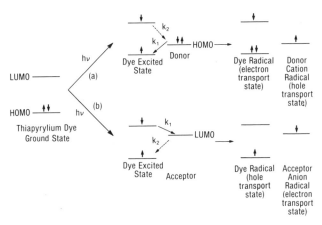

Fig. 16. Orbital energy-level diagram for the charge-generation process at the (a) thiapyrylium dye–donor and (b) thiapyrylium dye–acceptor interface. The highest occupied molecular orbital (HOMO) of the donor and the lowest unoccupied molecular orbital (LUMO) of the acceptor are such that the rate constants for recombination k_2 are much less than the rate constant for charge generation k_1.

TABLE II *Singlet–Triplet Splittings for Pyrylium and Pyrylium-Related Dyes[a,b]*

Dye	Singlet absorption (eV)	Singlet emission (eV)	Triplet emission (eV)	Singlet–triplet splitting (eV)
(structure: φ, φ, N, φ)	3.99 (310)	3.47 (356)	2.63 (471)	0.84
(structure: CH_3, CH_3, N, φ, N, φ)	3.64 (340)	3.07 (404)	2.59 (479)	0.48
(structure: CH_3, CH_3, N, φ, +N—H, φ; ClO_4^-)	2.76 (450)	2.45 (507)	2.16 (573)	0.29
(structure: φ, φ, +O, φ; ClO_4^-)	3.10 (400)	2.74 (451)	2.37 (523)	0.37
(structure: CH_3, CH_3, N, φ, +O, φ; ClO_4^-)	2.32 (535)	2.02 (612)	Unobservable	—
(structure: φ, φ, +S, φ; ClO_4^-)	3.10 (400)	2.32 (535)	2.31 (538)	0.01
(structure: CH_3, CH_3, N, φ, +S, φ; ClO_4^-)	2.09 (590)	1.96 (630)	Unobservable	—

[a] From Thompson and Wilt (1981).
[b] Wavelength is indicated in parentheses.

absolute value of the reduction potential of the acceptor should be less than or equal to that of the dye but not so much less that recombination with the trapped hole on the dye is competitive.

Table I shows that for most donors and a TNF acceptor these conditions are satisfied. They are not satisfied, however, for boron diketonates, although presumably they could be modified to increase their electron affinity.

One consequence of the above model is that ϕ_0 is independent of the donor or acceptor used. Thus ϕ versus E for all donors should saturate at the same value of ϕ_0 at high fields, but the shape of ϕ versus E will vary with the donor or acceptor redox potential.

B.   One- and Two-Dimensional Onsager Models

A crucial test of the Onsager model for a three-dimensional isotropic charge distribution is the ratio of slope to intercept of the ϕ-versus-E plot at low fields (Braun and Chance, 1974). From Eq. (14) to first order in E,

$$\Phi(E) = \Phi_0 \exp\left(\frac{-\omega}{kT}\right) + \Phi_0 \left(\exp\frac{-\omega}{kT}\right) \frac{er}{2kT} \frac{\omega}{kT} E, \qquad (19)$$

so that the slope m/intercept b ratio (which is independent of r) is

$$m/b = (er\omega/2(kT)^2) = 3.76 \times 10^{-5} \quad \text{cm/V} \qquad \text{for} \quad \kappa = 3.0. \qquad (20)$$

Neither the undoped nor amine-doped thiapyrylium assembly follows the low-field Onsager prediction (solid lines in Figs. 13 and 14). This could be a result of bulk charge trapping at low fields. (Note, for example, that ϕ is thickness-dependent below 5×10^4 V/cm in the amine-doped film.) But it also could be due to pronounced deviations from a charge-generation process that is three-dimensionally isotropic. This should be especially noticeable in the undoped thiapyrylium assembly where the structure (Fig. 2) indicates a quasi-one-dimensional stacked array of molecules within the dye layer. Charge generation in one (or two) dimensions is considerably different from that in three dimensions. In one or two dimensions, the probability of a charge diffusing away from a given point (and not returning) is *zero* (Feller, 1957). As a consequence of this, as $E \to 0$, $\phi \to 0$; there is no free diffusion current at zero applied field, whereas in three dimensions the Onsager model predicts a finite current at zero field which from Eq. (19) is

$$\phi(r, T) = \phi_0 \exp(-\omega/kT). \qquad (21)$$

An Onsager model in one dimension has been worked out (Haberkorn and Michel-Beyerle, 1973), which when averaged over random orientations of the thiapyrylium stacks in the applied field (Young, private communication) becomes for a low field

$$\phi = \phi_0 \tfrac{1}{2}(\omega/er)^{-1}E \, \exp(-\omega/kT), \tag{22}$$

with a zero intercept.

The low-field data for the undoped thiapyrylium assembly (Fig. 13) is replotted as ϕ versus E in Fig. 17. Least-squares analysis indicates a slope/intercept ratio of 10^{-2}, almost three orders of magnitude larger than that of the three-dimensional model. Based on Borsenberger's estimate of $\phi_0 = 4 \times 10^{-2}$, the pair thermalization distance r from a numerical solution of Eq. (22) is 56 Å. Viewed in this way, the insertion of a high concentration of dopant could change the dimensionality of the charge-generation problem, for if the amine occupies sites above and below the thiapyrylium assembly layer, the charge-generation reaction

$$\text{dye biradical} + D \rightarrow \text{dye}^- + D^+ \tag{23}$$

would occur three-dimensionally as shown in Fig. 18. No theory has been worked out for such a complex charge distribution in which the hole on D^+ can execute a three-dimensional random walk among D sites while the electron executes only a one- or two-dimensional random walk on

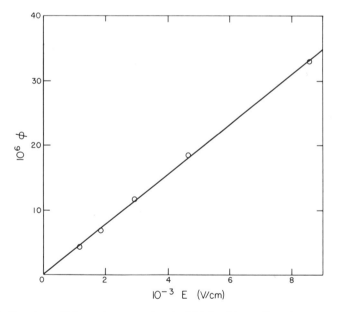

Fig. 17. Quantum efficiency ϕ versus electric field for thiapyrylium dye (I) complexed with Lexan polycarbonate. Slope/intercept ratio $S/I \sim 10^{-2}$, three orders of magnitude larger than predicted by the three-dimensional model.

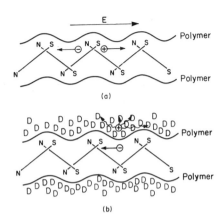

Fig. 18. Comparison of a one-dimensional photogeneration process with three-dimensional photogeneration. In (a) the electron–hole pair will be preferentially generated along the one-dimensional array of thiapyrylium molecules. $\phi(E \rightarrow 0)$ is zero for this process. In (b) the electron–hole pair is generated in three dimensions at the donor (D)–dye interface. $\phi(E \rightarrow 0)$ is >0 for this process.

thiapyrylium assembly sites. Qualitatively, however, the overall effect would be to introduce a zero-field generation term not present in the strictly one- or two-dimensional random walk.

An alternate model considers the donor as surrounding individual filaments of the thiapyrylium assembly in which the filament diameter is small compared to the singlet exciton diffusion length (i.e., the distance the exciton can diffuse in its lifetime). From the transmission electron micrograph (Fig. 7), this length need not be more than 70 Å. Once the exciton is at the surface, a process such as that described by Eqs. (17) and (23) for charge-carrier generation could occur. This model requires that three-dimensional exciton diffusion processes be allowed.

C.   Chemical Evidence for Low Dimensionality

Although there is no direct experimental evidence for the location of the dopant, there is some chemical evidence that dye layer separation can occur.

If the perchlorate anion of (I) is replaced by the much larger anion (XII), the thiapyrylium assembly still forms in polycarbonate (II) with only a small shift in λ_{max} toward the blue by 20 nm (Fig. 19). This shift is qualitatively consistent with a decrease in the exciton bandgap of Fig. 8c, owing to the reduced layer–layer interaction along the b axis of the dye–polymer complex.

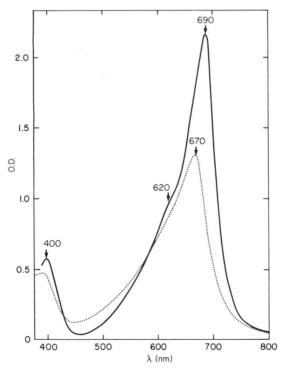

The fibrous morphology of the thiapyrylium assembly can be swelled by the insertion of very large anions; in fact, anions as large as (XIII) can be

Fig. 19. Optical density versus wavelength for thiapyrylium dye complexed with Lexan polycarbonate for different anions: ——, ClO$_4^-$, ---, sulfonate (XI).

inserted between the layers very readily, with little change in λ_{max} (Van Allan *et al.*, private communication). Given that such large inert structures are so readily incorporated, it is chemically feasible that neutral donors or acceptors might also be incorporated, in particular the amine (X).

Further work on the electron microscopy and x-ray diffraction of these films would be highly desirable to elucidate this point.

IV. SUMMARY AND CONCLUSIONS

We have reviewed some results indicating that the quantum efficiency ϕ for both electron and hole charge generation of thiapyrylium dye molecules is greatly enhanced when they are assembled into a quasi-one- or -two-dimensional morphology within a polycarbonate matrix. Moreover, when a donor or an acceptor is added to this structure, additional enhancement of orders of magnitude occurs in ϕ, approaching 0.6 at 10^6 V/cm.

There is nothing in either the chemistry or the physics of the thiapyrylium ring and its dyes suggesting that these assembly structures are unique to thiapyrylium. On the contrary, it is well known that cyanine dyes such as (XIV) and its relatives aggregate readily in solution (the J and

(XIV)

H aggregates) (Herz, 1977) and on the surface of silver halide grains. However, it is not a trivial matter to locate structures within a dye class that will easily form these low-dimensional assemblies within a polymer matrix (the normal state of affairs is the formation of a three-dimensionally ordered dye crystal, a state in which the polymer is excluded from the assembly) and at the same time have the correct morphology, the right oxidation and reduction potentials, and the proper donor or acceptor, all of which can be combined to form a state with high photogeneration efficiency.

In the absence of any thermodynamic data on the present system, it is difficult to understand how small structural changes of the molecule can effect such major structural changes in the assembly. In light of such effects, if the assembly can be considered to be in some thermodynamically stable (or metastable) state, the difference in free energy between it and the better known three-dimensionally crystalline ordered state of the dye must, in fact, be rather small.

ACKNOWLEDGMENTS

This work would not have been possible without the extensive involvement of many of my colleagues at the Eastman Kodak Company Research Laboratories. While it is not possible to give credit here to all these valued co-workers, I would like to thank certain individuals who were helpful in writing this paper: Calvin Salzberg who introduced me to aggregation phenomena; Don Vanas who instilled in me the notion of the importance of polymeric effects; Martin Berwick, Larry Contois, and Sal Marino for numerous discussions on the preparation aspects of the thiapyrylium–polymer complex; Fred Chen, Neil Haley, George Reynolds, and James Van Allan for their efforts in stimulating my thinking about dye chemistry; Douglas Smith for extended discussions on the x-ray structure; A. P. Marchetti and E. I. P. Walker for their elucidation of the optical properties; P. M. Borsenberger, William Mey, and Arun Chowdry for the electrical properties. Special thanks are due to G. L. Bottger, T. K. Dykstra, D. C. Hoesterey, and R. H. Young for critical reading of the manuscript; and, finally, I thank William Light and W. J. Dulmage for starting it all.

REFERENCES

Akamatu, H., Inokuchi, H., and Matsunaga, Y. (1956). *Bull. Chem. Soc. Jpn.* **29**, 213.

Albrecht, F. X., and Gruenbaum, W. T. Private communication.

Beal, C. N., Case, B. L., and Large, R. F. Private communication.

Berlinsky, A. J. (1976). *Contemp. Phys.* **17**, 331.

Borsenberger, P. M., Mey, W., and Chowdry, A. (1978a). *J. Appl. Phys.* **49**, 273.

Borsenberger, P. M., Chowdry, A., Hoesterey, D. C., and Mey, W. (1978b). *J. Appl. Phys.* **49**, 5555.

Braun, C. L. (1968). *Phys. Rev. Lett.* **21**, 215.

Braun, C. L., and Chance, R. R. (1974). *In* "Energy and Charge Transfer in Organic Semiconductors" (K. Masuda and M. Silver, eds.), p. 17. Plenum Press, New York.

Castro, G. (1971). *IBM J. Res. Dev.* **15**, 27.

Chang, J. C., and Mina, R. Private communication.

Chiang, C. K., Druy, M. A., Gare, S. C., Heeger, A. J., Louis, E. J., MacDiarmid, A. G., and Park, Y. W. (1978a). *J. Am. Chem. Soc.* **100**, 1013.

Chiang, C. K., Gau, S. C., Fincher, Jr., C. R., Park, Y. W., MacDiarmid, A. G., and Heeger, A. J. (1978b). *Appl. Phys. Lett.* **33**, 18.

Dahm, D. J., Horn, P., Johnson, G. R., Miles, M. G., and Wilson, J. D. (1975). *J. Cryst. Mol. Struct.* **5**, 27.

Dulmage, W. J., Light, W. A., Marino, S. J., Salzberg, C. D., Smith, D. L., and Staudenmayer, W. J. (1978). *J. Appl. Phys.* **49**, 5543.

Emerson, E. S., Conlin, M. A., Rosenoff, A. E., Norland, K. S., Rodriquez, H., Chin, D., and Bird, G. R. (1967). *J. Phys. Chem.* **71**, 2396.

Feller, W. (1957). *In* "An Introduction to Probability Theory and Its Applications," Vol. I, p. 237. Wiley, New York.

Ferraris, J., Cowan, D. O., Walatka, Jr., V. V., and Perlstein, J. H. (1973). *J. Am. Chem. Soc.* **95**, 948.

Geacintov, N. E., and Pope, M. (1971). *Proc. Int. Conf. Photoconduct., 3rd* (P. J. Warter, ed.), p. 311. Pergamon, Oxford.

Ghosh, A. K., and Feng, T. (1978). *J. Appl. Phys.* **49**, 5892.

Griffiths, J. (1976). "Colour and Constitution of Organic Molecules." Academic Press, New York.

Groner, F. Private communication.

Gutman, F., and Lyons, L. E. (1967). "Organic Semiconductors," p. 405. Wiley, New York.

Haberkorn, R., and Michel-Beyerle, M. E. (1973). *Chem. Phys. Lett.* **23**, 128.

Halm, J. M. (1977). *Tappi* **60**, 90.

Herz, A. H. (1977). *Adv. Colloid Interface Sci.* **8**, 237.

Inokuchi, H., and Maruyama, Y. (1976). *In* "Photoconductivity and Related Phenomena" (J. Mort and D. M. Pai, eds.), p. 155. Elsevier, Amsterdam.

Jeszka, J. K., Ulanski, J., and Kryszewski, M. (1981). *Nature* **289**, 390.

Kamimura, H. (1976). *In* "Physics of Semiconductors" (F. G. Fumi, ed.), p. 51. North-Holland Publ., Amsterdam.

Large, R. F., and Beal, C. N. Private communication.

Lipari, N. O., Duke, C. B., Bozio, R., Gerlando, A., Pecile, C., and Padva, A. (1976). *Chem. Phys. Lett.* **44**, 236.

Lipari, N. O., Rice, M. J., Duke, C. B., Bozio, R., Gerlando, A., and Pecile, C. (1977). *Int. J. Quantum Chem. Quant. Chem. Symp.* **11**, 583.

Liu, K., Lignowski, J. S., and Ullman, R. (1967). *Biopolymers* **5**, 375.

Marino, S. J. Unpublished.

McGlynn, S. P., and Smith, F. J. (1965). *Mod. Quantum Chem.* **3**, 67.

McGlynn, S. P., Azumi, T., and Kinoshita, M. (1969). *In* "Molecular Spectroscopy of the Triplet State," p. 88. Prentice Hall, Englewood Cliffs, New Jersey.

McRae, E. G., and Kasha, M. (1964). *In* "Physical Processes in Radiation Biology," p. 23. Academic Press, New York.

Melby, L. R., Harder, R. J., Hertler, W. R., Mahler, W., Benson, R. E., and Mochel, W. E. (1962). *J. Am. Chem. Soc.* **84**, 3374.

Melz, P. J. (1972). *J. Chem. Phys.* **57**, 1694.

Merritt, V. Y., and Hovel, H. J. (1976). *Appl. Phys. Lett.* **29**, 414.

Morel, D. L., Ghosh, A. K., Feng, T., Stogiyn, E. L., Purwin, P. E., Shaw, R. F., and Fishman, C. (1978). *Appl. Phys. Lett.* **32**, 495.

Mort, J., and Chen, I. (1975). *Appl. Solid State Sci.* **5**, 69.

Mulder, B. J. (1968). *Philips Res. Rep. Suppl. 4*.

Nielsen, N. A. (1976). *Tappi Reprography Conf., San Francisco, California*.

Norland, K., Ames, A., and Taylor, T. (1970). *Photogr. Sci. Eng.* **14**, 295.

Onsager, L. (1934). *J. Chem. Phys.* **2**, 599.

Onsager, L. (1938). *Phys. Rev.* **54**, 554.

Pai, D. M., and Enck, R. C. (1975). *Phys. Rev. B* **11**, 5163.

Patsis, A. V., and Seanor, D. A. (1976). "Photoconductivity in Polymers." Technomic Publ., Westport, Connecticut.

Perlstein, J. H. (1977). *Angew. Chem. Int. Ed. Engl.* **16**, 519.

Perlstein, J. H., Ferraris, J. P., Walatka, Jr., V. V., and Cowan, D. O. (1972). *In* "Magnetism and Magnetic Materials" (C. D. Graham and J. J. Rhyne, eds.), p. 1494. American Institute of Physics, New York.

Perlstein, J. H., Van Allan, J. A., Isett, L. C., and Reynolds, G. A. (1978). *Ann. N.Y. Acad. Sci.* **131**, 61.

Pfister, G. (1977). *Phys. Rev. B* **16**, 3676.

Schaffert, R. M. (1971). *IBM J. Res. Dev.* **15**, 75.

Schaffert, R. M. S. (1975). *In* "Electrophotography." Wiley, New York.

Scher, H. (1976). *In* "Photoconductivity and Related Phenomena" (J. Mort and D. M. Pai, eds.), p. 107. Elsevier, Amsterdam.

Scott, J. C., Etemad, S., and Engler, E. M. (1978). *Phys. Rev. B* **17**, 2269.

Stucky, G. D., Schultz, A. J., and Williams, J. M. (1977). *Ann. Rev. Mater. Sci.* **7**, 301.

Street, G. B., and Clarke, T. C. (1981). *IBM J. Res. Develop.* **25**, 51.

Thompson, D. R., and Wilt, J. R. (1981). Unpublished.
Toombs, G. A. (1978). *Phys. Rep.* **40C**, 181.
Torrance, J. B. (1979). *Acct. Chem. Res.* **12**, 79.
Tredgold, R. H. (1977). *Adv. Phys.* **26**, 80.
Van Allan, J. A., Chang, J., Costa, L., and Reynolds, G. A. (1977). *J. Chem. Eng. Data* **22**, 101.
Van Allan, J. A., Reynolds, G. A., Noonan, J., and Perlstein, J. H. To be published.
Walker, I. P. Private communication.
Young, R. Private communication.

Chapter 3

Photophysical Processes, Energy Transfer, and Photoconduction in Polymers

Robert F. Cozzens
DEPARTMENT OF CHEMISTRY
GEORGE MASON UNIVERSITY
FAIRFAX, VIRGINIA

I. INTRODUCTION

Photochemistry is the branch of chemistry that deals with the processes following the absorption of a photon of electromagnetic radiation by a molecule. When the absorbed photon is in the visible or ultraviolet region of the spectrum, the absorbed energy results in an electronically excited species. The various processes involving the distribution, relocation, and ultimate fate of the excess energy associated with an excited or energy-rich molecule, following the absorption of a photon of electromagnetic radiation, are collectively referred to as *photophysical* processes. Processes resulting in a permanent chemical alteration of a molecule or its

93

chemical reaction with another molecule are called *photochemical* processes. It is possible for excess energy initially localized on a given chromophore to be transferred to another nearby chromophore either on the same or on a different molecule. These *energy-transfer* processes may be either *intramolecular* or *intermolecular* in nature and may occur in one concerted step over distances ranging from van der Waals contact radii to over 100 Å. Following several one-step transfers of energy, the initial packet of excitation energy may find itself considerably removed from the site of initial photon absorption. Subsequent photochemical and photophysical processes may therefore occur at distances remote from the site of the original electronically excited state. One of the many possible photophysical processes is one that leads to a separation of electrical charge, thus producing a positive (hole) or negative (electron) species. If either of these charges is able to migrate under the influence of an electric field, electrical conductivity results and the material is said to be a photoconductor.

All the photochemical and photophysical processes that can occur in small molecules may also occur in macromolecular systems. In addition, extensive intramolecular energy migration and interactions between nearby groups is also possible in macromolecules. A study of the general photochemistry of polymers should proceed first by way of the world of small molecules.

A. Electromagnetic Radiation and Photons

Electromagnetic radiation is best described as a wave propagating at the speed of light c (2.997 × 10^8 m/sec) having associated with it an oscillating electric field and magnetic field both mutually perpendicular to each other and to the direction of propagation. The frequency of oscillation of the electric and magnetic field vectors is denoted by ν and is related to the wavelength λ (the closest distance between two points of identical amplitude on the wave) by the relationship

$$\nu = c/\lambda. \tag{1}$$

The original, classical description of electromagnetic radiation was made by Maxwell in 1862 as a set of partial differential equations. The picture of electromagnetic radiation as simply a wave could be verified by its ability to undergo constructive and destructive interference with other waves of electromagnetic radiation. However, such a simple picture needed reinterpreting following observations by Hertz and further studies by Millikan (1916) on the photoelectric effect. Further anomalies developed following observations of the scattering of x rays shown by Compton (1923) and quantitative study of the light emitted from a blackbody radiator by Max Planck (1901). Explanation of the photoelectric effect by Einstein after the work of Planck led to the concept of electromagnetic radiation behaving not simply as waves but rather as discrete packets of energy called *photons*. Thus, electromagnetic radiation cannot be described simply as a particle or as a wave but rather as something having simultaneous characteristics of both. The energy of one photon was determined by Planck and confirmed by Einstein to be given by

$$\varepsilon = h\nu = hc/\lambda = h\nu, \tag{2}$$

where ν is the reciprocal of the wavelength, and is referred to as the wave number, and c is the speed of light. Planck's constant h has the value of 6.6256×10^{-35} J sec. Since chemists seldom deal with one of anything, but rather with bulk collections, a collection of Avogadro's number of photons is defined as an *Einstein* just as Avogadro's number of molecules is defined as a *mole*.

The electromagnetic radiation spectrum extends from very low-energy, long-wavelength, low-frequency photons in the radio region and beyond into the less understood area of "electric" waves, and up into the high-energy, short-wavelength, high-frequency region of x rays and beyond into gamma rays and the mysterious world of cosmic rays. In the midregion of the spectrum are the ultraviolet (200–400 nm) and visible (400–800 nm) rays which, when absorbed by a molecule, induce electronic transitions between molecular orbitals. Photons in the infrared and microwave regions induce vibronic and rotational transitions, respectively, when absorbed by molecules. The energy of photons in the visible and ultraviolet spectral regions fall in the 170–800 kJ/Einstein range, which is comparable in magnitude to the bond energies of typical chemical bonds, and thus these photons are capable of bringing about chemical transformations and therefore are of particular interest to the photochemist.

B. Electronic Excitation

Electrons in a given molecular orbital within a molecule may interact with an incident photon and be promoted to a more energetic molecular orbital. The photon is either absorbed or not absorbed; it is not possible for only a portion of the energy of a photon to be absorbed. One photon induces one transition within a molecule. Thus, before photochemistry can begin, a single photon must be absorbed. This is a statement of the *first law of photochemistry*. The resulting energy-rich molecule is referred to as being in an excited state. Photochemistry begins with this excited state.

An electron in a given orbital may be in one of two possible spin states; that is, it may have a spin quantum number of $+\frac{1}{2}$ or $-\frac{1}{2}$, which we will denote as an arrow pointing upward ↑ or downward ↓, representing the direction of the magnetic field vector associated with the spinning charged particle. The two electrons required to fill a given molecular orbital may have their spins paired (↑↓) or unpaired (↑↑). A system with a single electron (↑) in an orbital is a free radical and generally quite reactive. Molecules with all electrons in paired spin states (↑↓) are said to be in a *singlet* state, while those with two unpaired electrons (↑↑) are called *triplets*. The term "singlet" refers to the fact that only one wave function is required to describe the energy state, while the triplet system has three different wave functions each having the same energy and is thus a threefold degenerate state. Most organic molecules in their lowest electronic energy state or ground state are singlets, and thus the initial absorption of a photon is generally a singlet-state molecule.

Transitions between molecular orbitals that do not result in changes in electron spin are more probable or *allowed* than those requiring spin changes, which are quantum mechanically *forbidden*. The term "forbidden" should be thought of as meaning low probability rather than not occurring at all. Because of this quantum mechanical selection rule and the fact that most ground-state molecules are in the lowest singlet state, the absorption of a photon of appropriate energy will lead to an electronically excited singlet state S_1 (or S_2, S_3, . . . , S_n), whereas the ground singlet state is denoted by S_0. This absorption process is described by

$$S_0 + h\nu \rightarrow S_1. \tag{3}$$

Similarly, it is possible to absorb a photon of greater energy and excite a ground singlet molecule to an even higher singlet state,

$$S_0 + h\nu' \rightarrow S_2 \text{ (or } S_3, \text{ etc.).} \tag{4}$$

The probability of an incident photon of appropriate energy being absorbed by a molecule is directly proportional to the transition moment that quantum mechanically couples the two states and is thus related to the product of the wave functions that describe the states. This transition moment is indicative of the given molecule and is directly related to the molar absorption coefficient or extinction coefficient ε as used in the classical Beer–Lambert relationship

$$A = \log(I_0/I) = \log(1/T) = \varepsilon l c, \tag{5}$$

where l is the sample path length, c is the sample concentration, T is the transmittance, I_0 is the incident light intensity, I is the intensity of the transmitted light, and A is the absorbance or optical density. The molar absorptivity ε is a function of the wavelength of the absorbed photon, and a plot of ε versus λ is the absorption spectrum of the sample.

C. Photophysical Processes

Once a molecule has been excited to a higher electronic state by the absorption of a photon, it may undergo a variety of photophysical processes. In addition, any of the various excited electronic states may be involved in photochemical reactions resulting in permanent alterations in its chemical nature. First, consider only the photophysical processes. Because vibrational transitions take place at rates on the order of 10^{13} sec^{-1} and electronic transitions between higher-lying singlet states (S_2, S_3, etc.) and the first excited singlet (S_1) are quantum mechanically allowed, most organic molecules initially in higher singlet states are rapidly deactivated to the lowest vibrational level of the lowest excited singlet state in times short compared to chemical reactions. Therefore, interest should be concentrated on properties and transitions involving the lowest excited singlet (S_1) and triplet (T_1) states.

Once in the first excited singlet state a number of avenues for the loss of excess energy are presented to the molecule, ultimately leading it back to the ground excited state. The molecule may emit a photon of light of energy equal to the energy difference between the lowest vibrational level of the first excited singlet and some vibrational level of the ground singlet state. This process is termed *fluorescence,* and the emitted light is of equal or longer wavelength than the initially absorbed photon:

$$S_1 \rightarrow S_0 + h\nu' \quad \text{(fluorescence).} \tag{6}$$

It is also possible for an electron in the excited singlet molecule to undergo a spin transition, thus leading to an excited triplet molecule:

$$S_1 \rightarrow T_1 \quad \text{(intersystem crossing).} \tag{7}$$

This *intersystem crossing* process is quantum mechanically forbidden, however, because the energy difference between the two states is small, the process may be relatively efficient in many cases. In the presence of certain substances, such as heavy atoms, intersystem crossing is even more probable.

Once in the triplet state the molecule may emit a photon of electromagnetic radiation $h\nu''$ of lower energy than fluorescence and return to some vibrational level of the ground singlet state. This process, termed *phosphorescence,* is again forbidden because of the electron spin transition required, however, it is one of the few routes back to the ground state remaining for the energy-rich triplet molecule:

$$T_1 \rightarrow S_0 + h\nu'' \quad \text{(phosphorescence).} \tag{8}$$

Because phosphorescence is spin-forbidden, the time required for this light-emission process is much longer than that required for the shorter-wavelength fluorescence. Phosphorescence decay times may range from 10^{-4} sec to hours, whereas fluorescence decay is generally in the 10^{-9}–10^{-7} sec range. The difference in observed lifetimes between excited singlets and excited triplets as detected by light emission may thus be used to differentiate experimentally between these two processes. If the higher vibrational levels of the ground singlet state come close to or overlap the lower vibrational levels of the excited triplet state, it is possible for the triplet to undergo a radiationless transition and become vibrationally deactivated by an *intersystem crossing* process. The excess energy would ultimately be degraded to heat.

If two excited triplet molecules come into close proximity, they could mutually induce an electron spin transition causing one molecule to become an excited singlet while the other drops back to the ground singlet state:

$$T_1 + T_1 \rightarrow S_0 + S_1. \tag{9}$$

This radiationless process is called *triplet–triplet annihilation*. Although the process occurring in each molecule is quantum mechanically forbidden, the total process is allowed since the net electron spin is conserved. The resulting excited singlet may be subsequently involved in all the various processes in which any other S_1 molecule may be involved. If the S_1 molecule resulting from triplet–triplet annihilation were to undergo fluorescence, the emitted electromagnetic radiation would have the same spectral distribution as normal fluorescence, however, because it arises by way of a long-lived triplet, the decay rate of this fluorescence would appear exceptionally slow. This emission is thus referred to as *delayed fluorescence,* and its existence is indicative of triplet–triplet annihilation or some other process wherein an excited singlet is formed from a triplet species. A summary of photophysical processes is given in Fig. 1.

Two identical nearby molecules or two chromophores on the same molecule, one of which is in its ground singlet state while the other is an excited singlet, can couple to form an excited dimer and behave as a unit called an *excimer*. The excimer has a dissociated ground state and thus has a structureless emission spectrum and no absorption spectrum. Excimer formation and emission may be represented by the sequence

$$S_0 + h\nu \rightarrow S_1, \tag{10}$$
$$S_1 + S_0 \rightleftarrows (S_0S_1)^*, \tag{11}$$
$$(S_0S_1)^* \rightarrow S_0 + S_0 + h\nu'''. \tag{12}$$

If the excited singlet had been formed by triplet–triplet annihilation, then the resulting excimer emission would appear delayed. Excimer emission

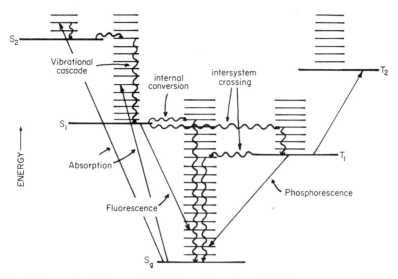

Fig. 1. Energy-level diagram indicating various photophysical processes possible for a typical molecule.

$h\nu'''$ is lower in energy and structureless compared to fluorescence $h\nu'$. If the two singlet chromophores involved in excimer formation are not identical species, then this special type of system is called an *exciplex*.

It is possible for an excited singlet or triplet molecule to come within close proximity of another molecule in the ground state and transfer its electronic excitation energy to the ground state system, resulting in a new excited species. Such an *energy-transfer* process may be represented by

$$D^* + A \rightarrow D + A^*, \tag{13}$$

where D and A represent an energy donor and acceptor molecule or chromophore, respectively, and the asterisk (*) denotes an electronically excited species, either singlet or triplet. For such an energy-transfer process to be allowed, and thus efficient, total electron spin must be conserved. Energy-transfer processes may occur over relatively large distances (up to 100–200 Å) in one concerted, radiationless step and should not be confused with the *trivial process* whereby the excited donor emits a photon that is reabsorbed by the acceptor molecule. The various factors governing the exchange of electronic excitation energy between two chromophores will be discussed in detail later.

D. Photochemical Processes

An electronically excited species, either in a singlet or triplet spin state, may undergo chemical reaction before it has time to loose its excess electronic energy by one of the many photophysical routes provided. This excited species may have an entirely different electron distribution than the ground-state system and thus may be involved in totally unique chemical reactions compared to the reactions common to the ground-state molecule. From the chemical point of view, it is best to think of the excited species as a chemically unique entity. The photochemistry of small molecules has been thoroughly reviewed by Calvert and Pitts (1966), Turro (1967), Chapman (1967), Noyes *et al.* (1967), and many others. The application of the basic rules of photochemistry, which were developed by studying small molecules, to polymeric systems is complexed by the properties and reactions unique to polymers. For instance, gas-phase reactions in polymers are out of the question, the purity of polymers is often dubious, the effect of chain coiling and entanglement is often variable and unknown, and the stability of free-radical fragments in polymers is generally greater than with small molecules as a result of decreased mobility. Numerous reviews on the photochemistry of polymers have been published, including those by Fox and Cozzens (1969a,b), Grassie (1956), and Labana (1976).

An electronically excited molecule, singlet or triplet, may undergo any of numerous photochemical reactions, including bond rupture forming free-radical fragments, intramolecular rearrangements, and oxidation leading to ketone, aldehyde, and peroxide groups, etc. One reaction unique to polymer molecules is depolymerization or unzipping following the initial formation of a free-radical fragment. The major requirement for unzipping is that the C–C bond in the polymer backbone be thermodynamically weaker than the bond between pendant groups and the main-chain carbons. Once a free radical is formed on the main chain, the polymer will split off as a monomer, unit by unit, and continue to depolymerize to a monomer. The *ceiling temperature* is the temperature above which the equilibrium between the monomer and polymer lies in the direction of the monomer. At temperatures above the ceiling temperature a main-chain polymer free radical formed by a photodissociation reaction will continue to unzip to monomer without additional input of energy.

In air, photooxidation commonly causes photochemical damage to many polymer systems and is one of the major causes of yellowing with age. It has been postulated that the formation of aromatic diones is a major contributor to increased visible light absorption with the aging of polystyrene base polymers and copolymers. In addition, end group diones may serve as a source of secondary chain-breaking processes. Another possible source of polymer yellowing is the formation of polyene structures which could absorb in the visible region of the spectra as a result of long-range conjugation and resulting π-electron mobility.

In nearly all cases of polymer photodegradation, the initial photon absorption is not by the basic structural unit of the polymer but rather by impurity groups, end groups, or chain irregularities which absorb at wavelengths longer than those predicted for the hypothetically pure, perfect polymer. As a result, photodegradation is induced in most polymers at surprisingly long wavelengths. Polymers such as polyethylene ($-CH_2CH_2-$), poly(vinyl chloride) ($-CH_2CHCl-$), etc., should absorb only in the far-ultraviolet region of the spectrum, yet photodegradation can be induced by near-ultraviolet radiation. Thus, impurities dominate the photophysics and photochemistry of most polymer systems.

II. EXPERIMENTAL TECHNIQUES

One of the most useful tools available for study of the photophysics of polymers is the spectrophosphorimeter. This instrument monitors the spectral distribution and intensity of light emitted from a sample while simultaneously exciting the sample with a predetermined wavelength of monochromatic light, or the reverse can be performed by observing the emission at one wavelength while varying the excitation wavelength. The emitted light intensity (fluorescence, phosphorescence, delayed fluorescence, excimer fluorescence, etc.) may be recorded as a function of either the wavelength of the exciting light or the wavelength of the emitted light. The resulting spectrum is referred to as an *excitation spectrum* or an *emission spectrum*, respectively. Several publications exist that review these techniques and catalog emission spectra of various molecules, such as those by Parker (1968) and Hercules (1966). The intensity of the emitted light at a fixed wavelength resulting from excitation by a fixed wavelength may be monitored as a function of time following the termination of excitation. The resulting intensity-versus-time plot is the *decay curve* of the species responsible for the emission. Kinetic information regarding the sum of the processes leading to emission may thus be obtained directly.

Phosphorescence, being quantum mechanically forbidden, has an apparent rate of decay much slower than that of the allowed fluorescence process. Fluorescence emission can be separated from phosphorescence by incorporating a mechanical chopper in the path of the excitation light, which operates out of phase with a similar chopper placed in the path of the emitted light. If the chopping rate is slow compared to the rate of fluorescence (and mechanical choppers will be since fluorescence decay is on the order of 10^{-9}–10^{-7} sec) but fast compared to phosphorescence decay, then the fluorescence spectrum is removed from the output, leaving only phosphorescence. By varying the chopper frequency it is possible to separate different emissions by their decay rates. Delayed fluorescence can be separated from normal or prompt fluorescence, even though they have the same spectral distribution, by use of this time-resolving technique.

Electron paramagnetic resonance spectroscopy (EPR or ESR) is another tool useful to the photochemist. The ability to detect molecules with a net electron spin not equal to zero permits the observation of triplet molecules as well as free-radical fragments. It is often possible for detailed EPR studies to describe the microscopic chemical environment of the observed species and thus predict its structure, as well as to monitor the kinetics of the excited state. Direct photolysis of a sample within the resonance cavity is possible, and real-time kinetic measurements may be made on triplet or free-radical species.

Because excited-state molecules are generally quite unstable and have short lifetimes, special handling techniques are required for their observation and study. Frequently, samples are maintained at liquid-nitrogen temperatures (77 K) during the course of observation either as neat samples or in a glass-forming solvent such as a 1 : 1 mixture of tetrahydrofuran and diethyl ether. Glassy systems are used to reduce the mobility of the species being studied, as well as that of impurities. In the case of polymer systems, the difficulty of finding a suitable glass-forming solvent is magnified by the poor solubility of polymers in most media. Many electronically excited molecules either react chemically with molecular oxygen or their natural lifetimes are considerably shortened in its presence; thus, rigorous exclusion of all molecular oxygen is often necessary. Ground-state molecular oxygen is a triplet species and can interact with other triplets leading to singlets without undergoing spin-forbidden processes and thus may quench or shorten the lifetime of many excited species. The possibility of energy transfer to some fortuitous impurity molecule, or a chromophore within a macromolecule, requires the utmost care in maintaining the purity of all samples being studied.

III. ENERGY TRANSFER

A. Intermolecular Electronic Energy Transfer

The radiationless transfer of electronic energy from one molecule to another nearby molecule has been observed for many systems involving both singlet and triplet states. Several comprehensive reviews on the subject of electronic energy transfer have been published including those by Bennett and Kellogg, (1967) and Wilkinson (1967). Quantum mechanical selection rules require that for an allowed process the net electron spin of the system must be conserved. This does not prohibit a given molecule from changing spin as long as the net spin for the entire process remains unchanged. A quantum mechanically forbidden process may still occur, even though forbidden, but with a reduced probability or with a requirement of closer physical proximity than that for an allowed process.

Radiationless energy transfer can be divided into two categories according to the particular mechanism involved, namely, *coulombic* and *exchange*. Energy transfer by a coulombic mechanism involves a quantum-mechanical coupling by a dipole–dipole interaction between the two molecules involved in the energy-transfer process. Higher-order interactions such as dipole–quadrapole and quadrapole–quadrapole are also possible but are generally of little importance in most systems. If the transition involving the energy-accepting molecule (A) is spin-allowed, the transfer may occur over a relatively long distance, up to about 50–150 Å. If the acceptor transition is spin-forbidden (i.e., requires an electron spin change such as S $\rightarrow$ T or T $\rightarrow$ S), the efficiency of energy transfer by a coulombic mechanism will be low. Energy transfer by an exchange resonance mechanism is a shorter-range phenomenon, generally taking place over distances of less than 20 Å, and is often referred to as a contact process. This mechanism is of particular importance in systems where dipole–dipole coupling is weak or where the transition taking place in the acceptor is spin-forbidden.

For weak dipole–dipole interactions between donor (D) and acceptor (A) molecules, Förster (1959) derived an expression for the rate constant for the transfer of electronic energy from D* to A given by

$$k_{DA} = 8.79 \times 10^{-25}(K\Omega/n^4\tau_D^0 R_0^6), \tag{14}$$

where

$$\Omega = \int_0^\infty f_D(\bar{\nu})\varepsilon_A(\bar{\nu}) \, d\bar{\nu}/\nu^4, \tag{15}$$

K is a steric orientation factor, n is the refractive index of the medium, τ_D^0 is the radiative lifetime of the donor, R_0 is the separation distance between the donor and acceptor, ε_A is the absorption coefficient of the acceptor molecule, and f_D is a function describing the emission spectrum of the donor normalized to one. The above rate equation represents a first-order process for a system of donors and acceptors held in a rigid medium at a fixed distance greater than 20 Å. In cases involving transitions that are spin-allowed for the corresponding radiative process for the excited donor (i.e., where Ω is large), R_0 values on the order of 50–100 Å are obtained from this equation. When transitions are not fully allowed, higher-order terms such as dipole–quadrapole interactions described by Dexter (1953) and vibronic interactions discussed by Katsuura (1965) become increasingly important.

The rate constant for energy transfer by an exchange mechanism has been derived by Dexter (1953) and is given by

$$k_{D-A} = K^2(2\tau/\hbar) \, e^{-2R/L} \int_0^\infty f_D(\bar{\nu}) \varepsilon_A'(\bar{\nu}) \, d\bar{\nu}, \tag{16}$$

where K and L are constants. In contrast to the Föster equation for coulombic interaction, this equation for exchange resonance interaction does not depend upon the oscillator strength of the acceptor. Because of this, spin-selection rules for the acceptor need not be adhered to as long as the net spin for the entire process is conserved.

Because of the close contact distances required for energy transfer by an exchange mechanism, it would be expected that the efficiency of this process would be diffusion-controlled in fluid solutions of D and A. A diffusion-controlled rate expression derived by Debye (1942) is given by

$$k_D = 8RT/3000n, \tag{17}$$

where n is the viscosity of the solvent, R is the gas constant, and T is the absolute temperature.

Long-range transfer by a coulombic mechanism requires spin-allowed transitions for both the donor and acceptor such as

$$\text{D*(singlet)} + \text{A(singlet)} \rightarrow \text{D(singlet)} + \text{A(singlet)}, \tag{18}$$
$$\text{D*(singlet)} + \text{A(triplet)} \rightarrow \text{D(singlet)} + \text{A*(triplet)}, \tag{19}$$

whereas a process such as

$$\text{D*(triplet)} + \text{A(singlet)} \rightarrow \text{D(singlet)} + \text{A*(singlet)} \tag{20}$$

is quantum mechanically spin-forbidden. However, if the donor phosphorescence process is of high efficiency, even though spin-forbidden, and if the triplet lifetime is relatively long, energy transfer by a coulombic process may occur by this otherwise spin-forbidden reaction. Observations of long-range coulombic energy transfer by this process have been observed by Ermolaev and Sveshrikova (1962, 1963).

B. Examples of Intermolecular Energy Transfer

Experimental observations involved in the detection of electronic energy transfer include depolarization of fluorescence, luminescence sensitization, and the quenching of emission and other spectroscopic techniques. Transfer of singlet excitation energy from 1-chloroanthracene to perylene in rigid glasses at 77 K, as well as in fluid solutions at room temperature, has been observed by Bowen and Livingston (1954) and Bowen and Brooklehurst (1955). By varying the concentration it was shown that this energy-transfer mechanism involved long-range dipole–dipole interaction between the two molecules. Singlet energy transfer between monomolecular layers of dyes separated by a uniform inert substrate has been observed by Zwick and Kahn (1962) and by Drevhage *et al.* (1963). These systems consisted of an ionic dye chromophore attached to a long hydrocarbon chain. The dye was spread on water along with a fatty acid. By carefully controlled dipping of a solid substrate it was possible to build up layers of dye chromophores separated by a known thickness of hydrocarbon; thus energy-transfer distances could be directly measured. Transfer distances of 30–100 Å were obtained, depending upon the specific dye chromophore, that agreed well with the value calculated by the Föster equation shown above.

Triplet-to-singlet energy transfer by a coulombic mechanism was first observed by Ermolaev and Sveshnikova (1962, 1963) involving triphenylamine as the donor and chrysoiden, chlorophylls *a* and *b*, and pheophytsins *a* and *b* as acceptors. The phosphorescence decay of the donor was observed as a function of acceptor concentration. Transfer distances of 55 Å were measured that compared favorably with a Föster theory prediction of 40 Å. Energy transfer between the triplet state of phenanthrene-d_{10} and the excited singlet state of rhodamine B was observed by Bennett *et al.* (1964). Their experimentally observed transfer distance of 47 Å distance was predicted by the Förster theory.

The effect of singlet energy transfer on the apparent lifetime of the excited donor molecule, D^*, in a solid system as a function of concentration was also studied by Bennett *et al.* (1964) using nonfluorescent dyes as acceptors and pyrene as the donor. The natural radiative lifetime of pyrene (260 nsec) decreased as a function of acceptor concentration in such a manner as to predict a transfer efficiency that was an R^{-6} function of the separation distance between D and A. This is in agreement with Förster's theory for dipole–dipole interaction and transfer of energy by a coulombic mechanism.

Transfer of triplet energy in a glassy solution detected by the sensitization of phosphorescence of the acceptor was first observed by Terenin and Ermolaev (1952). Their studies involved donors such as carbonyl compounds, amines, and various hydrocarbons, and acceptors such as naphthalene and its derivatives, biphenyl and quinoline. By observing the phosphorescence of the acceptor as a function of concentration it was determined that the energy-transfer distance was 12–13 Å, which agrees with a short-range exchange mechanism for this quantum mechanically forbidden process. Similar experiments by Inokuti and Hirayama (1965) involving triplet transfer from benzophenone to naphthalene or 1-bromonaphthalene resulted in a measured transfer distance of 13 Å. The sensitized emission spectrum of the acceptor was identical to the emission spectrum of the acceptor following direct excitation, thus the possibility of formation of a bimolecular complex by charge-transfer interaction was ruled out. Siegel and Goldstein (1965) studied the transfer of triplet energy from benzophenone or phenanthrene-d_{10} to naphthalene-d_8 and concluded that the relative orientation of the donor and acceptor molecules had little effect on the efficiency of triplet–triplet energy transfer by a short-range exchange resonance mechanism.

Transfer of electronic energy from an excited donor polymer molecule to a nonpolymer acceptor molecule has been reported for many systems. Quenching of triplet poly-1-vinylnaphthalene by piperylene has been reported by Cozzens and Fox (1969), indicating energy transfer from a naphthalene chromophore in the polymer to piperylene, a well-documented triplet quencher. Transfer of triplet energy from poly(methyl vinyl ketone) (PMVK) to naphthalene incorporated in a film of the polymer at 77 K has been reported by David *et al.* (1972). These workers observed the phosphorescence quenching of PMVK simultaneously with the sensitization of naphthalene phosphorescence. A study of the quenching and sensitization of emission as a function of naphthalene concentration indicated an effective transfer distance of 11.3 Å. Similar work by these authors (1970) has been performed using polyvinylbenzophenone as sensitizer. The transfer of singlet energy from poly(methyl vinyl ketone) to benzophenone in films of the polymer at 77 K has also been reported.

Triplet energy transfer between identical molecules was reported by Ermolaev (1963, 1964) in a glassy system at 77 K of naphthalene and benzophenone as a function of benzophenone concentration. The results indicated a migration of triplet energy between numerous benzophenone molecules before encountering a naphthalene molecule. Studies on the depolarization of phosphorescence as a function of concentration for several similar organic systems in rigid media by Chaudhuri and El-Sayed (1965) indicated a triplet–triplet energy transfer by an exchange mechanism, from molecule to molecule throughout the sample. This phenomenon of energy being transferred between several like molecules until it eventually finds an energy-trapping site is the model used to describe emission properties of single crystals of organic molecules.

C. Electronic Energy Transfer in Organic Crystals

Nonradiative transfer of energy from one chromophore to another in an organic crystal occurs by an exciton mechanism, as described by

Davydov (1962) and McClure (1960), resulting from strong quantum mechanical coupling between chromophores. By such a mechanism the energy may be transferred many times before becoming trapped or localized on some chromophore of slightly lower energy. An exciton model describes the excitation energy as being a delocalized (wavelike) packet of energy spread over many chromophores. Nieman and Robinson (1962) and Sternleiht *et al.* (1963) have observed the migration of triplet excitons in pure as well as in mixed crystals. The occurrence of delayed fluorescence in many organic crystals is the result of triplet–triplet annihilation following the collision of two triplet excitons migrating throughout the crystal matrix. In the case of a mixed crystal, or of a fortuitous impurity in an otherwise pure system, one can think of the crystal as a host–guest system where energy is delocalized (or rapidly hopping) over many host molecules, eventually becoming localized (or trapped) on a guest molecule that has a somewhat lower energy level. Further migration is thus prevented. Host-to-guest energy transfer in molecular crystals has been reviewed by Windsor (1965). As a result of this exciton mechanism, emission by trace quantities of appropriate impurities may be detected in many organic crystals. In many cases the efficiency of energy trapping is so great that *only* the impurity emission is observed even after exhaustive purification.

Observations of triplet exciton migration in benzophenone crystals doped with trace amounts of 1,2-benzanthracene were made by Hochstrasser (1964). Triplet exciton migration was found to be more efficient than singlet exciton migration in this system because of the long lifetime of the triplet and the high intersystem crossing efficiency from the singlet to the triplet state. The triplet exciton migration rate constant was found to be 10^{10} sec^{-1}. Similar work has been done by Korsunskii and Faidysh (1963) and Colson and Robinson (1968).

Singlet exciton migration in solid solutions of perylene in *N*-isopropylcarbazole has been studied by Klopffer (1969a–c). His studies involved measurements of the ratio of the fluorescence intensity of the host to that of the guest molecule as a function of the concentration of the guest molecule. The results indicated the average displacement of the trapped energy from the initial site of absorption to be 1350 Å. Based on a hopping (or three-dimensional random walk) model this displacement required an energy-transfer rate constant of 4×10^{13} sec^{-1}. It is obvious from this finding that the light emission from a crystal will often be dominated by impurities. Similar studies by Azume (1971) have resulted in experimental transfer rate constants for triplet excitons on the same order of magnitude as those reported for singlet migration.

Polymers can be thought of as having properties of order, somewhat like single crystals, while at the same time possessing a certain degree of disorder, like solutions of small molecules. Polymers in dilute solution may be thought of as one-dimensional single crystals; a relatively long-range displacement of energy can be expected, and energy trapping by impurities often dominates the photochemistry and photophysics of polymers.

D. Intramolecular Energy Transfer and Excimer Formation

Transfer of triplet electronic excitation energy from one chromophore to another, different, chromophore on the same molecule was observed by Lamola *et al.* (1963, 1965). The molecule studied involved a naphthalene group separated from a benzophenone group by a chain of one, two, or three methylenes (I). In this system, the two chromophores are free to

(I)

rotate with respect to one another. Breen and Keller (1968) and Keller (1968) observed intramolecular energy transfer in system (II). However,

(II)

in this case the absorption and emission spectra did not look like the sum of those of the independent chromophores, and thus a charge-transfer complex between the two chromophores was postulated. Sensitization of acceptor fluorescence and quenching of donor emission in small dimer molecules have been studied by DeMember and Filipescu (1968), Conrad and Brand (1968), and many others. Transfer of triplet excitation energy from one chromophore to another on the same small molecule has been less thoroughly studied, but cases have been reported by Keller and Dolby (1969), Leermakers *et al.* (1963), Cowan and Baum (1971), and many others.

The coupling between two nearby identical chromophores, one of which is in the excited singlet state and the other in the ground state, leads to the formation of excimers and excimer fluorescence. No change in the absorption spectrum upon formation of an excimer is detected, thus indicating that no stable ground-state complex exists. The field of excimer formation has been extensively reviewed in recent years by Förster (1969) and Birks (1970). Considerable work involving intermolecular excimer formation in solutions as a function of concentration has been reported by Smith *et al.* (1966), Birks *et al.* (1968), and Hirayama and Lipsky (1969). Because of the required alignment of the two interacting chromophores, temperature, stereochemistry, and solvent viscosity are all controlling factors in excimer formation.

The formation of intramolecular excimers has been reported by Vala and co-workers (1965) for polystyrene and poly-1-vinylnaphthalene in a fluid solution, and by Klopffer (1969a–c) for the 1,3-biscarbazolylpropane molecule (III). In this molecule, fluorescence of the carbazole

$$N—CH_2—CH_2—CH_2—N$$

(III)

chromophore was less intense than expected, and a new broad, structureless emission, red-shifted from that of the normal fluorescence, was defined as excimer fluorescence. The absorption spectrum was identical to the sum of the absorption spectra of the independent chromophores, eliminating ground-state complex formation. This work was extended by Klopffer (1969a–c) to include the polymer poly-N-vinylcarbazole (PVK) which has a molecular weight of about 9×10^5. Excimer fluorescence dominates the emission spectra of this polymer molecule. The concentration of excimer-forming sites in the solid phase was found to be 10^{-3} mole/base mole unit. In addition, the transfer of singlet energy in a solid film of the polymer to guest molecules (perylene, trinitrofluorenone, and hexachloro-p-xylene) was found to follow an exciton mechanism where the excimer-forming sites competed with the guest molecules as trapping locations for the migrating electronic energy. Thus, energy migrates along and between polymer chains in a manner analogous to that observed in molecular crystals. Excimer fluorescence emission and rates of decay of emission have been used in characterization of the stereochemistry and structure of polymers.

Intramolecular triplet energy transfer in polymers has been reported by Cozzens and Fox (1969) in glassy solutions of poly-1-vinylnaphthalene at 77 K. They observed the delayed fluorescence resulting from the triplet–triplet annihilation of two triplet packets of energy migrating along a polymer chain (see Fig. 2). Delayed fluorescence and phosphorescence

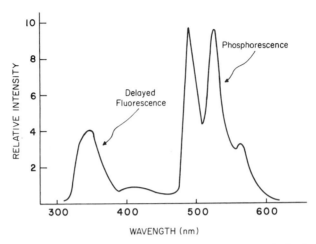

Fig. 2. Delayed emission spectra of poly-1-vinylnaphthalene in THF–Et$_2$O glass at 77 K excited at 290 nm.

intensities were monitored as a function of the intensity of the exciting light. The delayed fluorescence was biphotonic in nature, as would be expected for a triplet–triplet annihilation mechanism, whereas the phosphorescence was monophotonic. Delayed emission as a function of the concentration of the polymer was compared to that from 1-ethylnaphthalene solutions, a compound chosen as a model for a segment of the polymer chain. The effect of piperylene, a known triplet quencher, on the intensity of phosphorescence and delayed fluorescence has confirmed that the triplet migration is intramolecular in nature and leads to triplet–triplet annihilation when two triplets come close enough to interact. Triplet energy transfer along chains of polyvinylphthalimide in thin films leading to trapping by guest molecules has been reported by Lashkov and Ermolaev (1967). The observed emission was from acceptor molecules, such as anthracene and benzyl, and migration occurred along the polymer chain by way of triplet excitons.

Transfer of triplet energy along chains of polyvinylbenzophenone in solid films at 77 K has been reported by David *et al.* (1970). Energy transfer from a triplet state localized on a benzophenone chromophore in the polymer to a naphthalene molecule incorporated in the film exhibited a transfer distance of 36 Å. This is greater than predicted for a single-step process, indicating intrachain transfer prior to localization of the energy in the naphthalene trap.

After trapping energy migrating along a polymer chain by an external acceptor, the next step in the study of intramolecular energy transfer was to incorporate an intrachain energy trap. Fox and Cozzens (1969a,b) synthesized copolymers of polystyrene and poly-1-vinylnaphthalene with various ratios of monomers. Delayed emission spectra were obtained from rigid glass solutions of tetrahydrofuran and diethyl either (1 : 1) at 77 K. It was observed that the phosphorescence of the poly-1-vinylnaphthalene groups was sensitized, while the phosphorescence of the polystyrene chromophores was reduced in the copolymer compared to a mixture of the two polymers at the same concentration (see Fig. 3). This sensitization and quenching indicated an intramolecular transfer of energy along the polystyrene chain until a naphthalene chromophore was encountered, where energy became trapped. The delayed fluorescence from the naphthalene was weak in the low-vinylnaphthalene mole-fraction copolymers and increased as the percentage of vinylnaphthalene groups increased. Copolymers of this type can be considered analogous to one-dimensional crystals doped with guest acceptor groups. Energy may migrate randomly along the polymer chain until it encounters a chromophore that is lower in energy, at which point the migrating exciton becomes localized. The trapping chromophore may be a physical defect or a chemical impurity (such as an oxidation site) incorporated onto the chain. The efficiency of triplet quenching in copolymer films at 77 K by an intrachain group compared to that of mixed polymers was studied by George (1972), and little difference was observed; at least in the case of poly(styrene-*co*-vinylnaphthalene).

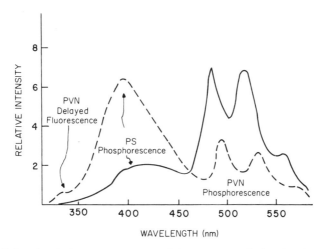

Fig. 3. Delayed emission spectra. ——, poly(styrene-*co*-1-vinylnaphthalene); ---, polystyrene mixed with poly-1-vinylnaphthalene. Both systems have a molar vinylnaphthalene concentration of 1.02%. $\lambda_{ex} = 260$ nm; 77 K; VN = 10^{-3}.

Once localized, a migrating energy packet may become involved in various photophysical and photochemical processes. The importance of the photochemical processes to the overall photostability of the polymer system is obvious. The possibility of photostabilizing a polymer by the incorporation of intrachain energy traps has been investigated (Fox *et al.,* 1976). The quantum yield for chain scission upon photolysis for copolymers of styrene and/or methyl methacrylate with traces of copolymerized vinylnaphthalene is lower than in a system with added 2-methylnaphthalene rather than copolymerized vinylnaphthalene. This indicates that intrachain trapping of migrating energy is a possible mechanism for polymer photostabilization. The efficiency of intrachain energy transfer within a vinyl aromatic polymer molecule is high, and the migration distance is limited by the concentration of traps within the chain. In the case of polystyrene, intrachain migration has been observed by Hirayama (1963, 1965) to occur over distances in excess of 140 styrene units. The necessity for orbital overlap between adjacent chromophores on a polymer chain as required for efficient energy-transfer limits the type of polymers in which intramolecular energy transfer is important. The experimental techniques of detection are generally limited to luminescence, and therefore only polymers with high quantum yields for emission lend themselves to easy study, even though the intramolecular energy-transfer process may occur in many systems where it is experimentally unobserved.

Excimer formation by two neighboring chromophores on a polymer chain has been extensively studied. In glassy dilute solutions of most vinyl aromatic polymers at 77 K, excimer fluorescence is weak or nonexistent. In fluid solutions at room temperature nearly all the emission is excimer formation to occur, as proposed by Vala *et al.* (1965) and by Hiraymaa (1963, 1965), the singlet energy migrating along a polymer chain must encounter a site suitable for excimer formation. In fluid solutions, where molecular geometry is constantly changing, there are many possible excimer sites at a given instant compared to the low-temperature glassy system. The extent to which intrachain excimers compete with physical or chemical defects as an energy trap has been studied by Fox *et al.* (1972) for the poly-1-vinylnaphthalene, poly-2-vinylnaphthalene, and polystyrene systems. The thermodynamic stability of the excimer in these polymers was determined by monitoring the relative amount of normal fluorescence to excimer fluorescence as a function of temperature. In fluid solutions of these polymers, the extent of excimer formation decreases with decreasing temperature as a result of reduced molecular motion, as shown in Fig. 4. The excimer trap depth appeared to be less than that for the corresponding dimeric compound and could compete as a trapping site

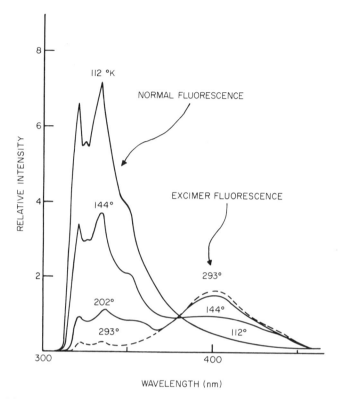

Fig. 4. Fluorescence spectra of poly-2-vinylnaphthalene in THF–Et$_2$O at 10^{-4} base molar naphthalene taken at various temperatures.

with other shallow traps within the chain. However, excimer formation is not a barrier to the migration of energy. In poly-*N*-vinylcarbazole it has been shown by Kloppfer that excimer-forming sites compete with impurities and defects as traps for migrating energy and that the excimer site has a concentration on the order of 10^{-3} mole/base mole. Emission studies by Johnson (1975) involving poly-*N*-vinylcarbazole show two unique types of excimer sites, one a deep trap and the other a shallow trap. Time-resolved fluorescence studies on poly-*N*-vinylcarbazole by Hoyle *et al.* (1978) confirm the existence of two excimer sites, one of which appears to exist prior to excitation and is assumed to be a function of the rigidity of the polymer backbone, while the other is formed after excitation as the molecule randomly moves into the appropriate configuration. The range of singlet energy migration in poly-*N*-vinylcarbazole films has been shown by Okamoto *et al.* (1974) to be limited to 400–900 chromophores as a result of excimer site trapping.

In polymer films where coiling and tangling of polymer chains tends to bring chromophores closer together, interchain transfer would be expected to compete with intrachain migration. The range of energy migration in films will be greater than in dilute solutions, and the importance of traps is increased. The process of triplet energy migration and trapping at 77 K in films of various vinylnaphthalene polymers and copolymers with styrene and methyl methacrylate has been studied by Fox *et al.* (1971) and Fox *et al.* (1972a,b). It was observed in these systems that the nature of the delayed emission spectra was determined by energy traps, both intrinsic and extrinsic to the polymer chain. Triplet migration in these polymers is quite efficient, with its range limited by the concentration of effective traps. Trapped triplet energy is emitted as trap phosphorescence or delayed exciplex fluorescence. An exciplex is an excimer in which the two chromophores are not identical. The delayed exciplex emission results from complexation between an excited singlet formed by triplet–triplet annihilation and a neighboring ground singlet impurity chromophore. Only emission from the trapping impurity was observed in neat films of vinylnaphthalene polymers. This is the situation common to many organic molecular crystals. The fortuitous traps in these vinyl aromatic polymers appear to be oxidation products either incorporated into the polymer during synthesis or formed by degradation following polymerization. Recent work involving poly-α-methylstyrene polymers with naphthalene end groups indicates extremely efficient energy transfer in undiluted films at 77 K leading to sensitized phosphorescence from the naphthalene. Studies by Powell (1971) on the fluorescence emission and decay times of solid films of polyvinyltoluene doped with known amounts of p-terphenyl- and diphenylstilbene indicate the migration of singlet excitons in the polymer with both excimer-forming sites and p-terphenyl competing as energy traps. The exciton undergoes a three-dimensional random walk in these solid films until it encounters a trapping site, involving both intrachain and interchain transfer of singlet energy.

Observation of the migration of singlet energy along chains of poly(methyl vinyl ketone) and poly(methyl isopropenyl ketone) in dilute fluid solutions at room temperature has been made by Somersall and Guillet (1972). Studies were made on the efficiency of intermolecular quenching by biacetyl of the fluorescence of extrachain incorporated naphthalene. Energy is apparently transferred intramolecularly between neighboring carbonyl groups. Measurement of the efficiency of migration in various solvents indicate that the process is not totally diffusion-controlled, at least in low-viscosity solvents.

Fluorescent lifetime and depolarization measurements made by North and Soutar (1972) on systems of dye groups such as rhodamine B and dichlorofluorescein attached to poly-*N*-vinylcarbazole in dilute solutions indicate the transfer of singlet energy along the polymer chain to the trapping dye group. Poly-*N*-vinylcarbazole in dilute solutions was found to display a higher degree of fluorescence depolarization than most other polymers, indicating a high efficiency of intramolecular singlet energy transfer. All these conclusions were based on the assumption that all unbound dye molecules had been extracted from the polymer sample.

E. The Importance of Energy Transfer in Polymers

The phenomenon of nonradiative energy transfer in polymer systems may be of considerable importance in understanding and controlling the photodegradation of polymers. The high mobility of singlet and triplet energy within vinyl aromatic polymers (and perhaps others also) allows migrating energy to find reactive sites along a polymer chain within the radiative lifetime of the excited state. Thus, "weak links" incorporated in a polymer chain become important. Conversely, it is possible to protect photochemically polymer systems by incorporating appropriate energy traps along the chain, perhaps in the form of end groups, which radiate the energy rather than undergoing chemical degradation. It has been found in our laboratory that certain polymers are photostabilized by the intrachain incorporation of trapping groups such as the poly(methyl methacrylate-*co*-styrene) system.

The application of the theory of intramolecular energy transfer to biological systems is becoming more important. Biological processes such as vision, photosynthesis, and energy transport within DNA molecules are only three of many processes requiring energy transfer from one molecular site to another in a living system. The regularity of chromophores along certain protein molecules lends some of these proteins to basic energy-transfer study, and some work along these lines has been done, for example, by Longworth and Battista (1970).

Intramolecular as well as intermolecular energy transfer is important in the photoconductivity of polymers because it allows energy to migrate to an appropriate trapping site where an electron and a hole are generated. The mobility of the resulting charge carrier is analogous to the mobility of an exciton.

One of the major difficulties encountered in the study of energy transfer in polymer systems is the purity and regularity of the polymer molecules and the synthesis of new, unique systems with desired traps incorporated onto the chain. The problem of purity is one that is difficult to overcome. The efficiency of energy migration and ultimate trapping is so high in many systems that trace concentrations of impurities dominate the emission spectra and perhaps the entire photochemistry of the system. The future of these studies depends largely upon the availability of suitable materials for study.

IV. POLYMER PHOTOCONDUCTIVITY

In order for any material to conduct electricity it is necessary that charge carriers be present (positive and/or negative) and that these carriers be free to move under the influence of a polarizing electric field. The conductivity σ of a system is proportional to the concentration of charge carriers c, their mobility μ, and the magnitude of charge on the carrier n and thus may be represented by

$$\sigma = nc\mu. \tag{21}$$

A photoconductive material is one in which the conductivity increases upon illumination. For commercial application in an electrophotographic process a photoconductive material should exhibit an increase in conductivity of at least three orders of magnitude upon irradiation by light at the wavelength and intensities encountered in the application. The increase in conduction originates from an increase in the concentration of charge carriers upon the absorption of photons. The photogeneration of charge carriers is one of the numerous photophysical processes in which an electronically excited state may become involved. All the various photophysical processes compete for the absorbed excitation energy along with various photochemical processes that may permanently alter the photoconductive material. The interest in developing commercially acceptable polymeric photoconductors stems largely from the potential polymers have for ease of fabrication, custom chemical modification to incorporate desirable properties, and relatively low cost.

A. Historical Development of Photoconductivity

The most interesting and thoroughly studied polymeric photoconductor is poly-*N*-vinylcarbazole, hereafter referred to as PVK, and its derivatives, copolymers, and extrinsically doped mixtures. The first report on the photoconductivity of this polymer was by Hoegl *et al.* (1958) who proposed its practical use as an electrophotographic agent. They reported that the sensitivity of this photoconductor could be enhanced by the addition of various organic compounds including dyes, Lewis acids, aldehydes, and ketones. Continuing work by Hoegl (1965) expanded the list of sensitizing additives that could be incorporated in a solid matrix of PVK, all of which were present at relatively low concentrations. Similar studies on doped PVK were made by Hayaski *et al.* (1966) and also by Landon *et al.* (1967) who proposed that the charge-transfer band for the polymer–dopant complex was active in enhancing photoconduction. Since then it has been generally accepted that, in the case of organic electron acceptor dopants, the formation of charge-transfer complexes provides a mechanism for photoconduction different from that provided by dye sensitization or a simple extension of the absorption spectrum of the sample into the visible, but rather that these complexes are an integral part of the photoconduction process. In the case of charge-transfer complex formation much higher concentrations of molecular dopants are possible than with simple dye sensitization; concentrations are encountered up to and including equal molar quantities of electron acceptor and donor chromophores.

The subject of polymer photoconduction has been extensively reviewed in recent years. Seanor (1972) has reviewed the electrical properties of several groups of polymers, and books by Gutmann and Lyons (1967) and Katon (1968) review the semiconductor properties of organic polymers. Photoconduction is extensively surveyed in books edited by Frisch and Patsis (1972) and Patsis and Seanor (1976) and in another edited by Mort and Pai (1976) as well as in a general review of photoconductive polymers by Pearson (1977), and by Hatano and Tanikawa (1978).

A study of the photoconductivity of polymers can be divided into two general areas, both of which independently contribute to the photoconduction of the materials, namely, photogeneration of charge carriers and charge transport or mobility in a polarizing field. The basic observation of photoconduction by measuring the current passing through a sample under an electric field upon irradiation with the appropriate wavelength of light is a measure of a combination of these two processes and in and of itself does not resolve them. Photogeneration of charge carriers is wavelength-dependent and results from the separation of an electron from some chemical entity or group, leaving a positively charged hole. Charge mobility results from the separation of this hole and electron and the transfer of an electron from a neighboring group to a positive hole in the case of *hole migration* or the transfer of an electron to a neighboring neutral group in the case of *electron migration*. These processes are represented in Fig. 5. In neither case is there any mass movement, but simply electron transfer which should not be confused with ionic conduction.

B. Photogeneration of Charge Carriers

The first step in the photogeneration of charge carriers is the absorption of a photon of light, followed by various photophysical processes includ-

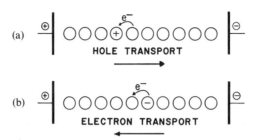

Fig. 5. Schematic representation of hole migration (a) and electron migration (b) in a conductive polymer in an electric field.

ing energy migration to a trapping site. Most of the understanding of carrier generation in organic materials originates from studies on molecular crystals. Application to the processes involved in polymer photoconductivity is considerably more complex because of the lack of long-range order and purity in polymeric systems. The photogeneration processes in polymers may be grouped into four general categories: (a) intrinsic photoconduction of neat or pure polymers (as if there were such things), (b) dye sensitization, (c) charge-transfer trapping, and (d) photoinjection of charge at an interface of two materials such as polymer and electrode.

Charge generation in poly-N-vinylcarbazole has been extensively studied. Work by Pai (1970) and Regensburger (1968) is of particular interest, and the importance of impurities and the role of surface states and contaminants has been well established by Pfister and Williams (1974). This latter work showed that the effect of impurities dominated the photoconduction of even the purest polymers, leading one to question any study on intrinsic photoconduction. The spectral dependence of the photogeneration of charge carriers in PVK films has been found in our laboratory to change with aging of the polymer, showing a continuing increase in photoresponse in the 350–450 nm region while in the 250–300 nm region there is an initial drop in photoresponse followed by little change upon continued aging. Upon aging under ambient laboratory conditions yellowing of the film is noted.

The generally accepted mechanism for the photogeneration of charge carriers in PVK involves the localization of migrating excitation energy on a chromophore followed by electron transfer to a neighboring group, resulting in an excited-state charge-transfer complex. If the coulombic attraction between the electron and hole is on the order of kT, then a mobile charge carrier is formed. All the photophysical processes competing for the electronic excitation energy combine to reduce the quantum yield for carrier generation, and the various photochemical processes result in impurity formation and permanent alteration of the chemical structure of the polymer, as indicated by yellowing. It has been demonstrated by Comizzoli (1972) that the fluorescence of PVK films is reduced in the presence of an electric field, indicating that migration of the carriers in the field prevents germinate pair recombination and reversion of the mobile charge-transfer state to an excited singlet state. This indicates that carrier production occurs by way of the excited singlet state and not the excimer state.

In many cases it is noted that the photogeneration of charge carriers is dependent on the magnitude of the electric field used in making measurements. Pai (1969) observed a dependence of quantum yield on electric field that approximated the function $\exp(\alpha E^{1/2})$. At higher fields the quantum yield approached 0.1, meaning that only 10% of the absorbed photons result in charge-carrier formation. This field dependence may be due to the Poole–Frenkel effect, or field-assisted thermal ionization, as suggested by Hughes (1971a,b) who studied carrier generation in PVK by x-ray excitation. The general Onsager (1938) theory for germinate recombination of electron–hole pairs may be used to explain the field dependence of carrier formation. With this theory it is possible to calculate the initial carrier-separation distance in a photoconductor, which for PVK films is found to be about 60 Å.

The carrier-generation process in PVK films that have not been exhaustively purified is dominated by impurity trapping and thus is extrinsic in nature. Because of the great variety of impurities and defects possible in these polymer films studies on extrinsic effects should be limited to purified polymers into which known dopants have been introduced; even then, questions about purity may always be raised. The light emission and photocurrent of PVK films doped with dimethyl terephthalate, a weak electron acceptor, was studied by Yokogama *et al.* (1975) who found that the fluorescence of the carbazole–terephthalate exciplex was quenched by an electric field, indicating a field-assisted thermalization of the exciplex forming free-charge carriers. The initial excitation energy migrates among the carbazole chromophores to an exciplex-forming site where dissociation into an excited charge-transfer state, $D^+ + A^-$, occurs, where D represents the electron donor chromophore (carbazole) and A is the electron acceptor (terephthalate).

The most commonly studied doped PVK system involves the electron acceptor 2,4,7-trinitro-9-fluorenone (TNF). Hoegl and Neugebaur (1960) reported that the addition of TNF up to 10 mole % in PVK greatly increased the photoresponse of the polymer. Shattuck and Vahtra (1966) reported that higher concentrations of TNF, up to a 1 : 1 molar ratio, produced a highly sensitivity photoconductive material suitable for elec-

trophotographic applications. This system was evaluated more extensively by Schaffert (1971). It is reasonably well established that photogeneration occurs by way of energy trapping by the charge-transfer complex
formed between the TNF and carbazole chromophores. Melz (1972) studied photogeneration in this system using transient electrostatic discharge
techniques and concluded that the carrier-generation efficiency was
strongly field-dependent. The initial carrier separation distance from Onsager theory calculated by the method of Batt *et al.* (1968) as applied to
single-crystal anthracene is found to be about 38 Å for 1 : 1 doped TNF–
PVK, nearly half the value obtained for the undoped polymer. Larger
separation distances are found for lower concentrations of TNF. The recombination of charge carriers in the 1 : 1 doped TNF–PVK system has
been studied by Hughes (1972) using a pulsed ruby laser in order to
produce a high concentration of charge pairs. The results showed that the
fraction of initial charge carriers that succeeded in escaping recombination was independent of carrier mobility and was directly related to the
electric field and distance separating the polarizing electrodes. These results are in good agreement with the diffusion-controlled recombination
theory formulated by Langevin (1903).

The region of spectral sensitivity for photogeneration of charge carriers
may be extended by dye sensitization. Dye-sensitized photoconduction
has been reviewed by Reucroft (1975), and Mylnikov and Terenin (1968)
have studied dye sensitization of metal polyacetylenides. Confusion
still exists as to the mechanism of dye-sensitized photoconduction in
polymers. The question is whether the light-absorbing dye transfers its
excitation energy to the photoconductive polymer which in turn uses it to
generate charge carriers or whether the carriers are formed by the dye
molecules themselves and then are injected into the photoconductive
matrix. Studies that will help resolve this question have been made by
Weigl (1972), but further work needs to be done.

C. Carrier Mobility

Several experimental techniques may be used to measure the mobility
of charge carriers in nearly insulating materials. All these methods involve
transient response time measurements and are thus able to measure carrier mobility independent of carrier concentration. Transient time or
time-of-flight measurements involve the injection of a charge sheet at an
interface and noting the time required for the charge to reach another
interface a known distance (film thickness) away. Charge carriers may be
injected into a polymer film at an interface with another organic layer, a
metal electrode, or an inorganic semiconductor. The first reported mea-

surement of charge drift mobilities in an organic polymer was made by Vannikov (1967) on films of poly(vinyl acetate). Similar studies were made on films of poly(vinyl chloride) and poly(acetonitrile-*co*-vinylpyridine) by Kryszewski *et al.* (1968). In all these studies, the carrier mobility in organic polymers was found to be quite low, generally less than 10^{-4} cm^2/V sec.

Charge injection into PVK films from a layer of vacuum-deposited amorphous selenium has been studied by Regensburger (1968), and similar techniques are being used by numerous workers at the Xerox Corporation. An injection barrier of about 6 V/μm with a spectral response corresponding to selenium was observed. Photoinjection from gold, copper, and aluminum has been reported by Lakatos and Mort (1968) and Mort and Lakatos (1970). This work involved hole injection at the metal–polymer interface and resulted in an observed photoemission threshold energy for the valence band edge of poly-*N*-vinylcarbazole of about 6 eV. Transient photocurrent pulses generally exhibit long tails, indicating a dispersion of charge carriers as they traverse the thickness of the polymer film. Pai (1969) investigated the transient photoresponse in PVK films and found the pulse decay to be dependent in an exponential manner upon the film thickness, temperature, and applied electric field.

Mobility measurements in films of the PVK–I_2 charge-transfer complex have been made by Herman and Rembaum (1966). Measurements of charge mobility in 1:1 PVK–TNF films by Schaffert (1971) show that charge carriers are predominately electrons, while in the undoped polymer conduction is attributed to a combination of holes and electrons. Similar conclusions were reported by Seki and Gill (1971) who found an exponentially field-dependent electron drift mobility of about 10^{-7} cm^2/V sec at a field of 10^5 V/cm. Mobility measurements as a function of acceptor doping concentration over a wide range were made by Gill (1972).

The effect of steric interactions on the mobility of holes in various carbazole-containing polymers has been studied by Limburg *et al.* (1975) and Froix *et al.* (1975). Hole mobilities for three carbazole polymers were measured by the xerographic discharge technique as a function of applied electric field. The highest mobilities were found for polymers that sterically exhibited the highest degree of carbazole ring overlap. This generalization is also supported by mobility measurements made by Okamoto *et al.* (1973a,b) on chloro- and bromosubstituted poly-*N*-vinylcarbazole, poly-1-vinylpyrene, and poly-9-vinylacridine.

The photoelectric properties of copolymers and alternating copolymers of *N*-vinylcarbazole have been studied by Okamoto *et al.* (1973a,b), and Pittman and Grube (1974) studied poly(*N*-vinylcarbazole-*co*-vinylferrocene). All these materials exhibited lower photoconductivity than pure, undoped PVK. It may be concluded that the poorer photoconductivity of these systems results from the inability of carbazole groups to interact because of steric hindrance, thus leading to reduced carrier mobility. The effect on intrinsic photogeneration of charge carriers in these systems is unknown. Yanus and Pearson (1974) have synthesized poly-2-vinyl-fluorenone and reported charge mobilities of less than 10^{-8} cm^2/V sec at a field of 10^5 V/cm, a much lower mobility than that observed for carbazole polymers.

The molecular-weight dependence of photoresponse in PVK has been measured by Tsuchihashi *et al.* (1975) who found photoconductivity independent of molecular weight at $\sim 10^5$, with some dependence at lower weights. The dependence at lower molecular weights may well be the result of end group involvement. Recent studies on systems, such as *N*-isopropylcarbazole (a model for the PVK unit) doped into an otherwise nonphotoconducting polymer matrix such as poly(methyl methacrylate) should eventually help to clarify the steric and structural dependence of photoconductivity.

REFERENCES

Azume, T. (1971). *J. Chem. Phys.* **54**, 85.
Batt, R. H., Braun, C. L., and Hornig, J. J. (1968). *J. Chem. Phys.* **49**, 1967.
Bennett, R. G. (1964). *J. Chem. Phys.* **41**, 3037.
Bennett, R. G., and Kellogg, R. E. (1967). *In* "Progress in Reaction Kinetics" (G. Porter, ed.), Vol. 4, p. 215. Pergamon, Oxford.
Bennett, R. G., Schwenker, R. P., and Kellogg, R. E. (1964). *J. Chem. Phys.* **41**, 3040 (1964).
Birks, J. B. (1970). "Photophysics of Aromatic Molecules." Wiley (Interscience), New York.
Birks, J. B., Srinivasan, B. N., and McGly-n, S. P. (1968). *J. Mol. Spectrosc.* **27**, 266.
Bowen, E. J., and Brocklehurst, B. (1955). *Trans. Faraday Soc.* **51**, 774.
Bowen, E. J., and Livingston, R. (1954). *J. Am. Chem. Soc.* **76**, 6300.
Breen, D. E., and Keller, R. A. (1968). *J. Am. Chem. Soc.* **90**, 1935.
Calvert, J. G., and Pitts, J. N., Jr. (1966). "Photochemistry." Wiley, New York.
Chapman, O. L. (1967). "Organic Photochemistry." Dekker, New York.
Chaudhuri, N. K., and El-Sayed, M. A. (1965). *J. Chem. Phys.* **42**, 1947.
Colson, S. D., and Robinson, G. W. (1968). *J. Chem. Phys.* **48**, 2550.
Comizzoli, R. B. (1972). *Photochem. Photobiol.* **15**, 399.
Compton, A. H. (1923). *Phys. Rev.* **21**, 715.
Conrad, R. H., and Brand, L. (1968). *Biochemistry* **7**, 777.
Cowan, D. O., and Baum, A. A. (1971). *J. Am. Chem. Soc.* **93**, 1153.
Cozzens, R. F., and Fox, R. B. (1969). *J. Chem. Phys.* **50**, 1532.
David, C., Demarteau, W., and Geuskens, G. (1970). *Eur. Polym. J.* **6**, 537.
David, C., Putman, N., Lempereur, G., and Geuskens, G. (1972). *Eur. Polym. J.* **8**, 409.
Davydov, A. S. (1962). "Theory of Molecular Excitons" (translated by Kaska and Oppenheimer). McGraw-Hill, New York.
Debye, P. (1942). *Trans. Electrochem. Soc.* **82**, 205.

DeMember, J. R., and Filipescu, N. (1968). *J. Am. Chem. Soc.* **90**, 6425.
Dexter, D. L. (1953). *J. Chem. Phys.* **21**, 836.
Drevhage, K. H., Zwick, M. M., and Kuhn, H. (1963). *Ber. Bunsenges. Phys. Chem.* **67**, 62.
Ermolaev, V. L. (1963). *Sov. Phys.—Usp.* **6**, 333.
Ermolaev, V. L. (1964). *Opt. Spectrosc.* **16**, 299.
Ermolaev, V. L., and Sveshnikova, E. B. (1962). *Bull. Acad. Sci. USSR Phys. Sci. (Engl. Transl.)* **26**, 29.
Ermolaev, V. L., and Sveshnikova, E. B. (1963). *Dokl. Akad. Nauk. SSSR* **149**, 1295.
Förster, Th. (1959). *Discuss. Faraday Soc.* **27**, 7.
Förster, Th. (1969). *Angew. Chem. Int. Ed. Engl.* **8**, 333.
Fox, R. B., and Cozzens, R. F. (1969a). Photochemistry, *Encycl. Polym. Sci. Technol.* **2**, 760.
Fox, R. B., and Cozzens, R. F. (1969b). *Macromolecules* **2**, 181.
Fox, R. B., Price, T. R., and Cozzens, R. F. (1971). *J. Chem. Phys.* **54**, 79.
Fox, R. B., Price, T. R., Cozzens, R. F., and McDonald, J. R. (1972a). *J. Chem. Phys.* **57**, 534.
Fox, R. B., Price, T. R., Cozzens, R. F., and McDonald, J. R. (1972b). *J. Chem. Phys.* **57**, 2284.
Fox, R. B., Price, T. R., and Cozzens, R. F. (1976). *In* "Ultraviolet Light Induced Reactions in Polymers" (S. S. Labana, ed.), Am. Chem. Soc. Symp. Ser. 25, p. 242. American Chemical Society, Washington, D.C.
Frisch, K. C., and Patsis, A. (eds.) (1972). "Electrical Properties of Polymers." Technomic, Westport, Connecticut.
Froix, M., Williams, D. J., and Foedde, A. O. (1975). *Bull. Am. Phys. Soc.* **20**, 473.
George, G. A. (1972). *J. Poly. Sci. Polym. Phys. Ed.* **10**, 1361.
Gill, W. D. (1972). *J. Appl. Phys.* **43**, 5033.
Grassie, N. (1956). "Chemistry of High Polymer Degradation Processes." Butterworths, London.
Gutmann, F., and Lyons, L. E. (1967). "Organic Semiconduction." Wiley, New York.
Hatano, M., and Tanikawa, K. (1978). *Prog. Organ. Coatings* **6**, 65.
Hayashi, Y., Kuroda, M., and Inami, A. (1966). *Bull. Chem. Soc. Jpn.* **39**, 1660.
Hercules, D. M. (Ed.) (1966). "Fluorescence and Prosphorescence Analysis, Principles and Applications." Wiley (Interscience), New York.
Herman, A. M., and Rembaum, A. (1966). *J. Appl. Phys.* **37**, 3642.
Hirayama, F. (1963). Thesis, Univ. of Michigan.
Hirayama, F. (1965). *J. Chem. Phys.* **42**, 3163.
Hirayama, F., and Lipsky, S. (1969). *J. Chem. Phys.* **51**, 1939.
Hochstrasser, R. M. (1964). *J. Chem. Phys.* **40**, 1038.
Hoegl, H. (1965). *J. Phys. Chem.* **69**, 755.
Hoegl, H., and Neugebauer, C. A. (1960). U.S. Patent 3,162,532, June 13.
Hoegl, H., Sus, and Neugebauer, C. A. (1958). U.S. Patent 3037861, September 8.
Hoyle, C. E., Nemzek, T. L., Mar, A., and Guillet, J. E. (1978). *Macromol.* **11**, 429.
Hughes, R. C. (1971a). *Chem. Phys. Lett.* **8**, 403.
Hughes, R. C. (1971b). *J. Chem. Phys.* **55**, 5442.
Hughes, R. C. (1972). *Appl. Phys. Lett.* **21**, 196.
Inokuti, F., and Hirayama, F. (1965). *J. Chem. Phys.* **43**, 1978.
Johnson, G. E. (1975). *J. Chem. Phys.* **62**, 4697.
Katon, J. E. (1968). "Organic Semiconducting Polymers." Dekker, New York.
Katsuura, K. (1965). *J. Chem. Phys.* **43**, 4149.
Keller, R. A. (1968). *J. Am. Chem. Soc.* **90**, 1940.

Keller, R. A., and Dolby, L. J. (1969). *J. Am. Chem. Soc.* **89**, 2768; **91**, 1293.

Klopffer, W. (1969a). *J. Chem. Phys.* **50**, 1689.

Klopffer, W. (1969b). *Chem. Phys. Lett.* **4**, 193.

Klopffer, W. (1969a). *J. Chem. Phys.* **50**, 2337.

Korsunskii, V. M., and Faidysh, A. N. (1963). *Sov. Phys.—Dokl.* **8**, 564.

Kryszewski, M., Szymanski, A., and Swiatek, J. (1968). *J. Polym. Sci. Part C* **16**, 3915.

Labana, S. S. (ed.) (1976). "Ultraviolet Light Induced Reactions in Polymers," Am. Chem. Soc. Ser. 25. American Chemical Society, Washington, D.C.

Lakatos, A. I., and Mort, J. (1968). *Phys. Rev. Lett.* **21**, 1444.

Lamola, A. A., Leermakers, P. A., Byers, G. W., and Hammond, G. S. (1965). *J. Am. Chem. Soc.* **85**, 2670; **87**, 2322.

Landon, M., Lee-Doller, E., and Weigl, J. (1967). *Mol. Cryst.* **3**, 241.

Langevin, P. (1903). *Ann. Chim. Phys.* **28**, 289,443.

Lashkov, G. I., and Ermolaev, V. L. (1967). *Opt. Spectrosc.* 462.

Leermakers, P. A., Byers, G. W., Lamola, A. A., and Hammond, G. S. (1963). *J. Am. Chem. Soc.* **85**, 2670.

Limburg, W. W., Yanus, J. F., Goedde, A. O., Williams, D. J., and Pearson, J. M. (1975). *J. Polym. Sci.* **13**, 1133.

Longworth, J. W., and Battista, M. (1970). *Photochem. Photobiol.* **11**, 207.

McClure, D. S. (1960). *Solid State Phys.* **8**, 1.

Melz, P. J. (1972). *J. Chem. Phys.* **57**, 1694.

Millikan, R. A. (1916). *Phys. Rev.* **7**, 355.

Mort, J., and Lakatos, A. I. (1970). *J. Non-Cryst. Solids* **4**, 117.

Mort, J., and Pai, D. M. (eds.) (1976). "Photoconductivity and Related Phenomena." Elsevier, Amsterdam.

Mylnikov, V., and Terenin, A. (1968). *J. Polym. Sci. Part C* **16**, 3655.

Nieman, G. C., and Robinson, G. W. (1962). *J. Chem. Phys.* **37**, 2150.

North, A. M., and Soutar, I. (1972). *Faraday Trans I* **68**, 1106.

Noyes, W. A., Hammond, G. S., and Pitts, J. N. (eds.) (Series, 1967–present). "Advances in Photochemistry." Wiley, New York.

Okamoto, K., Kato, K., Marao, Kusabayaski, S., and Mikawa, H. (1973a). *Bull. Chem. Soc. Jpn.* **46**, 2883.

Okamoto, K., Kusabayaski, S., Yokoyama, M., and Mikawa, H. (1973b). *Int. Conf. Electrophotogr., 2nd, Washington, D.C.* See SPSE Publ. 1974.

Okamoto, K., Kato, K., Marao, S., Kusabayaski, S., Mikawa, H. (1974). *Bull. Chem. Soc. Jpn* **47**, 749.

Onsager, L. (1938). *Phys. Rev.* **54**, 554.

Pai, D. M. (1969). *J. Chem. Phys.* **50**, 3568.

Pai, D. M. (1970). *J. Chem. Phys.* **52**, 2285.

Parker, C. A. (1968). "Photoluminescence of Solutions." Elsevier, Amsterdam.

Patis, A. V., and Seanor, D. A. (1976). "Photoconductivity in Polymers, An Interdisciplinary Approach." Technomic Pubs. Corp., Westport, Connecticut.

Pearson, J. M. (1977). *Pave Appl. Chem.* **49**, 463.

Pfister, G., and Williams, D. J. (1974). *J. Chem. Phys.* **61**, 2416.

Pittman, C. V. and Grube, P. L. (1974). *J. Appl. Polym. Sci.* **18**, 2269.

Planck, M. (1901). *Ann. Phys.* **4**, 553.

Powell, R. C. (1971). *J. Chem. Phys.* **55**, 1871.

Regensburger, P. J. (1968). *Photochem. Photobiol.* **8**, 429.

Reucroft, P. J. (1975). *Polym.-Plast. Technol. Eng.* **5**, 199.

Schoffert, R. M. (1971). *IBM J. Res. Dev.* **15**, 75.

Seanor, D. A. (1972). *In* "Polymer Science" (A. D. Jenkins, ed.), p. 1233. North-Holland Publ., Amsterdam.

Seki, H., and Gill, W. D. (1971). *Proc. Int. Conf. Conduct. Low-Mobil. Mater., 2nd* p. 409. Taylor and Francis, London.

Shattuck, M. D., and Vahtra, U. (1969). U.S. Patent 3,484,237, June 13, 1966; December 16.

Siegel, S., and Goldstein, L. (1965). *J. Chem. Phys.* **43,** 4185.

Smith, F. J., Armstrong, A. T., and McGlynn, S. P. (1966). *J. Chem. Phys.* **44,** 442.

Somersall, A. C., and Guillet, J. E. (1972). *Macromolecules* **5,** 410.

Sternleiht, H. G., Nieman, G. C., and Robinson, G. W. (1963). *J. Chem. Phys.* **38,** 1326.

Terenin, A. N., and Ermolaev, V. L. (1952). *Dokl. Akad. Nauk SSSR* **85,** 547.

Tsuchinashi, N., Enomoto, T., Tanikawa, K., Tajivi, A., and Hatano, M. (1975). *Macromol. Chem.* **176,** 2833.

Turro, N. J. (1967). "Molecular Photochemistry." Benjamin, New York.

Vala, M. T., Jr., Haebig, J., and Rice, S. A. (1965). *J. Chem. Phys.* **43,** 886.

Vannikov, A. U. (1967). *Sov. Phys.—Solid State* **9,** 1068.

Weigl, J. W. (1972). *Photochem. Photobiol.* **16,** 291.

Wilkinson, F. (1964). *Adv. Photochem.* **3,** 241.

Windsor, M. W. (1965). *In* "Physics and Chemistry of the Organic Solid State" (D. Fox, M. Labes, and A. Weissberger, eds.), p. 343. Wiley (Interscience), New York.

Yanus, J. F., and Pearson, J. M. (1974). *Macromolecules* **7,** 716.

Yokogamo, M., Endo, Y., and Mikawa, H. (1975). *Chem. Phys. Lett.* **34,** 597.

Zwick, M. M., and Kuhn, H. (1962). *Z. Naturforsch.* **17A,** 411.

Chapter 4

Photovoltaic Phenomena in Organic Solids

Vingie Y. Merritt
IBM CORPORATION
CORPORATE HEADQUARTERS DIVISION
IBM JOURNAL OF RESEARCH AND DEVELOPMENT
WHITE PLAINS, NEW YORK

I. INTRODUCTION

Perhaps more than any other single factor, the increasingly recognized need to develop inexpensive and potentially renewable energy sources has caused a tremendous expansion of research efforts in photovoltaics. The

127

economic appeal of being able to efficiently convert a free and abundant supply of solar energy directly into electricity is obvious. Although the inorganic semiconductor materials silicon, gallium arsenide, sulfide salts of cadmium and copper, and various alloys of these materials form the basis for the vast majority of research and development work in this area, this chapter will discuss efforts that have concentrated on the use of organic photovoltaic materials.

Several features of organic materials make them attractive candidates for use in photovoltaic devices. They are potentially inexpensive and readily available. The actual fabrication of devices can be rather simple and inexpensive compared with techniques presently used in most inorganic systems (although there have been continued advances in these areas). For example, well-known spin-coating or dip-coating techniques might be used. Many organic materials can also be used in thin-film form; as a result, material costs can be reduced considerably. Organic materials can, in most cases, be easily derivatized; thus, the possibility exists for tailoring specific features such as absorptivity, reflectivity, tensile strength, elasticity, adhesion, resistivity, density, dielectric constant, etc.

To date, the overwhelming barriers to the successful utilization of organic devices have been their extremely complex photophysics (and photochemistry) and especially their extremely poor conversion efficiencies [1–22], despite early theoretical predictions of efficiencies of a few percent [23]. However, in the past few years, there have been reports of efficiencies close to 1% with various intensities and device configurations [24–27]. Figure 1 shows the power-conversion efficiencies of common inorganic and organic photovoltaic systems. Obviously, organic materials are at present nowhere nearly competitive with the traditional inorganic materials, where not only are the physics of the materials well understood but solar conversion efficiencies as high as 22% have been attained [28, 29]. Indeed, even for organic solar cells for which the fabrication and material costs can be considered negligible, the conversion efficiencies must still be ~10% to be competitive with today's fossil fuel source electricity, and problems of dc-to-ac conversion and power storage remain. It is hoped, however, that a better understanding of the principles of photovoltaic behavior with respect to organic materials will both give further insight into the requirements for truly efficient organic photovoltaic materials and stimulate much-needed additional research in this area.

A. The Photovoltaic Effect—A Simple Definition

In this chapter we discuss only solid-state systems, and the term *photovoltaic effect* will refer to the development of a voltage across an electro-

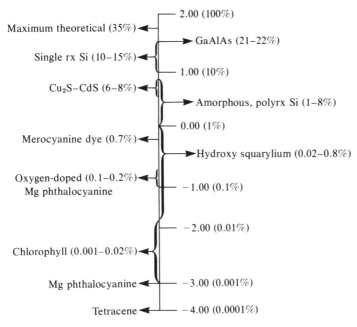

Fig. 1. Comparison of power-conversion efficiencies for organic and inorganic photovoltaic devices (log scale).

static potential barrier under the influence of light. (A separate and abundant body of literature exists on *photogalvanic* effects, which are due to electrochemical potential barriers, mostly in solid–liquid and liquid–liquid organic systems [30]. These effects are not discussed here.) The photovoltaic effect considered here is to be distinguished from the Dember effect [1, 31–43] and the photomagnetoelectric effect (PME) [44–50]. The Dember effect can be described as the development of a voltage difference across a sample along the direction of light propagation, and results from varying degrees of light penetration, light absorption, and resultant carrier generation, and from differences in the diffusivity of carriers (mobilities). The PME effect involves the development of a potential (Hall voltage) in a sample at mutual right angles to the direction of light propagation and an applied magnetic field. Neither of these phenomena will be discussed further.

In the classical photovoltaic effect, the electrostatic potential gradient exists in the dark and results from the presence of an interfacial region where the net majority-carrier density has been reduced (depleted) from the bulk equilibrium value. This *depletion region* (or space-charge layer) will be discussed in greater detail in Section II.

B. Photovoltaism versus Photoconductivity

The existing potential gradient assists in the generation, separation, and migration (if applicable) of the *charged carrier species* produced by light. It is this "built-in" field that distinguishes *photovoltaism* from *photoconductivity,* where externally applied fields are required to produce current. Normally, in the absence of such an applied field, photogenerated species in a photoconductor simply recombine, giving no net current flow. Thus, any photovoltaic material must be photoconductive, but the converse is not necessary. The special considerations involved in finding a photoconductor that will be a good photovoltaic material will be discussed.

A simple four-step description of the photovoltaic process might then be, as follows:

(1)   photogeneration of charged or neutral *carrier species,*
(2)   *charge* separation or generation from neutral carrier species,
(3)   charge transport, and
(4)   charge collection to yield current.

II. THEORETICAL ASPECTS

A. Barrier Contacts—Dark Considerations

The creation of the depletion layers can be explained either by using the concepts of semiconductor band theory (for example, see Refs. [1, 2, 37–41, 42b, 51–85] or by relying solely on thermodynamic principles [86]. Although some workers object to describing organic materials, even approximately, as semiconductors with any degree of long-range delocalized band structure, the approximation is useful for an introduction to the principles of photovoltaic behavior in organic materials. Once one is familiar with these concepts, it is then appropriate to redefine and reevaluate actual photovoltaic results observed in organics in light of the much more complicated and nonbandlike electronic character of most of these materials. It is the purpose of this chapter first to provide a simplified introduction to photovoltaic effects in certain "well-behaved" organic systems (direct generation of free charge carriers and bandlike semiconductor behavior) and then to explore and explain observed deviations from the ideal case.

1. Intrinsic and Extrinsic Semiconductors and Work Functions

Figure 2 shows ideal energy-level diagrams for intrinsic and extrinsic semiconductors. The energy of the highest normally occupied ground

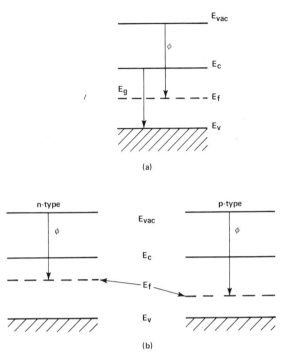

Fig. 2. Energy-band pictures for (a) intrinsic and (b) extrinsic (n- and p-type) semi-conductors.

vibrational state (or molecular orbital) is represented in this figure as the upper valence band edge E_v. (All energy levels are given relative to a reference vacuum level E_{vac}.) The lowest-energy excited (conduction) state available is represented by the lower conduction band edge E_c. The Fermi level E_f can be thought of as a thermodynamic potential or the chemical potential per electron. It is determined by the Fermi–Dirac distribution function (see Refs. [1, 31, 42c]) and is the energy at which the probability that a given energy level is occupied is $\frac{1}{2}$. For $p \ll N_v$,

$$E_f = E_v - kT \ln(p/N_v),$$

where p is the concentration of holes in the valence band and N_v is the effective density of states in the valence band. In other terms, E_f can be considered as the limiting level to which available energy bands are normally filled with electrons. In a pure intrinsic semiconductor (no states within the bandgap E_g), E_f is situated exactly halfway between E_v and E_c. The work function (ϕ) of a pure single-crystal material is simply $(E_{vac} - E_f)$.

For extrinsic, impure, and polycrystalline semiconductors the picture is far more complex. For n-type semiconductors (where the majority carriers are electrons in the conduction band) donor levels are present (excess electrons can be thermally excited into low-lying conduction levels), and the effective (quasi or steady-state) Fermi level lies closer to E_c. For p-type semiconductors (where the majority carriers are holes in the valence band) vacancies are created in the normally occupied valence levels because of the presence of localized acceptor states within the bandgap, and the effective Fermi level lies closer to E_v (see Fig. 2) [42, 75a, 87–90]. Although 99.99999% purities have been achieved for traditional inorganic semiconductor materials, such purity levels in most organic materials have not as yet been attained. As a result, numerous impurity states (donor or acceptor) exist within the bandgap.

One can think of the value $(E_v - E_{vac})$ as being related to the ionization potential I_p of the material (the energy needed to remove an electron from the material), whereas the value $(E_c - E_{vac})$ is related to the electron affinity χ (the energy needed to add an electron to the material) (see Fig. 3). Work functions can be calculated for pure intrinsic single-crystal semiconductors by $(I_p - \frac{1}{2}E_g)$ or by $(\chi + \frac{1}{2}E_g)$, where E_g can be estimated from the onset of absorption in the electronic excitation spectrum. The work functions can also be measured by means of external photoemission, where ϕ is given by a known collector work function plus the measured saturation voltage (see, for example, Refs. [91–94]).

The theoretical "flatband" pictures (shown in Figs. 2 and 3), even for absolutely pure intrinsic semiconductors, do not address deviations in carrier densities and energy-level distributions that occur at surfaces or defect sites; see, for example, Ref. [95]. Dangling bonds, surface lattice vacancies or distortions, etc., can produce variations in the density of states (band bending) even in the absence of any impurities or contact effects (discussed next).

It must also be remembered that polycrystalline values of E_v, E_c, and E_g (and thus E_f) are not the same as those for their single-crystal counterparts because of polarization energies P due to additional atomic interactions and lattice effects; see Fig. 3c and Refs. [94, 96–99].

All these factors make quantitative determination of work functions for organic materials extremely difficult. The work functions of about 40 organic conductors have been estimated from the data of various workers [1, 6, 16, 17, 24b, 94, 100–102] and are listed in Table I. In addition, the most recently published preferred work functions for common electrode materials can be found in Ref. [103]. The great majority of monomeric organic photoconductors and dyes studied have estimated work functions between 4.0 and 5.0 eV; for polymeric materials, the range may be much broader.

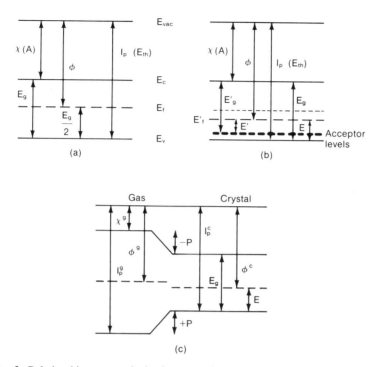

Fig. 3. Relationships among ionization potentials, electron affinities, work functions, energy gaps, and polarization energies: (a) intrinsic semiconductor, (b) extrinsic semiconductor, and (c) comparisons of energy levels for gas- and crystal-phase semiconductors. (From Ref. [94].)

Most treatments of the theory of photovoltaic effects have extensive discussions and mathematical derivations for homo- or heterojunction devices, which involve the contact of two semiconductor materials (with equal or unequal bandgaps, respectively) but different conductivity types (pn, np, etc.) [1, 2, 10, 42, 62, 75c,d, 88b, 104–116]. After a brief introduction to these concepts, this chapter will concentrate on Schottky barrier devices, which involve contact of a semiconductor with a metal. This is because most of the detailed work on organic photovoltaic effects has been on devices of this type.

2. Contact Potentials

When two semiconductors with different work functions are brought into contact there will be a diffusion of majority and minority carriers across the interfacial region until a single effective Fermi level E_f' is attained. At equilibrium, the current contributions to each side of the junction (see Fig. 4) must be equal. The current contributions to this dark

TABLE I *Work Functions of Some Organic Photoconductors*

Material	ϕ	Ref.
Anthracene	4.71	[94]
1,2-Benzanthracene	5.24	[1^a, 100]
Benzidene TCNQ	4.81, 4.97	[94]
β-Carotene	4.0–4.75, 4.3	[1^a, 100]
Chlorophyll	$\approx$4.5	[1^a]
Chrysene	4.1–4.2	[1^a]
Coronene	4.1–4.5	[1^a]
Crystal violet	$\approx$3.3–4.1	[1^a]
Cu Phthalocyanine	4.56	[94]
Dibenzophenathiazine DDQ	4.58, 4.63	[94]
Dibenzophenathiazine$_2$ DDQ	4.60, 4.62	[94]
3,3-Diethylthiacarbocyanine	$\approx$4.05	[1^a]
Erythrosin	4.4, $\approx$4.5, 4.6	[100, 101b]
Hydroxy squarylium	4.15, 4.2	[24b]
Indanthrone	4.76, 4.94	[101b]
Isoviolanthrone	4.62	[94]
Malachite green	3.4–4.2, 4.9	[100]
Methylene blue	4.1, 4.7	[100]
Methyl squarylium	4.5	[24b]
Merocyanine	4.65	[100]
Mg Phthalocyanine	4.75, 4.85, 4.96	[17]
Naphthacene	3.8–4.6	[1^a]
Naphthalene	3.8–4.4	[1^a]
Pentacene	4.33	[101a]
Perylene	4.54, 4.63	[94]
Phenosafranin	4.35, $\approx$4.5	[100]
Phenothiazene	3.6–4.0	[1^a]
Phthalocyanine (metal-free)	4.08, 4.41, 5.15	[94, 100]
Pinacyanol	4.0, $\approx$4.1–4.2, 4.5	[1^a, 100]
Polyethylene	4	[102]
Pyranthrene	4.47, 4.54	[94]
Pyrene	$\approx$4.0–4.2	[1^a]
Quaterylene	3.88 (intrinsic), 4.46 (extrinsic)	[101a]
Rhodamine B	4.2, 4.5	[100]
Rhodamine 6G	4.65	[100]
TCNQ	5.01	[94]
Tetracene	4.30, 4.51	[94, 101a]
Tetrathiatetracene	4.24, 4.32	[94, 101a]
Violanthrene	4.48, 4.53, 4.76	[94, 101a]
Violanthrone	4.60, 4.66	[94]
Zn Phthalocyanine	5.15, 5.2	[100]

^{a} Table 6.7.

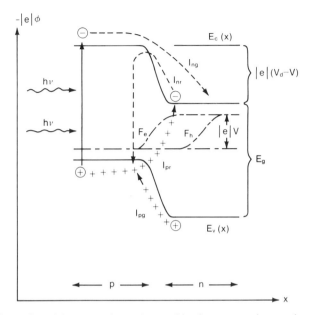

Fig. 4. Illustration of the generation and recombination currents in a p-n junction device. I_{pg}, Generation current due to electrons in the p-type material; I_{pr}, recombination current due to electrons from the n-type material diffusing into the p-type material; I_{nr}, recombination current due to holes from the p-type material diffusing into the n-type material; F_e and F_h, potential steps for electrons and holes; V_d, diffusion voltage. (From Ref. [63].)

"equilibrium" situation result from *generation currents* I_g and *recombination currents* I_r. Generation currents are due to thermal generation of *minority* carriers on each side of the junction and are defined as being negative. Recombination currents are due to the diffusion of *majority* carriers into neighboring regions and are defined as being positive. The generation currents are essentially determined by the temperature T of the device and can be considered to be independent of any potential field eV across the device. The recombination currents, however, are affected by such fields. The net dark current at equilibrium

$$I_{de} = I_r - I_g = 0, \tag{1}$$

where for the p-n junction $I_r = I_r(\text{holes on the p side}) + I_r(\text{electrons on the n side})$ and $I_g = I_g(\text{electrons on the p side}) + I_g(\text{holes on the n side})$. This diffusion of carriers causes a net depletion of majority carriers on each side of the interface and leads to an effective "bending" of the E_c and E_v bands (since E_g must be maintained) by an amount equal to the diffusion potential $|e|V_d$. It is to be noted that, for these p-n junctions, the maximum value of $eV_d = E_g$, which would be achieved for heavily doped materials where the Fermi level is "pinned" to E_v or E_c.

The analogous situations involving contacts of n- and p-type semiconductors with metals are shown in Figs. 5a–5d. Here again, the net equilibrium dark current is given by Eq. (1), but I_r and I_g are considerably simplified since only one semiconductor is involved. When a semiconductor is in contact with a metal, the dark diffusion process depends on the relative (to E_{vac}) values of the work functions for the metal ϕ_m and the semiconductor ϕ_{ns} (n-type) or ϕ_{ps} (p-type). When $\phi_m > \phi_{ns}$, electrons diffuse into the metal, lowering the effective Fermi level (relative to E_v and E_c) and creating a region depleted of majority carriers near the interface (Fig. 5a). This is depicted by bending of the bands near the interface. In the bulk of the material the relative level of E_f to E_v and E_c remains unchanged. Now, if electrons in the bulk of the semiconductor are to further diffuse into the metal, they must first overcome the diffusion potential $|e|V_d$, which is theoretically determined by $\phi_m - \phi_{ns}$ (or $|\phi_m - \phi_{ns}|$). See, for example, Ref. [116].

The electronic diffusion potential in Schottky barriers is still limited by E_g, since if $\phi_m - \phi_s > E_g$, electrochemistry (ionization of the metal and the semiconductor) would occur and an electrochemical potential (ionic conduction, photogalvanic effect) would result. As for the minority carriers, it is easier for them to diffuse into the metal since holes "flow" up a potential gradient. The result is a rectifying (or blocking) contact. When $\phi_m < \phi_{ns}$, electrons diffuse into the semiconductor, raising the effective Fermi level and creating a region with an excess of majority carriers (accumulation region) in the semiconductor near the interface (Fig. 5b). This accumulation of majority carriers is easily accommodated by the metal, and it is in fact easier for such carriers to flow into the metal; i.e., there is an ohmic contact.

By a similar analysis of p-type Schottky devices, when $\phi_m > \phi_{ps}$ an ohmic contact results. Thus we have the following general rules:

$$\phi_{ps} > \phi_m \quad \text{or} \quad \phi_{ns} < \phi_m = \text{Schottky barriers,} \tag{2}$$

$$\phi_{ps} < \phi_m \quad \text{or} \quad \phi_{ns} > \phi_m = \text{ohmic contacts.} \tag{3}$$

By proper choice of the electrode materials (ϕ values), one can in theory manipulate the diffusion voltage V_d (see also Section II.B.1). Table I listed estimated work function values for some common organic conductors, and Ref. [1] tabulates most of the well-known organic conductors according to n- or p-type conductivity. In addition, the most recent preferred work function values for common electrode materials can be found in Ref. [103]. Theoretically, the higher the V_d the better, since the maximum photovoltage will be limited by this value. The use of some electrode

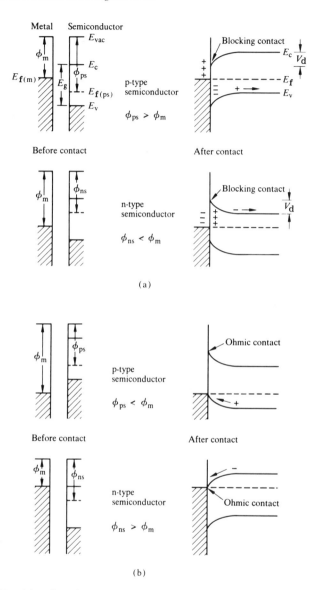

Fig. 5. Band bending that occurs in Schottky barrier devices. (a) Blocking contacts formed when p- and n-type semiconductors make contact with metals with work functions less than or greater than those of the semiconductor, respectively. (b) Ohmic contacts formed when the work functions of the metals are greater than (p-type) or less than (n-type) those of the semiconductors.

materials may not be practical, however, either because of cost or other device-fabrication considerations (deposition requirements, thermal stability, tendency toward oxidation or reduction, etc.). The extent of this band bending, the *depletion depth,* will be discussed in Section III.C. Other methods for manipulating the barrier potential are discussed in the following sections.

3. Injection Potentials

Also to be considered is the ability of majority carriers in the metal to further diffuse into the semiconductor. We will define two *injection potentials,* one for electrons ϕ_{ie} and one for holes ϕ_{ih}:

$$\phi_{ie} = (E_f - E_c) + |e|V_d = (\phi_m - E_c) + |e|V_d, \tag{4}$$

$$\phi_{ih} = (E_v - E_f) - |e|V_d = (E_v - \phi_m) - |e|V_d. \tag{5}$$

The second equalities in Eqs. (4) and (5) follow from the single Fermi level at equilibrium being essentially equal to ϕ_m, since the metal can be considered an infinite source and sink for carriers, which would further lower these injection barriers. It can thus be seen that, if $|e|V_d \leq (\phi_m - E_c)$ or $(E_v - \phi_m)$, the electrodes can be ohmic or injecting; they inject electrons if $\phi_m \approx E_c$ and holes if $\phi_m \approx E_v$. For $|e|V_d > (\phi_m - E_c)$ or $(E_v - \phi_m)$ no injection occurs [114a]. These electrode injection effects can contribute significantly to observed photocurrents (see Refs. [1, 37, 39–42, 60–62, 74, 87, 95, 104–106, 114] and later discussions on space-charge-limited currents).

Remember that the band pictures described so far do not account for any localized deviations in energy levels due to defects or surface states. Also remember that these electrostatic potentials exist in the dark.

4. Dark I–V Rectifying Behavior and Applied Voltages

Referring again to the p-n junction case, if the electrostatic potential of the p side is raised by V relative to the n side (forward bias), e.g., by means of an applied voltage V_{appl}, we no longer have the previous equilibrium situation (Fig. 6). There evolves a new equilibrium potential barrier $|e|V$ due to further diffusion and recombination of majority carriers:

$$|e|V = |e|(V_d - V_{appl}), \tag{6}$$

where a forward bias is defined as positive and a reverse bias as negative. The recombination current I_r increases, while I_g remains constant. The following relationships have been derived [63] for *ideal* junctions or Schottky devices:

$$I_r = I_g \exp(|e|V/kT), \tag{7}$$

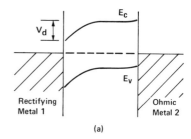

(a)

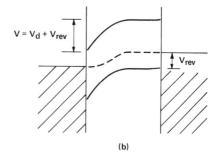

(b)

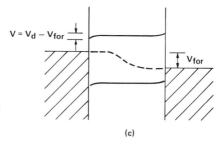

Fig. 6. Effect of bias voltage on a Schottky barrier (p-type organic): (a) unbiased, (b) reverse bias, (c) forward bias.

(c)

where k is the Boltzmann constant, V is the effective voltage across the device, and the temperature T is given in absolute degrees. Note that if $V_{\mathrm{appl}} = 0$, $V = V_{\mathrm{d}}$. The *net* dark current I_{d} across the junction (or barrier) is then given by

$$I_{\mathrm{d}} = I_{\mathrm{r}} - I_{\mathrm{g}} = I_{\mathrm{g}}[\exp(|e|V/kT) - 1]. \tag{8}$$

In terms of the net dark current density J_{d} ($=I_{\mathrm{d}}/A$, where A is the junction area),

$$J_{\mathrm{d}} = J_0[\exp(|e|V/kT) - 1], \tag{9}$$
$$J_0 = I_{\mathrm{g}}/A = A^*T^2 \exp(-V_{\mathrm{d}}/kT), \tag{10}$$

where e is the electron charge, A^* is the Richardson constant (1.2×10^6 A/m² K^2), J_0 is the saturation current density under reverse bias (determined by the physical and chemical properties of the conductor), and V_d is the diffusion voltage.

For most organic and many inorganic materials, particularly at low applied voltages (≤ 0.4 V), a diode factor B must be added to fit experimental data to theory:

$$J_d = J_0[\exp(|e|V/BkT) - 1].\qquad(9')$$

The value of B ranges from 1 to 3 and is usually taken as 2 [6, 63].

From these equations it can be seen that the dark current–applied voltage curve should show rectification (the device behaves as a diode). The applied voltage can be used to alter the effective voltage across the device. When a reverse-bias voltage V_{rev} is applied to a contact region (V_d), the effective barrier voltage V will be enhanced:

$$V = V_d + V_{rev} = V_d - (-V_{appl}).$$

If a forward-bias voltage V_{for} is applied, V may be smaller than V_d, completely eliminated ($V_{for} = V_d$), or even have a sign opposite that of V_d; i.e., the contact becomes nonblocking, injecting, or ohmic;

$$V = V_d - V_{for} = V_d - V_{appl}.$$

It can also be seen that if $V \approx BkT$, where T is now the ambient device temperature, majority carriers in the semiconductor have enough thermal energy to overcome the electrostatic potential barrier and carriers in the electrodes may also have enough energy to overcome the injection potentials discussed previously.

Electron tunneling or field emission can occur through thin surface barriers, such as between electrodes and a material (carrier injection, see Fig. 7a) or through thin potential barriers between E_v and E_c (external field emission, see Fig. 7b) such as those created by an applied voltage.

5. "Electrode Fields"

In cases where the effective dark barrier gradient across the sample is extremely small, a dominant applied field can be created across the sample by the use of two electrodes with very different work function values [24b, 27]. This principle has recently been used in the design of higher photovoltage outputs from organic photovoltaic devices [12, 16, 17, 26, 27]. The dark I–V and photo I–V behavior of such "modified" Schottky devices can be greatly affected by such "electrode fields," particularly in cases where the quantum efficiency of charge generation and transport is

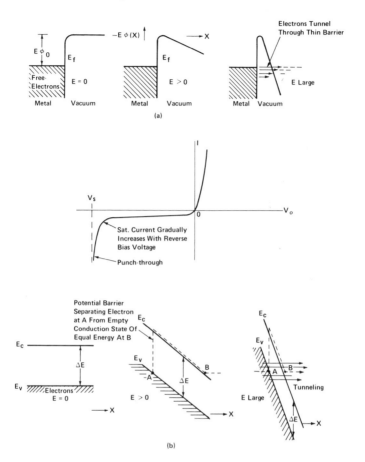

Fig. 7. Deviations from perfect diode behavior caused by (a) field emission due to externally applied voltages and (b) internal field emission within the bulk semiconductor. In each case the series is for increasing applied voltage (right to left).

field-enhanced. Whether the dominant influence on charge transport is an electrode-induced field or an actual Schottky (contact) depletion field gradient will likely depend on the height, width, and gradient of any active barrier region (see later discussion) in relation to sample thickness and to any "external" field, as well as on the actual charge-generation mechanism involved. In cases where V is essentially determined by V_{appl} (from electrodes) one would not expect to be able to easily detect a small gradient within the sample. Here, the field across the sample would be essentially uniform, and both it and the photovoltage would be determined by the difference in work function values between the two electrodes. If, in addition, ϕ_{m1} or ϕ_{m2} were $\approx E_c$ or E_v (respectively) electrode injection mechanisms might dominate. A later section will discuss how one can probe for the existence of finite depletion regions within a sample.

6. MIS (MOS) Cells

Another variation of the typical Schottky device that has recently been used with organic photovoltaic devices is the MIS (MOS) structure, in which a thin interfacial insulating (oxide) layer is inserted between the Schottky barrier metal electrode and the semiconductor. This very thin layer allows tunneling currents to pass through it and presumably helps either to decrease current losses due to surface recombination or to increase V in some fashion. It may also affect the value of B. The derivation of the equation expressing the dark I–V characteristics of such devices can be found in Refs. [24b, 27, 117–121];

$$J_d = J_0 \exp(-\chi^{1/2}d)[\exp(eV/BkT) - 1], \tag{11}$$

where χ is the electron affinity and d is the thickness of the insulating (oxide) layer.

7. Double-Barrier Contacts

The separation and collection of photogenerated carriers in a Schottky photovoltaic device is most efficient if there is one blocking (barrier) contact and one completely ohmic contact. Quite often with organic materials a truly ohmic contact is not present at the "collecting" electrode; one must at least hope for an injecting contact. Very often *two* blocking contacts exist, and the two barrier voltages are subtractive (Fig. 8). The sign of any net photovoltage is determined by the dominant barrier contact (as well as by the type of minority carrier, where it is generated, and the configuration of the external circuit). If $V_1 = V_2$, $V = 0$. Normally, any additional barrier contact will reduce the total effective barrier voltage V, and thus the maximum photovoltage possible from the device. There have been mathematical treatments of the double-barrier case [12, 16, 18]. Any nonohmic behavior at the "collecting" electrode can cause the rate of flow of photogenerated charge carriers out of the device to be less than their rate of generation (dI_g/dt), giving rise to space-charge-limited (SCL) effects. These are particularly important in materials with high bulk resistivities and for high light intensities, where dI_g/dt is large. These effects can significantly alter the behavior from that expected for ideal photovoltaic devices. These factors are discussed in greater detail in Sections II.B and III.

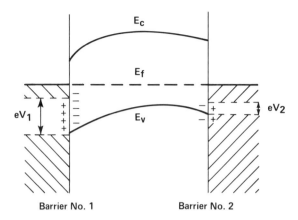

Fig. 8. Band pictures from double-barrier contacts. The total effective voltage of the barrier $V = V_1 - V_2$ and $V_{ph} \approx 0$ when $V_1 \approx V_2$. (From Ref. [16b].)

8. Doping

Since theoretically V (and V_d) are determined by the difference in work functions, anything that might alter the total number of free charge carriers or the net current, and thus the effective Fermi levels, will also influence the effective barrier height (and the net photocurrent). We have already mentioned that heavy doping maximizes V (by "pinning" E_f to E_c or E_v). Doping can also be used to create changes either in surface layers or the bulk material. Conductivity mechanisms can be altered, for example, by actual chemical changes, by the addition of new carrier species, by changes in the character or distribution of trapping or defect sites, by changes in the dielectric constant ε or by other alterations in the electronic properties of the bulk materials. Figure 9 and Table II summarize changes observed in the photovoltaic parameters of Ga/OHSq and phthalocyanine

TABLE II *Photovoltaic Measurements for Undoped and Doped OHSq Devices*[a]

Dopant	Electrodes	Intensity (mW/cm²)	V_{oc} (V)	J_{sc} (μA/cm²)	FF	η (%)
None	Ga/In-O	0.86	0.20	57	0.30	0.004
Methoxy-DEASP			0.42	140	0.30	0.021
None	Ga/In-O	94	0.51	330	0.27	0.048
Bromine			0.59	790	0.46	0.23
None	Ga/Sn-In-O	94	0.52	0.55	0.27	8.2×10^{-5}
Bromine			0.52	2.5	0.27	3.7×10^{-4}

[a] He/Ne light source.

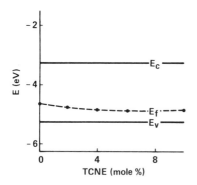

Fig. 9. Position of Fermi level in phthalocyanine doped with tetracyanoethylene (TCNE). (From Ref. [42c].)

Schottky barrier cells on doping (OHSq with bromine vapor and the hole transport material methoxy-DEASP[24b]). See also Refs. [1,25a,39,42b,c, 75a,c, 90, 95–99,122–133]. These effects will be seen both in the dark and under illumination. For successful photovoltaic applications, doping should enhance or at least preserve a high $I_{\mathrm{ph}}/I_{\mathrm{dark}}$ ratio, otherwise the maximum photovoltage will be very small (to be derived).

B. Effect of Light on Barrier Contacts

Figure 10 illustrates what happens when light falls on the barrier or junction region; see also Fig. 4. The effect of light will of course depend on the exact carrier-generation mechanism and the particular device configuration involved, but we use the example of a p-type organic material Schottky barrier cell in which $I_{\mathrm{ph}} \gg I_{\mathrm{d}}$. For simplicity of discussion at the moment, we will assume that each electron–hole pair is generated by a single absorbed photon and separated within the barrier region. (If charge

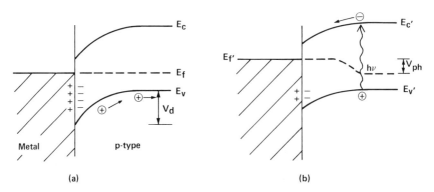

Fig. 10. Effect of light on a Schottky barrier. (a) In the dark, (b) under illumination.

generation occurs via an excitonic mechanism *only* at the electrode surface or within the barrier region, the observed effects will be similar.) The observed photocurrent will then be the difference between current due to species generated by light I_L and current (opposite sign) due to recombination. The change in hole (majority carriers) concentration in the valence band will be insignificant; however, the change in electron (minority carriers) concentration in the conduction band will be substantial. Thus, the observed photocurrent can be considered as due solely to the generation of minority carriers.

Recall that in the dark there is buildup of negative charge (see Fig. 5) in the organic material near the interface. Excess holes generated in the depletion region by light recombine with these negative charges, while the photogenerated electrons flow toward the surface under the influence of the barrier potential. Thus, at equilibrium (in the light) a new effective Fermi level is attained, and the net change in potential is the *observed* V_{ph}.

1. Photovoltage versus Barrier Voltage

Assuming no current losses due to recombination or trapping, the equations for V_{ph} for a Schottky barrier device are

$$V_{ph} = \left(\frac{kT}{e}\right) \ln \left[1 + \frac{I_{ph}}{I_d} \right] \tag{12}$$

$$= \left(\frac{kT}{e}\right) \ln \left\{ 1 + \frac{I_{ph}}{I_0} [\exp(eV/BkT) - 1] \right\}. \tag{13}$$

In theory, V_{ph} can be no greater than V and, since the maximum value of the potential eV is E_g, it can be seen that, if $I_{ph} \gg I_d$, the maximum possible photovoltage (in eV) is $\approx E_g/2$ or (in V) $V/2e$ (for B = 2). If $I_{ph} \approx I_d$,

$$V_{ph} = \left(\frac{kT}{e}\right) \left(\frac{I_{ph}}{I_0}\right) \exp\left(\frac{V}{BkT}\right). \tag{14}$$

Thus, we can see the need to minimize I_0 and maximize V (E_g) and I_{ph}/I_d for the best theoretical photovoltages.

Experimentally observed photovoltages (photo-emfs) can be used to estimate effective barrier heights. Table III shows V_{ph} values for violanthrene [94] and chlorophyll [18] Schottky barrier sandwich cells as a function of electrode work function values. Note that changes in the sign of the photovoltage can be used to differentiate between rectifying and ohmic or injecting contacts. (A *neutral* contact results if $\phi_m = \phi_s$ and no current or, at best, a very small forward or reverse current results.)

Although quantitative agreement between the theoretical V and V_{ph} is in most cases poor, qualitatively the results are consistent with the previ-

TABLE III *Effect of Substrate Metals on Photovoltages Obtained from Violanthrene [94]
and Chlorophyll [16b] Schottky Barrier Cells[a]*

Illuminated contact	Violanthrene: Metal ϕ (eV)	V_{ph} (V)	V_b (theory) (V)
Pb	4.1	+0.06	+0.39[b]
Al	4.3	+0.20	+0.19[b]
Zn	4.3	+0.12	+0.19[b]
Sn	4.5	±0.05	≈0[c]
C (graphite)	4.7	−0.03	−0.21[d]
Au	5.1	−0.05	−0.61[d]
Pt	6.3	−0.12	−1.81[d]

Illuminated contact (ϕ) (eV)	Chlorophyll: Nonilluminated contact (ϕ) (eV)	Sign of V_{ph}	$\Delta\phi$ Electrodes (eV)	$\Delta\phi$ Barrier/ organic (eV)
Al (4.1)	Hg (4.5)	−	−0.4	−0.4
Al	Al (4.1)[e]	+[e]	0.0[e]	−0.4 (−0.4)[f]
Al	Au (4.8)	−	−0.7	−0.4
Al	Ni (5.0)	−	−0.9	−0.4
Al	Cu (4.6)	−	−0.5	−0.4
Au (4.8)	Al (4.1)	+	+0.7	−0.4 (frontwall)[g] +0.4 (backwall)
Cr (4.3)	Hg (4.5)	−	−0.2	−0.2
Cr	HgIn (3.6?)	+	+0.7	−0.2 (−0.9)[f,g]
Cr	Ga (3.8)	+	+0.5	−0.2 (−0.7)[f,g]

[a] For violanthrene, $\phi \approx 4.49$; for chlorophyll, $\phi \approx 4.5$; both are p-type materials.

[b] Rectifying contact.

[c] Neutral contact.

[d] Ohmic or injecting contact.

[e] Varying amounts of aluminum oxide are indicated on the two Al surfaces; therefore the work functions are unknown.

[f] In frontwall configuration the sign of the photovoltage will be shown; in backwall configuration, the sign is reversed.

[g] Presence of two barrier contacts indicated; one tends to offset the other.

ously discussed barrier theory. It is extremely difficult (if not impossible) to determine quantitatively the exact effects of recombination or trapping due to high bulk resistivities, defects, surface states, numerous impurity states, etc., on the different values of the actually observed photovoltages. Extensive treatments have been given by numerous workers in the field of organic photoconductivity.

2. Photocurrent and the Barrier Region

The photovoltaic behavior of an organic device is a function of the total amount of light absorbed within the active region and how effectively this absorbed light generates photocurrent. The reader is referred to numerous excellent and detailed treatments of photoconductivity mechanisms and principles in organic systems; for example, see Refs. [1, 14, 23, 37–42, 51–74, 89a, 90, 95, 106, 113–114, 122–168]. A simplified discussion of some key aspects is provided here.

Under illumination, the generation current discussed previously increases. In the simplest, near-ideal situation, the net observed equilibrium current density is given by

$$J = J_d - J_{ph} = J_0[\exp(|e|V/BkT) - 1] - J_{ph}. \tag{15}$$

The relationship between I_{ph} and V_{ph} has been treated in Eqs. (12) and (14). Sections III.B and III.C discuss relationships between the photovoltaic parameters V_{oc} and I_{sc} (open-circuit voltage and short-circuit current).

a. Bandgaps. In ideal intrinsic conduction processes, mobile (free) carriers are generated only by absorbed photon energies equal to or greater than the bandgap (or the barrier height). Thus, for organic materials with $E_g \leq 2$ eV, thermal or optical excitation of electrons from the ground state to the excited state can occur (using normal wavelengths) by an intrinsic conduction mechanism.

The bandgaps can be estimated from the thermal activation energies ΔE (temperature dependence of dark conductivities),

$$\sigma = \sigma_0 \exp(-\Delta E/kT), \tag{16}$$

and the optical excitation energy, from photoconduction and absorption spectra, or from the optical activation energy ΔE_{opt} (long-wavelength limit of photoconductivity) [42, 104, 169]. Remember though that extrinsic conduction can occur for materials with bandgaps >2 eV (see discussion on carrier generation).

Obviously, since smaller bandgaps (or barrier heights) favor a greater percentage of photons having enough energy to generate free carriers, the maximum intrinsic photocurrent is favored by a small E_g (or barrier height). The desired balance between this factor and those (discussed previously) that favor a high E_g (maximum V_{ph}, minimum J_0, heavy doping) has led to calculations [1, 37, 38, 63] of optimum bandgaps for solar cell applications, generally between 1.5 and 2.0 eV. A theoretical study of the effect of differing barrier heights on the power-conversion efficiency of organic Schottky barrier cells of 1 : 1 polyvinylcarbazole–trinitrofluorenone (PVK–TNF) [23] found a maximum theoretical value of 2% for a 2-eV bandgap (10^6 V/cm field); see Fig. 11 and also Ref. [25b].

   b.   Carrier-Generation Mechanisms.   What complicates the picture considerably for organic materials is that several complex current-generation mechanisms exist whereby free carriers can be produced by photons with energies $<E_g$, for example, from the dissociation or ionization of excitonic species (loosely bound electron–hole pairs). These excitons can "migrate" through the sample via energy (rather than free-electron) transfer. Free charge carriers are then produced, e.g., after interaction with electrode surfaces, defects, traps, etc., by collision with other excitons or additional photons or phonons or by any combination of these mechanisms. The actual charge carriers can be created homogeneously or heterogeneously. Although neutral in character, and therefore theoretically immune to field effects, it has been found that local field effects can affect excitonic diffusion, as well as the dissociation mechanisms. It is also possible for multiple excitonic species to exist (i.e., excitons where the electron–hole pairs are bound to differing degrees) in addi-

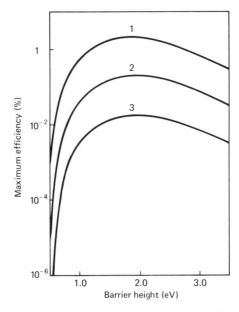

Fig. 11. Theoretical maximum photovoltaic conversion efficiency for 1:1 poly-vinylcarbazole–trinitrofluorenone (PVK–TNF) as a function of the barrier height in a photocarrier-generation limitation scheme. Curve 1: $g = 10^{-1}$; $E = 10^8$ V/m; $d = 0.008$–0.027 μm. Curve 2: $g = 10^{-2}$; $E = 10^7$ V/m; $d = 0.08$–0.27 μm. Curve 3: $g = 10^{-3}$; $E = 10^6$ V/m; $d = 0.7$–2.6 μm. (From Ref. [23].)

tion to intrinsically generated free carriers. Each species present has its own respective diffusion length, mobility, etc. Excitonic mechanisms are discussed in detail in Refs. [1, 106, 172–194].

Thus, for materials with $E_g > 2$ eV, extrinsic carriers can be generated by thermal activation of defects, by electrode-injection mechanisms, by ionization of excitons, by photoemission from electrodes [1, 170], or by detrapping of carriers held in states within the bandgap. Any one of these mechanisms could promote free carriers into the conduction band. Many of the photocurrents and photovoltages reported for the typical "insulating" organic photoconductors can be attributed to such extrinsic mechanisms and to Dember effects; see, for example, the photoconductivity and photovoltaic literature on aromatic hydrocarbon systems: anthracene [98, 134–147], pentacenes [148–150], perylenes [51a, 109, 124, 146, 147], naphthacenes [12, 19, 152], tetracenes [7, 9, 12, 153–160], miscellaneous [19, 51, 61, 94, 161–168].

c. Quantum Efficiency. Normally, if photocurrent is to result, not only must the excited carriers be energetic enough to overcome the effective dark barrier potential eV, but the carrier mobilities and lifetimes must also be such that carrier transport occurs through the entire sample under the existing potential gradient. In the simple idealized case (intrinsic generation of free carriers by a single photon) the photocurrent I_{ph} is given by

$$I_{ph} = \Phi e v \mu \tau E / d = geN, \qquad (17)$$

where g is the gain or photocarrier generation efficiency, N is the number of absorbed photons per unit area, Φ is the quantum efficiency of carrier generation, v is the sample volume, τ is the carrier lifetime, μ is the carrier mobility, E is the field, and d is the sample thickness [1, 42c, 104, 106, 190, 195, 196]. It is important to remember that Φ, μ, and τ refer to net values for all carrier species involved. For organic systems, the following points should be noted. First, even for a single carrier species, there is generally no single value for τ but rather a dispersion of values [1, 42]. Also, in high-field cases where μ is very small and the recombination kinetics very high, τ is essentially limited by the dielectric relaxation time $\tau_{rel} = \varepsilon \varepsilon_0 / \sigma = 10^{-12} \varepsilon / \pi \sigma$ (SCL conditions, see later section). More importantly, charge-carrier diffusion (and generation) is generally field-assisted.

What we do see from Eq. (17) is that high field strengths maximize I_{ph}. In conventional photoconductivity experiments this is achieved with an externally applied field. In photovoltaic applications, this can be achieved by the use of very thin films, since $E = V/d$. If the effective barrier voltage is approximately 0.5 V and $d = 100$ nm, $E = 5 \times 10^4$ V/cm. Thus, thin films favor high I_{ph} both because E is high and because recombination and trapping in high bulk-resistivity organic materials is minimized. However, one must still consider the resulting losses in absorbed photon energy for these films.

   d. Space-Charge-Limited Currents. In many organic systems, high bulk-resistivity values cause the photocurrents, particularly at high photon densities (intensities), to become space-charge-limited; see for example Refs. [1, 37, 39–42a, 60–62, 74, 87, 95, 105, 106, 114a]. The photocurrent in such cases can be estimated from Child's law:

$$I_{ph} = 9\varepsilon\varepsilon_0\mu V_{mp}^2/8d^3, \tag{18}$$

where V_{mp} is the photovoltage at maximum power output. Note that a necessary condition for SCL behavior is at least one ohmic electrode (otherwise saturation occurs). Figure 12 illustrates an example of power-conversion efficiency versus V for thick films (0.1–10 μm) of 1 : 1 PVK–TNF (SnO/Au cell excited with 600-nm monochromatic light), where there is a Child's law limitation [23].

   e. Theoretical Radiant-Energy-Conversion Efficiency. The efficiency of radiant-energy conversion η is a function of the radiant power density P, the absolute temperature T, the photovoltage at maximum power output V_{mp}, and the saturation dark current I_0. Recall from Eq. (10) that $I_0 (=J_0A)$ is in turn a function of the bandgap and the Schottky barrier potential eV_b. Thus,

$$\eta = \left[\frac{V_{mp}^2 I_0(e/kT)}{P} \right] \exp \frac{eV_{mp}}{kT} \tag{19}$$

and

$$(1 + eV_{mp}/kT) \exp(eV_{mp}/kT) = 1 + I_{ph}/I_0. \tag{20}$$

   3. The Active Photovoltaic Region

   a. Barrier Depth W. The distance to which the barrier potential gradient (diffusion potential or space-charge region) extends into the material helps define the active region of a Schottky barrier or junction device. There are differences in the profile and extent of these potential gradients in many of the organic systems studied (linear or nonlinear, extending throughout the entire sample thickness or existing only near the interface). The gradients are likely to be very dependent on individual trap distributions, defect densities, etc.

   The barrier (depletion) depth W in Schottky barrier devices is given by [42, 87, 171]

$$W = \sqrt{2\varepsilon\varepsilon_0 V/eN_b}, \tag{21}$$

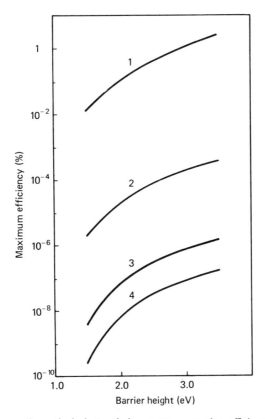

Fig. 12. Maximum theoretical photovoltaic energy-conversion efficiency for 1 : 1 PVK–TNF as a function of barrier height where there is a Child's law limitation. Curve 1: $d = 0.1$ μm; $E = 7.2 \times 10^6$–2.8×10^7 V/m. Curve 2: $d = 1$ μm; $E = 5.1 \times 10^5$–2.5×10^6 V/m. Curve 3: $d = 5$ μm; $E = 7.5 \times 10^4$–4.8×10^5 V/m. Curve 4: $d = 10$ μm; $E = 3.2 \times 10^4$–2.4×10^5 V/m. Compare with Fig. 11. (From Ref. [23].)

where ε is the dielectric constant, ε_0 is the permittivity of free space, V is the effective barrier voltage, and N_b is the bulk free-carrier density (dark conductivity) of the material.

First, it is important to realize that V is a dynamic parameter. That is, as the photovoltage builds up after the light is turned on, W varies. We will discuss this in reference to changes in the photoresponse spectrum with time, particularly when two barriers exist. It is also important to remember that, if significant electrode fields are present, W may be overshadowed.

Equation (21) indicates that extremely high values of N_b (e.g., in metals $N_b \sim 10^{20}$–10^{22} carriers/cm^3) will cause W to be so small (e.g., ~0.1–0.3

nm) that no appreciable depletion region exists. Thus, organic materials with extremely high intrinsic dark conductivities, even if $I_{ph} \gg I_d$, are not expected to be good photovoltaic materials. In most "impure" organic systems N_b is replaced by the concentration of donors or acceptors.

In p-n junction devices [87]

$$W = \left[\frac{2\varepsilon\varepsilon_0 V}{e} \left(\frac{N_a + N_d}{N_a N_d} \right) \right]^{1/2}, \tag{22}$$

where N_a and N_d refer to the concentration of acceptors on the p side and donors on the n side of the junction. Note that if $N_a \gg N_d$ (p$^+$-n junction) or if $N_a \ll N_d$ (p-n$^-$ junction), W is very small. The presence of very high densities of trapping or recombination sites within W can also compress the effective depletion region considerably.

Materials with extremely low conductivities would yield large W values (limited of course by the actual sample thickness), but the potential gradient across this region would be significantly reduced and would be inefficient in separating electron–hole pairs or in assisting the migration of charged species to the electrode surfaces. Keep in mind that in such cases (V_d very small) dark currents and photocurrents can arise from numerous extrinsic mechanisms. Recall that in cases where exciton ionization or dissociation gives charge species, the potential barrier in theory affects only already created charged species and that these may be created homogeneously or heterogeneously. Actually, in these cases, an image-charge dipole created at the interface [rather than W as in Eqs. (21) and (22)] affects the collection efficiency of that charge [27, 37]; see also Eq. (47) in Section III.I. Remember also that L_{ex} may be much greater than W.

There are reported examples in which evidence exists for finite depletion regions W less than the sample thickness t, as well as for only completely depleted regions [12, 17, 18, 20, 24b, 27, 197–199]. Each case must be considered separately and depends on the charge-generation mechanism(s) involved. We discuss methods for probing for the existence of W (and $W + L$) and for estimating its width in Sections III.H.1 and III.I.

Doping, application of bias voltages, electrode fields, or interfacial insulating layers can all be used to manipulate W.

b. Effective Diffusion Lengths L. The diffusion length L for a carrier is given by [42c]

$$L = \sqrt{D\tau} = \sqrt{(kT/e)\mu\tau}, \tag{23}$$

where D is the diffusion coefficient. For many organic materials diffusion or mobility is field-assisted; i.e., $\mu = f(E)$. Also recall that an average value for τ must be used [1, 42]. If E is very high (particularly likely in thin-film devices), the *effective* diffusion length L' may *appear* to be relatively independent of τ:

$$L' = \mu Et, \qquad (24)$$

where t is the transit time for carriers. If space-charge-limited conditions exist, recall that τ is defined by $\varepsilon\varepsilon_0/\sigma$ (see Section II.B.2.d).

In the case of excitonic mechanisms, the excitonic diffusion length L_{ex} may be significantly greater than L for the free carriers, particularly in high-resistivity materials. In fact, in organic materials L(holes) is generally greater than L(electrons) $\approx$ 1–10 nm, while $L_{ex} \approx$ 10–100 nm. As mentioned previously, it is difficult to access the field dependency of such mechanisms, since the final charge carriers can be created heterogeneously and from excitons bound to differing degrees. Methods for measuring diffusion lengths are discussed in Sections III.H.1 and III.I.

    c. Active Device Region $(W + L)$. The active region of a true Schottky barrier or junction device consists of the depletion depth W plus the effective diffusion length L for the carrier(s) involved (see Fig. 13). Any carriers created outside this region will simply recombine or be trapped before being collected, unless an externally applied field or an electrode field is present. Exclusive of these cases, one wants to maximize $(W + L)$ for efficient photovoltaic application. Techniques for estimating W, L, and $(W + L)$ are discussed in Section III.H.1. Of equal importance to maximizing $(W + L)$ is the necessity for similarly maximizing the number of photons absorbed within the active region of sufficient energy to overcome eV. These two aspects define the actual active photovoltaic region.

    d. Energy Absorption Depth d. Since conversion efficiency is a function of radiant power density, the proportion of incident power actually

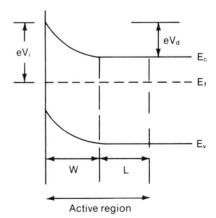

Fig. 13. The active region of a barrier contact is defined by the depletion depth W plus the effective diffusion length L of the material. Here the diffusion potential eV_d (barrier potential) and the electrode injection potential eV_i are also shown.

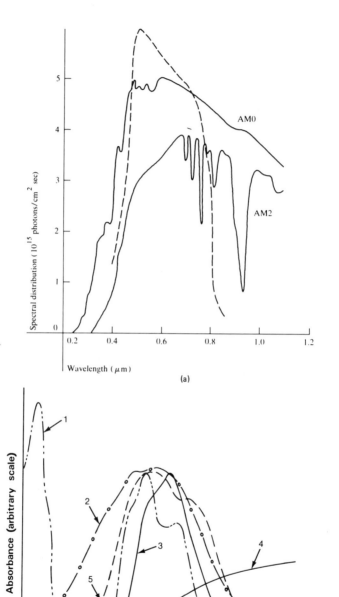

(a)

(b)

absorbed within the active region is important. For any particular material, the light absorption depth d is determined by the material's unique distribution (with wavelength) of absorption coefficients α:

$$AB = 1 - \exp(-\alpha t) = 1 - (RF + TR), \qquad (25)$$

where AB is the amount of absorbed light, t is the sample thickness, RF is the reflectance, and TR is the transmittance. Thus,

$$t = [\ln(1 - AB)]/\alpha. \qquad (26)$$

For example, a 300-nm-thick material with an average α value of 10^5 cm^{-1} would absorb about 95% of the light incident on it.

Obviously, the ideal situation would entail an α value such that nearly 100% absorption occurred within a thickness equal to the active barrier dimension, where the shape of the absorption spectrum essentially matched that of the desired incident power spectrum. For example, Fig. 14a shows a plot of the absorption spectrum of the organic photoconductive dye hydroxy squarylium (OHSq) (corrected for reflection and in the amorphous state) compared to standardized solar radiance spectra under outerspace conditions (AM0) and under normal cloudy terrestrial conditions (AM2). The amorphous state of this material has near-ideal absorption properties for photovoltaic applications [24]. (The microcrystalline state may be useful as an infrared-absorbing photoconductor.) Other potentially useful organic systems include thin films of photoconductive dyes such as the phthalocyanines, cyanines, other "squarylium" dyes, the merocyanines [16, 18, 24b, 26, 27], and highly absorbing (light) polymeric photoconductors such as the charge-transfer salts of polyacetylene [132, 133], polypyrrole [200], poly(p-phenylene) [201a], polynaphthalene [201b], polytetrathiafulvalene, and polypyrazoline systems, etc.; see Fig. 14b. Thin films of amorphous solid solutions of these materials may also be useful [24b].

Another intriguing possibility related to maximizing absorption in thin-film devices is the use of internal reflection via mirrors [237] or substrate topography [238]. Very little has been done on the use of antireflective or selectively absorbing topcoats (but see Ref. [24b]).

Fig. 14. (a) A comparison of the adsorption spectrum of amorphous hydroxy squarylium [24b] (---) with solar spectral radiance. AM0, extraterrestrial conditions; AM2, average cloudy terrestrial conditions. (b) Shape of absorption spectra for (1) amorphous metal-free phthalocyanine [209], (2) *trans*-polyacetylene [210], (3) Mg phthalocyanine [17], (4) AsF$_5$-doped polyacetylene [210], and (5) a merocyanine [26]. The absorbance scale is arbitrary, and the figures are meant only to indicate the breadth of absorption.

III. EXPERIMENTAL ASPECTS

A. Dark Behavior—Rectification

The equation for dark current I_d from a Schottky device was given by Eq. (2), where V is now the applied voltage. This equation explains the rectifying behavior observed in Schottky diodes (recall the discussion on the manipulation of V_b with bias voltages). If one applies an appreciable reverse bias to the cell, the term $\exp(eV/kT)$ becomes negligible compared to 1 and the current I approaches the asymptotic value $-I_0$. If an appreciable forward bias is applied, then $I \rightarrow \exp(eV/kT)$. A typical dark current–applied voltage curve for an ideal Schottky diode is shown as a solid curve in Fig. 15a.

Figure 15b shows part of a simplified equivalent circuit for a nonideal diode solar cell. The series (or spreading) resistance R_s is due to the bulk

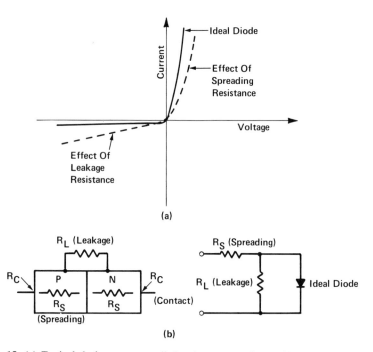

Fig. 15. (a) Typical dark current–applied voltage curve for an ideal Schottky diode is shown as a solid curve. The effects of series and spreading resistances are shown as dotted curves. (b) Simplified equivalent circuits for nonideal diode solar cells which show contact resistance R_c, series (spreading) resistance R_s, and shunt (leakage) resistance R_L (R_{sh} in text). $I = I_s [\exp(qV/kT) - 1]$. For reverse bias $I \approx -I_s$; for forward bias $I \propto \exp(qV/kT)$.

resistance of the (organic) material plus any contact resistance(s). For any particular organic material, R_s is dependent not only on sample thickness and device area but also on the applied voltage (because of resulting effects on trapping and/or recombination) and the light intensity (space-charge-limited effects). If contact resistance R_c is nonnegligible, this will also be included in R_s.

The shunt (or leakage) resistance R_{sh} is due predominantly to surface phenomena. Leakage currents are bilateral and tend to vary erratically. We see the effects of R_s and R_{sh} as dotted lines in Fig. 15a. In organic materials, the extremely high bulk resistivities can cause significant deviations from the ideal I–V curve. See for example, Figs. 16a–16f, which show typical dark I–V curves for tetracene [12], metal-free [22] and Mg phthalocyanine [17], chlorophyll [16, 18], hydroxy squarylium [24], and doped polyacetylene [132, 133] Schottky barrier cells. Note the distinct difference between these and the dark I–V curve for a "modified" Schottky cell made from a merocyanine dye in which an oxide layer is present between the organic and metal layers [17, 26]; see Fig. 16g.

B. Photocurrent–Photovoltage Curves

Let V be the voltage drop across a variable load resistance R_L, and I (J) be the current (current density) through it. The photovoltaic cell generates current $I_{ph}(J_L)$, and this photocurrent is partially offset by a diode forward current $I_d(J_D)$ (dark recombination current density); see Fig. 17a. This diode is shunted across the generator as shown in Fig. 17b. The total circuit current I (J) measured at the ammeter is now given *not* by

$$I = I_{ph} - I_d = I_{ph} - I_0[\exp(eV/kT) - 1] \tag{27}$$

but by

$$I = I_{ph} - I_0[\exp(eV^*/kT) - 1] - V_{sh}/R_{sh}, \tag{28}$$

where the voltage drop across the cell V^* is given by

$$V^* = V + IR_s + (R_s/R_{sh})V. \tag{29}$$

If $R_{sh} \gg R_s$ (and hopefully this is the case), we can ignore the V_{sh}/R_{sh} terms in Eqs. (28) and (29).

It should be noted that even when corrected for R_s and R_{sh}, Eq. (27) fails to agree with observed experimental behavior for voltages ≤ 0.4 V. In such cases an empirical factor B is used [63, 202] and kT is replaced by BkT in the above equations. The larger the value of I_0 (e.g., $>10^{-6}$–10^{-5} A/cm²), the larger the value of B. Generally $B \approx 2$, but for organics it is usually in the range of 2–3.

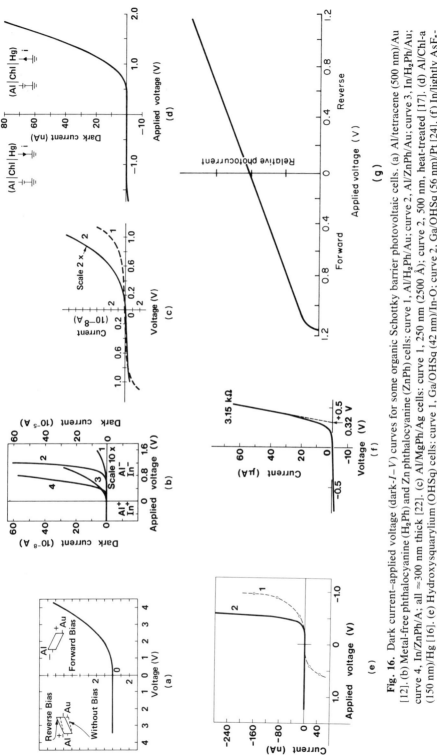

Fig. 16. Dark current–applied voltage (dark-*I–V*) curves for some organic Schottky barrier photovoltaic cells. (a) Al/tetracene (500 nm)/Au [12]. (b) Metal-free phthalocyanine (H₂Ph) and Zn phthalocyanine (ZnPh) cells: curve 1, Al/H₂Ph/Au; curve 2, Al/ZnPh/Au; curve 3, In/H₂Ph/Au; curve 4, In/ZnPh/A; all ≈300 nm thick [22]. (c) Al/MgPh/Ag cells: curve 1, 250 nm (2500 Å); curve 2, 500 nm, heat-treated [17]. (d) Al/Chl-a (150 nm)/Hg [16]. (e) Hydroxysquarylium (OHSq) cells: curve 1, Ga/OHSq (42 nm)/In-O: curve 2, Ga/OHSq (56 nm)/Pt [24]. (f) In/lightly AsF₅-doped *trans*-polyacetylene/Au [133]. (g) Modified Schottky barrier cell made from a merocyanine dye in which an oxide layer is present between the organic and metal layers [17, 26].

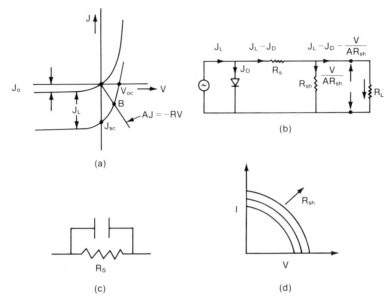

Fig. 17. (a) Current–voltage characteristics of a typical p-n junction or Schottky barrier device in the dark (J_D) and under illumination (J_L). The open-circuit voltage V_{oc} and the short-circuit current density J_{sc} are the intercepts on the voltage V and current density J axes. A and R are the junction area and load resistance. (b) Equivalent circuit of a typical well-behaved solar cell under actual operating conditions. R_{sh}, Shunt resistance; R_s, series resistance. (c) Alteration in the equivalent circuit for the high-resistivity organic device which behaves as a capacitor. (d) Family of I–V curves obtained for increasing R_{sh} values (direction of arrow).

For high-resistivity organic materials one should also consider additional alteration of the equivalent circuit of the series resistance part of Figs. 17b and 17c, where now the photovoltaic cell is considered to be a capacitor in parallel with a variable series resistance, which is a function of both the potential across the cell and the intensity L of incident light.

C. Definitions of V_{oc} and I_{sc}

Referring to Figs. 17b and 17c, if R_L is made much much larger than R_s (assuming negligible shunt current), essentially no current flows through the circuit (*open-circuit* condition) and the maximum output voltage (open-circuit voltage V_{oc}) is obtained.

If instead, R_L is made much much smaller than R_s (same assumption applicable to the shunt current), all available current flows through the circuit (*short-circuit* condition) and the maximum output current (short-circuit current I_{sc}) is obtained.

The mathematical equations relating V_{oc} to I_{sc} have been derived elsewhere [63, 202]:

$$V_{oc} = (kT/e) \ln[(I_{sc}/I_0) + 1], \tag{30}$$
$$I_{sc} = V_{oc}/R, \tag{31}$$

where R is the total internal resistance of the device. If $I_{sc}/I_0 \gg 1$,

$$V_{oc} \approx (kT/e) \ln(I_{sc}/I_0). \tag{32}$$

If intermediate values of R_L are used, intermediate values of I and V are obtained:

$$I = V_{oc} - V/R. \tag{33}$$

The shape of the output photovoltage versus output photocurrent curve is critically dependent on R ($R_s + R_c$, etc.) and on the diode characteristics (I_0, etc.) and is essentially a replot of the lower right quadrant of the photo-I–V curve shown in Fig. 17a. [The curve factor (fill factor FF) is discussed in the next section.]

Serious measuring problems can arise if $R_{sh} \lesssim R_s$ in Figs. 17b and 17c, since the current flow will now be distributed between the circuit branches. Readings at the voltmeter or ammeter will be smaller than the true output values for the cell. For example, if $R_{sh} = R_s$, the ammeter and voltmeter readings will be 50% of their true values. If the device shorts out, $R_{sh} \ll R_s$. A variable R_{sh} will give a family of similar-shaped photo-I–V curves (at constant intensity), but the FF will remain essentially constant; see Fig. 17d.

D. Output Power Measurements and the Fill Factor *FF*

The output power IV of the device can be calculated from the experimental values for I_{sc} and V_{oc}. Consider an *ideal* generator (constant-current generator). The dotted line in Fig. 18 shows the output photo-I–V curve that would be obtained. The maximum output power P_{max} is simply the product of the current and voltage at the maximum output power (I_{mp}, V_{mp}) [203, 204] or

$$P_{max} = I_{mp}V_{mp} = I_{sc}V_{oc} \qquad \text{(ideal case).} \tag{34}$$

For nonideal generators recall that Eq. (33) holds. Let us take the example of a generator of voltage V_{oc} with an internal resistance R (generator of constant internal resistance). Since as $I \rightarrow 0$, $V \rightarrow V_{oc}$ and $I_{sc} = V_{oc}/R$,

$$P = IV = I_{sc}[V - (V^2/V_{oc})]. \tag{35}$$

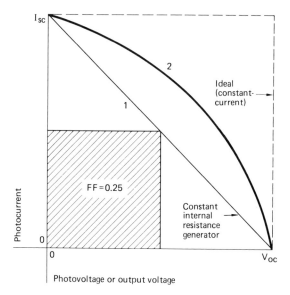

Fig. 18. Output current–voltage ($I–V$) curves for an ideal constant-current generator (dashed line), for a generator of constant internal resistance (curve 1), and for a typical photovoltaic device (curve 2). The curve factor (fill factor) FF is defined in the text.

The maximum output power occurs when [203, 204]

$$P_{max} = dP/dV = 0. \tag{36}$$

Here (see curve 1 in Fig. 18)

$$V_{mp} = V_{oc}/2, \qquad I_{mp} = I_{sc}/2 = V_{oc}/2R, \tag{37}$$

and

$$P_{max} = I_{sc}V_{oc}/4. \tag{38}$$

Most photovoltaic devices are generators of variable current (Fig. 18, curve 2). We now define a curve factor, the *fill factor FF* as

$$FF \equiv P_{max}/I_{sc}V_{oc}. \tag{39}$$

Thus, for an ideal generator $FF = 1$, whereas for a generator of constant internal resistance (curve 1) $FF = 0.25$. Typical values for photovoltaic devices range from 0.2–0.95; see for example Ref. [63]. A reasonably good value is anything above 0.75.

For most organic materials, high-bulk resistivities, large trap densities, nonideal contacts, recombination, short diffusion lengths, etc., result in field-dependent and space-charge-limited effects that cause the photovoltage and photocurrent outputs to be load-dependent (or load-limited). Again, the shape of the photo-I–V curve and the FF values reflect these effects (see, e.g., Refs. [157, 158]). Most organic materials have fill factors ≈ 0.25. Figures 19a–19e give examples of photo-I–V curves and FF values for some well-studied organic systems.

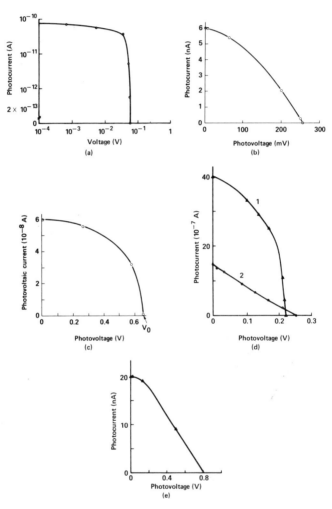

Fig. 19. Photocurrent–photovoltage (photo-I–V) curves for various organic Schottky barrier devices. (a) SnO_2 (NESA)/1 : 1 PVK–TNF (4.8 μm)/Au illuminated with 600-nm light [23]; (b) Al/Chl-a/Hg ($FF = 0.33$) [16]; (c) Al/tetracene/Au ($FF = 0.51$) [12]; (d) In/ZnPh/Au (curve 1; $FF = 0.48$) and In/H_2Ph/Au (curve 2; $FF = 0.25$) [22]; (e) Al/H_2Ph/Au ($FF = 0.30$) [22].

In a later section, the effects of space-charge-limited currents, trapping densities, recombination, etc., on the dependence of FF on the intensity of light incident on the device will be discussed.

In terms of I_{sc}, V_{oc}, and V_{mp} [20, 63, 203–205],

$$FF = (V_{mp}/V_{oc}I_{sc})\{I_{sc} - I_0[\exp(eV_{mp}/kT)]\}. \tag{40}$$

The fill factor can thus be determined easily by considering the ratio of the largest rectangular area that can be drawn under the output photo-I–V curve (maximum output power) to the area defined by $V_{oc}I_{sc}$.

E. Double Barriers and FF Values

The fill factor can be used to detect field-dependent photovoltaic responses such as those due to the presence of two contact barriers. For example, two Schottky devices made with similar thicknesses of amorphous hydroxy squarylium (OHSq) sandwiched between the electrode couples, Ga/Pt and Ga/In-O (a Schottky barrier in both cases occurs at the Ga–organic interface), gave substantially different fill factors (0.4 compared to 0.2, respectively) but essentially the same V_{oc} when measured at the same intensity [24b] (Fig. 20). This was attributed to a second barrier contact in the case of the In-O electrode. Differences are also observed in the intensity dependences; this is discussed in a later section.

F. Fill Factor as a Function of Electrode Effects

Earlier we alluded to the detrimental effects of poor "ohmic" contacts (or for that matter of two barrier contacts) on the performance of photovoltaic devices. The "ohmicity" of the nonbarrier electrode plays a crucial role in the value of the total series resistance R_s (which includes contact resistance R_c) and the fill factor FF. In particular, this electrode, depending on the type of semiconductor (carriers) and the external circuit, serves as an infinite reservoir of compensating charge carriers as current is drawn off into the external circuit. If appreciable recombination (or other deactivation) of the free carriers occurs at this electrode, the output photocurrent will be considerably reduced. See for example, Figs. 20 and 21.

During a detailed investigation of photovoltaic effects in backwall-illuminated hydroxy squarylium (OHSq) Schottky barrier cells [24b], attempts were made to improve the transmittance of light through the semi-transparent ohmic Pt electrodes (Ga/OHSq/Pt cells) in order to increase the *engineering conversion efficiency* η_e (see next section). It had also been

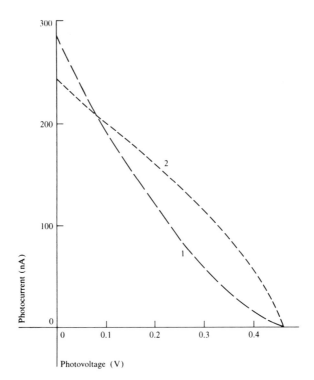

Fig. 20. Photo-*I–V* curves for single- and double-barrier cases. Curve 1 is for a Ga/OHSq (42 nm)/In-O cell where a second barrier contact is suspected at the organic–In-O interface (*FF* = 0.20). Curve 2 is for a well-behaved single-barrier Ga/OHSq (56 nm)/Pt cell (*FF* = 0.37) [24b].

observed that the use of "ohmic" In-O electrodes had shown evidence of double-barrier effects.

Previous work [207] had indicated that Bi_2O_3 (insulator) substrates had been successfully used to provide electrically continuous (i.e., complete homogeneous coverage by the metal to give a uniform sheet resistivity near that of the bulk material) rf-sputtered Pt thin films at much smaller thicknesses than normally needed (e.g., ≈2–3 nm as opposed to ≈ 10 nm). Therefore, in hopes of obtaining a similar benefit while maintaining the basic surface and electrical characteristics (work function) of Pt on a conductive substrate, "hybrid" electrodes of Pt and In-O were made. The photovoltaic behavior of Ga/OHSq Schottky devices were compared using "ohmic" electrodes of Pt, Pt–In-O, and In-O. Tables IV and V summarize the results. After correcting for differences in the transmittance values for the electrodes, the superior electrode was the hybrid Pt–In-O

TABLE IV *Transparencies and Sheet Resistances for In-O- and Pt-Coated Substrates*

Substrate	Transmission (%)	Sheet resistance ($\Omega/\square$)
In-O	95	191, 175
Pt/quartz	35	155
	40	262
Pt (3.5 nm)/Bi_2O_3 (10 nm)	49	155
Pt/In-O	43	76, 82
	45	78
	47	95

one. Note that the sheet resistivities for the hybrid electrodes (76–95 $\Omega/\square$) were nearly one-half to one-third of those for electrically continuous Pt on quartz or on Bi_2O_3 (155–262 $\Omega/\square$) and about one-half those for the In-O (175–191 $\Omega/\square$) electrodes. Although not completely understood, these results were interpreted as an indication of the enhanced "ohmicity" (less recombination, etc.) of the hybrid electrodes.

Poor contacts can also cause erratic and/or time-dependent photovoltaic parameters or gradual degradation of a device, as shown in Fig. 21.

G. Power-Conversion Efficiency

The *engineering* power-conversion efficiency η_e is defined as the ratio of the output power from the junction or barrier P_{out} (of area A) to the total power *incident* on the same device area A, P_{in}:

$$\eta_e \equiv P_{out}/P_{in}. \tag{41}$$

TABLE V *Comparison of Photovoltaic Devices of Solution-Cast OHSq on Pt, In-O, and Hybrid Pt–In-O Electrodes*

Electrodes	Electrode transmittance (%)	V_{oc} (V)	Intensity (mW/cm^2)	J_{sc} (μA/cm^2)	FF	η (%)	η_{corr} (%)
Ga/Pt	43	0.52	0.24	19	0.336	0.014	0.032
Ga/Pt, Bi_2O_3	49	0.42	0.68	33	0.30	0.006	0.012
Ga/Pt, In-O	47	0.38	0.68	210	0.36	0.42	0.086
Ga/In-O	>95	0.41	0.68	110	0.27	0.018	0.018

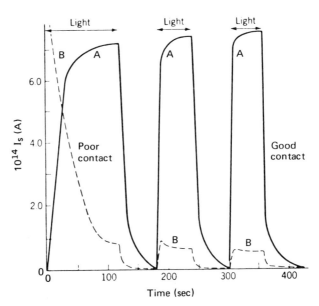

Fig. 21. Degraded performance of photoresponse (I_{sc}) for a sample with poor electrode contacts. The photocurrent obtained is considerably reduced and has a time-dependent response.

Thus,

$$\eta_e = (FF)J_{sc}V_{oc}A/P_{in} \qquad (42)$$

is also obtainable from the experimentally observable quantities I_{sc} and V_{oc}.

Often, in order to compare different materials, similar materials sandwiched between different electrode couples, or similar devices prepared with different electrode (semitransparent) thicknesses, a conversion efficiency based on the power *absorbed* by a device P_{ab} is used. This efficiency is akin to a quantum efficiency and will be termed a *corrected* conversion efficiency η_{corr}:

$$\eta_{corr} \equiv P_{out}/P_{ab}. \qquad (43)$$

If, for example, a chromium electrode is used that transmits only 50% of the light incident on it, $\eta_{corr} = 2\eta_e$. Although useful for comparison purposes, η_{corr} does not give a true picture of the actual device limitations; i.e., a real device cannot be made with a Cr electrode that is completely transparent. [See previous discussion of special "hybrid" electrodes designed to increase the percent of transmitted light through thin semitransparent electrode films while maintaining high sheet conductivities and the characteristics (predicted work function values) of the principal electrode material.]

H. Response Spectra

A response spectrum is a plot of any particular photovoltaic parameter (V_{ph}, V_{oc}, I_{ph}, I_{sc}, η, Φ, etc.) versus the wavelength of incident light. Recall that we previously described the optimal photovoltaic device as consisting of one blocking (barrier) and one ohmic electrode. We now discuss the relationships among the direction of incident illumination, sample thickness t, number of barrier contacts, carrier-generation mechanism(s), and the observed response spectra. See Refs. [6, 17, 18, 20, 22, 24b, 27, 61, 124, 206].

When discussing the direction of illumination, *frontwall* illumination refers to light incident on the same side as the barrier contact, whereas *backwall* illumination refers to light incident on the nonbarrier (ohmic or collecting) electrode.

We now discuss several sample situations in which the 100% absorption depth d_{100}, the sample thickness t, and the width of the active region ($W + L$) are varied. Unless otherwise noted, we assume a single-photon intrinsic photogeneration mechanism involving a separable electron–hole pair and a *single* barrier contact. The effects of changing these assumptions are treated individually.

1. The Cases $t = W + L$

Let us first discuss the optimal situation in which $t = t_{100} = W$ or ($W + L$). When $d_{100} = W$ or ($W + L$), the maximum possible photocurrent for any particular material will be obtained, since every potential photogenerated free charge carrier (limited only by Φ, α, and the incident power density L) falls under the influence of the electrostatic potential barrier and can be swept into the external circuit (neglecting electrode recombination). The response spectrum will resemble the absorption spectrum independent of the direction of illumination. Energetic bound electron–hole pairs created within a diffusion length L of the barrier region can diffuse into the barrier region defined by W and be separated, or they may diffuse toward the opposite electrode surface, where they may be deactivated. It is also possible for the electron–hole pairs to be separated within L at trapping, impurity, or defect sites (where localized potential wells may exist), and the resulting carriers can diffuse to the appropriate collecting electrode. In either case the shape of the response spectrum will again be independent of the direction of illumination. If $d_{100} > t$, the response spectrum will resemble the absorption spectrum independent of the direction of illumination; however, the photocurrent will fall off as $(d_{100} - t)$ increases, since the number of potential carriers available in the decreased thickness will decrease. Figure 22 shows frontwall response spectra that resemble the absorption spectra for various organic photovoltaic systems. For simple dissociative excitonic mechanisms, if *all* the *charged carriers* are created within $t = d_{100} = W + L$ (where L is now the diffusion length for the particular charge species), the shape of the response will be similar

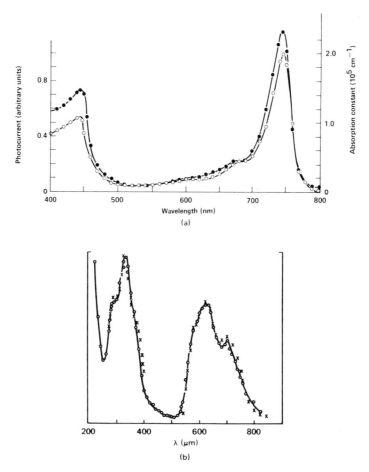

Fig. 22. Response spectra for organic photovoltaic devices in which the response essentially matches the absorption spectra. (a) I_{ph} of Al/Chl-a/Hg [16] (●, absorption spectrum; ○, spectral response); (b) capacitor photo-emf of amorphous H_2Ph (×) versus the absorption spectrum (○) [209, 211]; (c) I_{ph} of MgPh [209, 211]; (d) V_{ph} of MgPh.

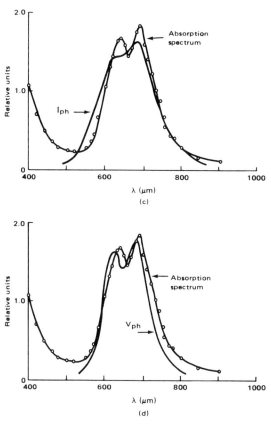

Fig. 22 (*Continued*)

to the absorption spectrum independent of the excitonic diffusion length L_{ex} as long as $L_{ex} > L$. If $L_{ex} < L$, a similar-shaped response will be obtained if $t = d_{100} = W + L_{ex}$. The *total* current densities may be considerably reduced, based on the proportion of excitonic particles yielding free charge carriers. If charges are created *only* at the barrier electrode surface, W is inconsequential and one wants $t = d_{100} = L_{ex}$. Excitonic diffusion lengths can be estimated from the slope of a J^{-1} versus α^{-1} curve. In fact, the presence of multiple or nonlinear slopes over the wavelength region of the response spectrum would indicate a variety of excitonic diffusion lengths and thus multiple or complex excitonic dissociation processes; see for example Ref. [27].

Later we will see how the principles of maximum response and similar frontwall/backwall response can be used to estimate values of $(W + L)$.

2. The cases $t > W + L$

In many practical situations with organic materials, the sample thickness is greater than the average active depth (estimated to be up to approximately 20 nm for free electrons or holes). Larger active depths tend to involve both excitonic diffusion lengths, which can be a few hundred nanometers, and one good ohmic contact. In such cases where $t > W + L$ (and L_{ex}) the relationships among d_{100}, t, and the direction of illumination become very important to the shape of the response spectrum. For all values of $d_{100} > W + L < t$ there is a "dead" region created within the device that will effectively degrade its photovoltaic response if charge carriers are produced in that region.

When $d_{100} = t > W + L$, the "inactive" region results from absorbed light creating bound electron–hole pairs that either recombine or deactivate before being separated or otherwise collected under the influence of the potential field. These, as well as any neutral excitonic species, are unaffected by the barrier potential. For frontwall illumination (through W) the response spectrum will resemble the absorption spectrum, although the details of sharp maxima and minima may be lost. However, for backwall illumination (light incident through the "dead" region and L) the sample itself will act as a light filter. Strongly absorbed light (absorbed near the incident surface) will be ineffective or much less effective in creating photocurrent, whereas weakly absorbed light (absorbed within $W + L$) will efficiently create photocurrent. As a result, the response spectrum will resemble the inverse of the absorption spectrum. Ghosh and his co-workers have calculated the relationships among spectral response (I_{ph}), d_{100}, t, and ($W + L$) for frontwall and backwall Mg phthalocyanine (MgPh) cells [17]; see Figs. 23a–23c. Their mathematical formula can accurately predict the shape of response spectra under these circumstances.

When $d_{100} < t > W + L$, inactive regions result from both absorption losses in any thickness $t - d_{100}$ and recombination losses as mentioned above. In any case where $d_{100} < t$, the bulk sample can act as a light filter when a backwall configuration is used (inverse relationship between the response spectrum and the absorption spectrum). Frontwall configurations will give response spectra resembling the absorption spectra, with a possible dampening of detailed features. Similar dependencies on the shape of the frontwall and backwall response spectra for d_{100} and t will also result if an exciton dissociates to give charge *only* at the barrier electrode surface if $d_{100} \leq t \leq L_{ex}$. For $t > L_{ex}$, see Section III.H.5. If multiple charge species or different charge-generation mechanisms are involved, a $W + L_{eff}$ (effective diffusion length) must be considered for each species. See later discussion on conduction mechanisms in Section III.H.5.

Lack of correlated effects can be due to several factors; among the more important are the presence of more than one barrier contact, a different carrier-generation mechanism, bias voltages, and poor electrode contacts.

3. Double Barriers

Recall the effect of a bias voltage on the contact barrier potential. The photovoltage and response spectrum will also be affected by any bias voltage due to a second potential barrier at the normally ohmic electrode. The total observed photovoltage is less than that theoretically calculated from a single barrier, since the two potentials are subtractive. The larger barrier potential is referred to as the *dominant* barrier. In relation to the normal absorption spectrum of the organic photoconductor, the response spectrum displays a complex relationship based on the relative barrier

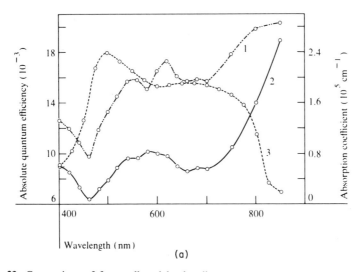

(a)

Fig. 23. Comparison of frontwall and backwall response spectra. (a) Backwall OHSq cells: curve 1, Ga/OHSq (42 nm)/In-O; curve 2, Ga/OHSq (56 nm)/Pt; curve 3, absorption spectrum of OHSq [24b]. (b) Experimentally obtained absorption spectrum of MgPh (part 1) and the frontwall (Al) and backwall (Ag) response spectra of an Al/MgPh/Ag cell (part 2) compared with the mathematically calculated frontwall and backwall spectra (part 3) [17]. (c) Experimental (solid triangles) and calculated (open triangles) action spectra of I_{sc} for In/H$_2$Ph (300 nm)/Au frontwall cell compared to the absorption spectrum (open circles). (d) Same as in (c) but for a backwall cell.

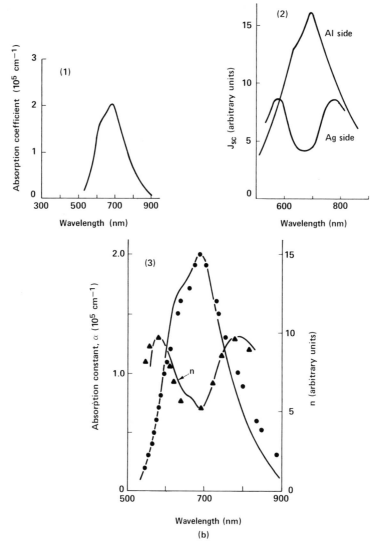

Fig. 23 (*Continued*)

heights, light penetration depths, diffusion lengths, etc. Tang and Albrecht [18] have treated this problem mathematically for double-barrier chlorophyll-a (Chl-a) devices; see Fig. 24. Very importantly, the spectrum (as well as other photovoltaic parameters) is also time-dependent, since one is dealing with space-charge effects. Each barrier, as it is being dynamically altered by the penetrating light, effectively biases the other barrier.

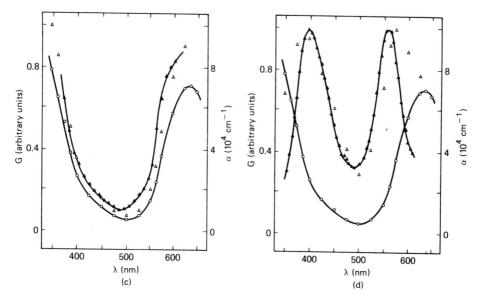

Fig. 23 (*Continued*)

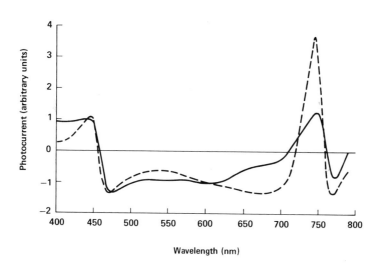

Fig. 24. Mathematical treatment of double barriers [18] and an example of experimental (solid curve) and calculated (dashed curve) response spectra for a Cr/Chl-a (150 nm)/Hg-In cell. The derived mathematical equation was $I_{ph} = A[(1 - e^{-k\delta_f}) - e^{-kl}(e^{k\delta_b} - 1)]$, where k is the wavelength-dependent absorption coefficient, δ_f and δ_b are the depths of the front- and back-barrier regions, and l is the film thickness. ---, calculated; ——, experimental; $hv \rightarrow$ Cr/Chl-a/Hr-In (150 nm).

Eventually, an equilibrium response is obtained. The direction of illumination with respect to the dominant barrier region is now analogous to the frontwall–backwall situation with single-barrier cells, only the response spectrum is far more complex. Figures 25a and 25b give a comparison of time-dependent (or -independent) response spectra for single- and double-barrier Chl-a cells [18]. In addition, Table VI gives initial and steady-state data on two Ga/OHSq Schottky barrier cells [24b] with single- and double-barrier situations.

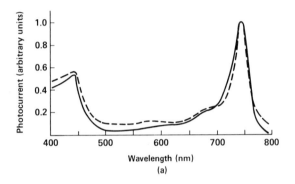

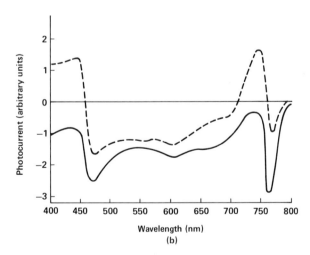

Fig. 25. Time-dependent response spectra caused by the internal biasing voltage in double-barrier cells. (a) Normal single-barrier Cr/Chl-a/Hg cell and (b) double-barrier Cr/Chl-a/Hg-In cell [18]. Solid curves, steady-state excitation; dashed curves, pulsed excitation.

TABLE VI *Steady-State Photovoltaic Responses (Open-Circuit Photovoltage) of Ga/OHSq/In-O and Ga/OHSq/Pt Schottky Barrier Devices*[a]

Device	Open-circuit photovoltage (V)	
	Initial response	Steady state
Ga/OHSq/In-O	0.553	0.434
Ga/OHSq/Pt	0.656	0.640

[a] Evidence exists for a double barrier in the In-O cell [24b].

4. *Applied Bias Voltages*

As mentioned in Section II.A.4, applied voltages influence the effective barrier potential and thus the photocurrent and photovoltage from a device. The response spectra will thus also vary depending on any applied bias; see Fig. 26.

5. *Conduction Mechanisms*

Far more complicated and apparently uncorrelated (with absorption spectra of pure organic materials) response spectra can be obtained if the process by which charges are created is more complex than our basic assumption given previously. Examples of such mechanisms can be grouped into extrinsic and intrinsic processes; some of these are listed in Table VII. The reader is referred to the vast literature on organic photoconductivity listed in Section II.B.2. In general, large current responses in regions where the pure material does not absorb, in infrared ranges where the material may absorb considerable thermal energy, or at the high-energy ultraviolet end of the spectrum, should all be suspect.

For organic materials in particular, excitonic conduction mechanisms can be extremely important. Since excitons can produce charge carriers by various proposed mechanisms involving energy transfer (at electrode surfaces; at defect, impurity, or other trapping sites; by collision of two excitons, an exciton, and a photon or phonon; etc.) in nonintegral multiples of the absorbed photon energies, the total quantum efficiency and the energy distribution of the final measured carriers can be very complex [1, 27, 106, 172–194]; see Section II.B.2.b and Fig. 27. Normally, one would not expect that diffusion of an uncharged exciton would be field-dependent. However, since the density of trapping sites and their related potential wells can be field-dependent, these may in turn affect the overall

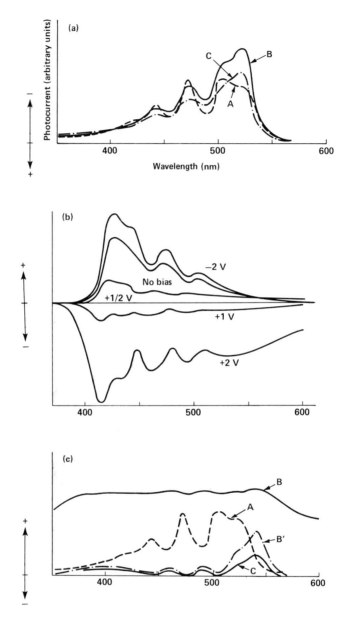

Fig. 26. Response spectra for Al/tetracence/Au as a function of the direction of illumination and applied bias voltages. (a) Frontwall configuration, (b) forward and reverse bias voltages applied, and (c) backwall configuration. Curve 1, −2 V; curve 2, no bias; curve 3, forward bias 1.4 V; curve 4, reverse bias 1.4 V; curve 5, no bias, photovoltaic mode. Curves A are the absorption spectra on Au. (From Ref. [12].)

TABLE VII   *Some Alternate Carrier-Generation Mechanisms*

Intrinsic

(1) Thermally stimulated current processes, e.g., thermal tunneling, thermal detrapping, previously trapped carriers, etc.
(2) Multiphoton absorption processes
(3) Energy transfer or excitonic conduction mechanisms
(4) Ionic conduction processes

Extrinsic

(1) Impurity conduction processes
(2) Electrode injection processes due to high fields, electrode photoemission, or photoionization
(3) Surface electrochemical or charge-transfer reaction processes

density of possible reaction centers that ultimately give rise to the charged species. Thus, diffusion of the excitonic species could be a function of the field. Once a charge carrier is created, it will of course be influenced by any potential field within a distance equal to the carrier's diffusion length. The observed response spectra in such cases are also very dependent on

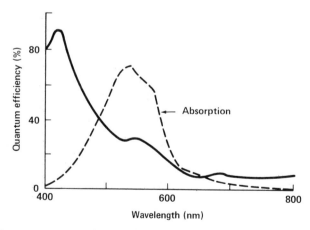

Fig. 27. Response spectrum for Al-Al$_2$O$_3$/merocyanine/Ag cell in which a complex excitonic conduction mechanism is proposed and where the quantum efficiency is energy- and field-dependent. (From a preprint of Ref. [27].)

whether the charges are created uniformly in the bulk of the material or in a heterogeneous fashion, such as only at a barrier (or injecting) electrode surface.

6. Other Effects

It is not unusual to see shifts by as much as 60 nm or other noncorrespondence to absorption spectra due to internal reflection, different recombination kinetics, photooxidation of localized ions or trapped carriers, charge transfer to impurities, mixing of excited states, reabsorption of energy by the sample, field-induced absorption, etc. [1, 8, 22, 157, 215, 223–226, 235, 236].

A summary of the considerations involved in interpreting response spectra for organic photovoltaic devices is given in Table VIII.

TABLE VIII *Considerations for Interpreting the Spectral Response of an Organic Photovoltaic Device*

(1)	Type of conduction process
	(a) Intrinsic/extrinsic
	(b) Electron–hole pairs/excitonic
(2)	Number of barrier contacts
(3)	Sample thickness relative to the
	(a) Penetration depth of light
	(b) Junction depth
	(c) Diffusion length
(4)	Direction of illumination
	Frontwall/backwall
(5)	Applied bias
(6)	Internal reflection effects

I. Measuring the Active Region

1. Capacitance Measurements

The presence of a separate and finite depletion region W within a sample can be probed by means of careful capacitance–voltage $(C-V)$ and capacitance–frequency measurements [7, 12, 17, 40, 42c, 197–199]. In a well-behaved Schottky barrier device,

$$C_d(V) = dQ/dV = \varepsilon\varepsilon_0(A/W), \qquad (44)$$

$$(d/dV)(1/C_d^2) = -2/Ne\varepsilon\varepsilon_0A^2, \qquad (45)$$

$$\frac{1}{C_{\rm d}^2} \frac{2}{Ne\varepsilon\varepsilon_0 A^2} (V_{\rm b} + V_{\rm appl}) = \frac{2}{Ne\varepsilon\varepsilon_0 A^2} (V_{\rm b} + V), \qquad (46)$$

where $C_{\rm d}$ is the differential capacitance, $V_{\rm appl}$ is the applied voltage, N is the number of charges entering the barrier capacitance, and A is the area. In a more detailed treatment [27],

$$V_{\rm b} = V_{\rm int} + (E_{\rm f} - E_{\rm v}) + (kT/e) - V_{\rm im}, \qquad (47)$$

where $V_{\rm int}$ is the intercept on the voltage scale (or diffusion potential) and $V_{\rm im}$ is the image potential; see Refs. [27, 38]. Great care must be taken, particularly when thin-film samples are involved, to make certain that the applied voltages do not (1) cause breakdown (punch through) of the films and (2) create such high internal electric fields (e.g., $\geqslant 5 \times 10^5$ V/cm) that W becomes insignificant with respect to the effective diffusion length L' [recall Eq. (24)]. The intercept of the tangent on the $1/C^2$ curve with the applied voltage axis can be used to determine $V_{\rm d}$; see Fig. 28. However, the frequency at which these measurements are made is *very* important in organic systems because of high bulk resistivities and space-charge-limited effects. In order to obtain meaningful data for many organic systems it is necessary to make capacitance measurements at frequencies below 1 Hz. In such cases, there are experimental problems associated

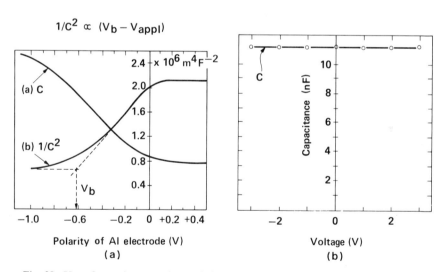

Fig. 28. Use of capacitance–voltage (C–V) measurements to determine the presence of a depletion region W and the value of the barrier (diffusion) voltage $V_{\rm b}$ ($V_{\rm d}$). The absence of a dependence of C on the applied voltage might indicate a completely depleted sample or problems associated with the frequency at which the capacitance was measured (see text). (a) Al/MgPh/Ag [17] and (b) Al/tetracene/Au [12].

with the type of capacitance bridge, the sampling times, and the sensitivities. Conventional small-signal $C-V$ measurements of Schottky barriers in these organics are difficult because the room temperature capacitance no longer depends on the applied bias; it is based on the capacitance of the *entire cell* rather than the barrier region alone. For thin-film samples, increased temperatures may be required in the 0.2–1.0 kHz range; e.g., see Ref. [199]. Even at higher temperatures, minimal applied voltages are required. Correctly calculated capacitances can be used to measure, or at least estimate, the widths of depletion regions W [20] from Eq. (44).

A new single-shot capacitor discharge technique has been reported [198] to be more successful for measuring capacitances and depletion depths for organic semiconductor Schottky barriers. In this technique, one looks at the integral capacitance C_i (rather than the differential capacitance C_d) as a function of applied bias V:

$$\frac{C_i}{C_0} = \frac{V_0}{V}\left[\left(1 + \frac{V}{V_0}\right)^{1/2} - 1\right], \tag{48}$$

Here

$$C_i = Q/V = \int_0^V C_d(V)\, dV/V, \tag{49}$$

$$Q/Q_0 = [(1 + V/V_0) - 1], \tag{50}$$

$$C_0 = Q_0/V_0, \tag{51}$$

$$Q_0 = AenW_0 = A(2e\varepsilon\varepsilon_0 NV_b)^{1/2}. \tag{52}$$

The subscript zero indicates the values at *zero applied bias, Q* is the amount of charge in the barrier region, and W_0 is the barrier width at zero applied field. For small biases ($V \ll V_0$),

$$C_d = C_i = C_0/2. \tag{53}$$

By curve-fitting a log–log plot of integral capacitance versus the applied bias to different values of C_0 and V_0 (simple translational scaling parameters if one plots C_i/C_0 versus V/V_0) one can determine W_0.

Using this method Popovic [198] determined a W_0 of 49 nm for a 150-nm Al/metal-free α-phthalocyanine/In-O Schottky barrier cell that showed no evidence of barrier existence when differential capacitance measurements were made between 0.01 and 2 kHz.

2. Photocurrent versus Thickness

One indirect experimental method for estimating the active depth ($W + L$) is based on the dependence of the output photocurrent on the sample thickness. For any particular device configuration, a plot of the short-circuit photocurrent I_{sc} versus the sample thickness will usually maximize near the optimal sample thickness. (The effects of depth of light penetration and absorption were discussed in Sections III.H.1 and III.H.2.)

Table IX gives data on the photoconductive dye OHSq [24] in two sandwich configurations (Ga/In-O and Ga/Pt). Here it is seen that the active region is considerably enhanced for the Pt device (single barrier contact). The optimal widths (or active regions) for various other organic materials have been estimated to be between 5 and 250 nm and depend considerably on the phase (crystallinity) of the material (see Section III.H), photocurrent-generation mechanism(s), device configuration, etc. See for example, Refs. [1, 20, 27]. Active regions defined primarily by excitonic diffusion mechanisms can be much larger than those defined by mechanisms involving free electrons and/or holes.

3. Response Spectra

Another indirect experimental technique for estimating active depths involves finding the sample thickness at which the response spectrum (also called the action spectrum) of a *single-barrier* device is the same whether the light is incident on the barrier electrode or the "ohmic" electrode (recall discussions in Section III.H).

4. Diffusion Lengths

An estimation of the diffusion length can be made by taking the slope of a J^{-1} versus α^{-1} plot as discussed in Section III.H.1. In addition, laser light mapping [208] has been used to estimate L by observing reductions in surface fields produced by surface stress.

TABLE IX *Short-Circuit Current Densities and Power Conversion Efficiencies of OHSq Cells as a Function of Film Thickness—An Estimate of the Optimum Cell Thickness [24b]*

Cell	Thickness (nm)	J_{sc} (mA/cm^2)	η (%)
Ga/In-O	10	0.18	0.002
	45	0.58	0.06
	90	0.0144	0.003
Ga/Pt	60	0.80	—
	90	1.244	0.09
	130	0.72	0.001

TABLE X *Typical Experimental Evidence for a Single Schottky Barrier or Junction Device*

---

(1) Dark current–voltage relationship:
 (a) Is there rectification?
 (b) Does I vary exponentially with V_{appl} (forward bias)?
(2) Capacitance–voltage relationship:
 (a) Does C vary with V_{appl} (see text for experimental precautions)?
 (b) Does the extrapolated value of V_b (C at $V_{appl} = 0$) roughly correspond to the predicted V_{ph}?
(3) Changes in photovoltaic behavior with the direction of light:
 (a) Does the sign of V_{ph} change with the direction of light?
 (b) Does the spectral response (zero applied voltage) for a frontwall cell resemble the absorption spectrum? Does it resemble the inverse of the absorption spectrum for a backwall cell (see text)?
(4) Changes in photovoltaic behavior with electrode material:
 (a) Does the amplitude and sign of V_{ph} roughly correlate with the difference in work function values between the electrodes or between the organic and one of the electrode materials?
(5) Does the relationship between V_{oc} and I_{sc} follow Eq. (30)?

Table X summarizes typical experimental evidence for the presence of a single Schottky barrier or junction.

J. Dependence of Photovoltaic Parameters on Light Intensity

1. Photocurrent

The reader is referred to an excellent discussion and review of the dependence of I_{ph} on incident light intensity L by Gutmann and Lyons (in particular, see Section 6.11, pp. 376–407 and Tables 6.12–6.16 in Ref. [1]). In general,

$$I_{ph} \propto L^n, \qquad n = f(N_f/N_t), \tag{54}$$

where N_f and N_t are the densities of free and trapped carriers above the Fermi level. The value of n thus depends on the number, distribution, and type (level) of traps present (or absent) on the carrier-generation and conduction mechanisms, and on diffusion and recombination processes in general; for example Refs. [1, 42c, 89b, 211a, 212–216]. For single-particle carrier schemes of organic insulators (in the dark), $0.5 \leq n \leq 1.0$. When $N_f \gg N_t$ (at high intensities), bimolecular recombination of the photogenerated free carriers dominates and $n = 0.5$. Where L is high enough for space-charge-limited (SCL) conditions to exist, saturation of I_{ph} with increasing L occurs; see for example, Figs. 29a and 29b. A linear dependence of

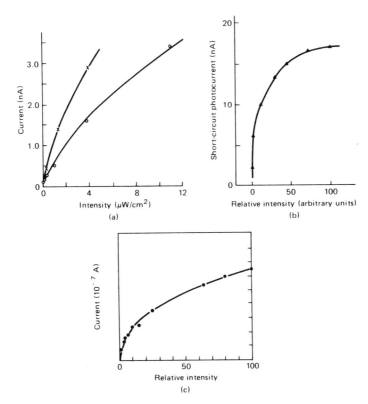

Fig. 29. Saturation in the photocurrent can occur at high intensities in materials with high bulk resistivities or poor "ohmic" contacts. (a) ×, octaethylporphin; and ○, tetraphenylporphin cells [20]; (b) Al/H₂Ph/Au [22]; (c) Al/MgPh/Ag [17].

I_{sc} on L ($n = 1$) has been observed in many organic Schottky barrier cells [8, 12, 16, 17, 22, 24b, 26, 27] and in all-organic [111, 112] and organic–inorganic p-n heterojunction devices [2, 75d]. In cases where two energetic particles are involved in creating free carriers, e.g., in photon–photon, exciton–exciton, and exciton–photon interactions, $n = 2$. There are numerous examples of superlinear dependence ($n > 1$) [217]. In addition, the dependence of I_{ph} on L can be different if an equilibrium (steady-state) condition has not been attained, since rise and decay times can be affected [124, 218]. It has also been shown that poor contacts or the presence of interfacial contamination at the barrier or junction, such as oxides on aluminum, can affect the value of n [22, 26, 27]. Figure 30 shows some typical I_{sc} (or I_{ph}) versus L plots for organic photovoltaic systems.

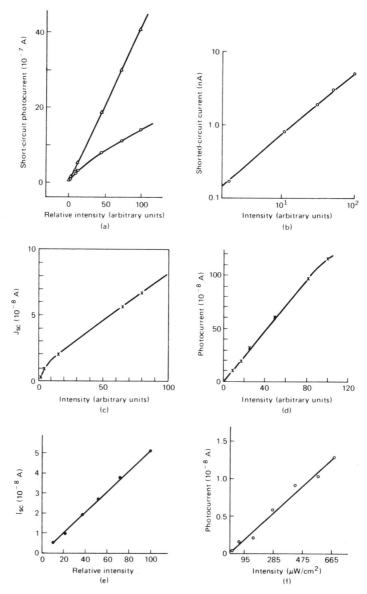

Fig. 30. Typical short-circuit and photocurrents as functions of the intensity of incident light for Schottky-type devices (a)–(d) and organic p-n heterojunction devices (e)–(f). (a) Al/H$_2$Ph/Au (O) and Al/ZnPh/Au ($\triangle$) [22]; (b) Al/Chl/Hg [18]; (c) Al/tetracene/Au [12]; (d) Ga/OHSq/In-O (backwall) [24b]; (e) merocyanine FX798 (p-type)/malachite green (n-type), $\lambda = 620$ nm; reading of 100 on intensity scale corresponds to 3.7×10^{15} quanta/cm^2 sec [42c]; (f) CdS/merocyanine A101, $\lambda = 558$ nm [42c].

2. Photovoltage

Since in an ideal Schottky barrier or p-n junction cell Eqs. (12)–(14) hold, and since $I_{ph} = f(L)$, we expect

$$V_{ph} \text{ (or } V_{oc}) \propto \ln L. \tag{55}$$

See for example Fig. 31, which shows the dependence of V_{oc} on L for various organic photovoltaic systems. Again, note that space-charge effects and/or field-dependent carrier generation schemes can complicate this dependence; in particular see Figs. 31c and 31d. In Fig. 31c the low light intensity gives an almost linear dependence [42c], as predicted in Ref. [1]; in Fig. 31d low-level light is also used [16].

3. Fill Factor FF

Recall that the fill factor has a complex dependence on the series resistance R_s, and thus on trapping effects, recombination, diffusion properties, etc., within the bulk of the organic material and at electrode surfaces or junction interfaces. Any dependence of these factors on light intensity L will also affect FF. For example, in most high-resistivity organic materials, as L increases, so do recombination processes, eventually leading to space-charge-limited conditions. In turn, one expects the value of FF to decrease with increasing intensity. In only one organic photovoltaic system has the change in FF with L been reported, the Ga/OHSq/Pt system [24], as shown in Fig. 32.

4. Power-Conversion Efficiency η

The ultimate dependence of η on L represents all the cumulative dependencies of I_{sc}, V_{oc}, and FF on L, the greatest effects being expected from I_{sc} and FF. For example, see Fig. 32 and Table XI, which summarizes the low-intensity (0.1–5 mW/cm²) and high-intensity (90–134 mW/cm²) photovoltaic behavior of Ga/OHSq/In-O and Ga/OHSq/Pt devices. The inferior

TABLE XI *Photovoltaic Properties of OHSq as a Function of Input Light Intensity*

Cell	Low intensity, 0.1–5 mW/cm²			High intensity, 90–134 mW/cm²		
	V_{oc} (V)	FF	η (%)	V_{oc} (V)	FF	η (%)
Ga/OHSq/In-O	0.37	0.31	0.06	0.55	0.20	0.02
Ga/OHSq/Pt	0.39	0.39	0.09	0.66	0.27	0.02

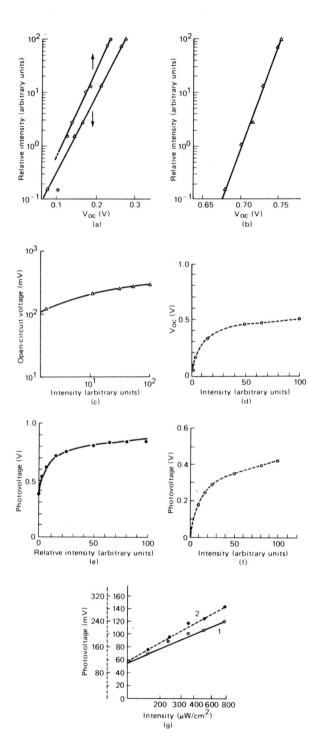

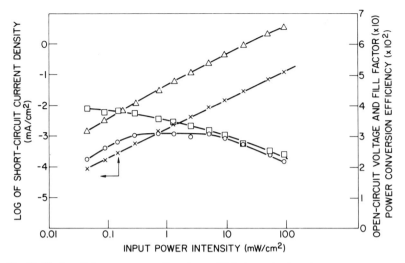

Fig. 32. Photovoltaic responses as functions of light intensity for a backwall Ga/OHSq (89 nm)/Pt Schottky barrier device. □, Fill factor (*FF*); ○, power-conversion efficiency AM0 light (η); ×, short-circuit current density (J_{sc}); △, open-circuit voltage (V_{oc}). (From Ref. [24b].)

high-intensity performance of the In-O cell (especially the decrease in *FF*) was attributed to space-charge effects arising from a double barrier.

5. Stability

In many organic systems irradiation of the devices causes irreversible degradation in the photovoltaic parameters as a result of thermal or chemical changes in the organic material. Photobleaching can occur, leading to optical absorption losses. Electrochemical reactions can also occur at electrode surfaces. One of the major challenges to the use (and study) of organic materials in photovoltaic devices remains their relative instability to long-term, high-intensity illumination.

K. Dependence of Photovoltaic Parameters on Temperature

1. Dark Currents and Photocurrents

Very fine reviews exist on the temperature dependence of the dark conductivity and photoconductivity of organic semiconductors. See for

Fig. 31. Typical photovoltages as functions of the incident light intensity for Schottky-type devices (a)–(f) and organic p-n heterojunction devices (g). (a) Al/H₂Ph/Au [22]; (b) Al/ZnPh/Au [22]; (c) Al/Chl/Hg [18]; (d) Al/tetracene/Au [12]; (e) Al/MgPh/Ag [17]; (f) Ga/OHSq/In-O backwall cell [24b]; (g) curve 1, CdS/merocyanine A101, λ = 558 nm; curve 2, AgI/rhodamine B, λ = 600 nm [42c].

example, Refs. [1, Chapters 6 and 10, Tables 4.8 and 8.2, and 37, 39–41, 42c]. In the brief discussion given here, thermopower effects (Seebeck and Peltier) [42c, 123] are excluded. The total observed temperature dependence for any particular organic photovoltaic device depends on the relative contributions of various effects. In fact, an observed temperature *independence* can be the result of canceling contributions, for example, of dn_c/dT and $d\mu/dT$ (to be discussed), of special trap distributions, of a dependence of Φ or τ [recall Eq. (17)] on temperature, etc.

a. Dark Conductivity—Intrinsic Generation and Recombination Currents. Remember that dark currents usually offset the observed photocurrents [see Eqs. (1), (7)–(10), and (16)]. Care must be taken to keep these contributions separated, since their temperature dependences can be very different.

Electrons can be thermally excited from valence states (impurity conductors are discussed later) into the conduction band when $kT \approx E_g/2$ (recall that E_f lies halfway between E_v and E_c in pure, intrinsic semiconductors). One obtains an exponential increase in the number of free carriers in the conduction band n_c with an increase in T [recall Eq. (16)];

$$\sigma = \sigma_0 \exp(-\Delta E/2kT) \quad [\text{or } (-\Delta E'/kT)],$$

where ΔE is the bandgap E_g of the material ($\Delta E'$ is an effective bandgap). In certain cases E_g (or E_g') can change with temperature.

This thermal *generation* current is partially offset by increased deactivation of free carriers by collisions of electrons with vibrating lattice atoms (phonon scattering), electrode surfaces, etc., and by increased recombination due to the diffusion of more electrons and holes across the junction or barrier. Because this rise in resistivity (mean free path of electrons decreases) is a fairly linear function of T, one still observes an approximately exponential increase in the conductivity with temperature. In the intermediate temperature ranges σ may appear to be relatively independent of T because of compensating factors. Eventually, at very high temperature, the conductivity is almost totally determined by phonon scattering and will begin to decrease with further increases in T. See, for example, Section 1.5–1.8 in [1] and Fig. 33.

b. Fermi–Dirac Distribution Function. The proportion of electrons (out of all possible) found at an energy level E is given by the Fermi–Dirac distribution function:

$$\frac{\rho}{\rho_0} = \frac{1}{1 + \exp[(E - E_f)/kT]}. \tag{56}$$

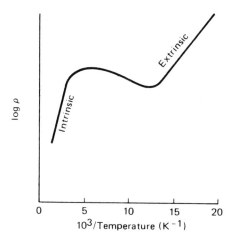

Fig. 33. Temperature dependence of the resistivity for a typical impurity semiconductor. (From Ref. [1, Fig. 1.13, p. 22].)

The shape of this distribution function is temperature-dependent. When normalized to $\rho_0 = 1$, at $T = 0$ K, the proportion of (or probability of finding) electrons in energy levels $E \leq E_f$ is 1, while for $E > E_f$ it is 0; i.e., the function is degenerate at E_f (Fig. 34). In fact, the Fermi level is defined as the energy at which the probability that a given energy level will be occupied (distribution function ρ) is exactly $\frac{1}{2}$. As the temperature is increased, a greater proportion of electrons occupy energy levels greater than E_f (see Ref. [1]); the curve rounds out as shown. At very high temperatures the classical Boltzmann–Maxwell distribution is obtained:

$$\rho = K \exp[(E_f - E)/kT], \tag{57}$$

where K is a proportionality constant.

   c.   *Dark Conductivity—Extrinsic Currents, Changes in E_f, and Trapping Effects.*   For impurity conductors, as $kT \rightarrow (E_c - E_i)$, where E_i is the energy of an impurity state, increased carrier densities (for n-type semiconductors) can be observed as a result of the excitation of electrons from localized donor impurity states (or traps) into the conduction band. For p-type semiconductors, as $kT \rightarrow (E_i - E_v)$, acceptor states localized in the bandgap near the valence band are filled with thermally excited electrons from valence states. In either case, the effective Fermi level is now raised relative to its lower-temperature value. For the case of donor impurities,

$$E_f = \frac{-E_d}{2} + \left(\frac{kT}{2}\right) \ln\left[\frac{2N_d}{c\pi^{1/2}(kT)^{3/2}}\right], \tag{58}$$

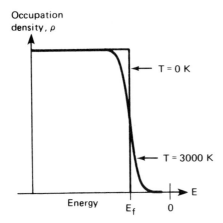

Occupation density, ρ

T = 0 K

T = 3000 K

Energy E_f 0   E

Fig. 34. The normalized Fermi–Dirac distribution function at $T = 0$ and 3000 K. (From Ref. [1, p. 11].)

where E_d is the energy of the donor and N_d is the concentration of donor levels. A similar expression can be derived for acceptor impurities. At absolute zero, all impurity semiconductors are insulators, since all bands are either completely full (E_v) or completely empty (E_c). At $T \le 100$ K, essentially all donor levels are emptied (or acceptor levels filled). Further increases in the temperature will produce no increase in the number of free carriers until $E_f \le E_g/2$, at which time the material behaves just like an intrinsic semiconductor.

These changes in E_f will in turn alter the contact (diffusion, barrier, or electrode injection) potentials in photovoltaic devices and, as a result, the observed photocurrents and photovoltages. The exact temperature dependence will be a complex function of the number, type, location, and energy (depth) of the traps (impurities, defects, etc.).

One can also obtain thermal detrapping of previously trapped carriers; however, when all trapping sites have been emptied, this current contribution will cease (see for example Ref. [153]). Any changes may be reversible or irreversible. Because both minority and majority carriers can be trapped differently, thermal release of these carriers can actually cause the dominant conductivity type (p or n) to change [234]. See Figs. 35a and 35b. Again, recall that we are referring to dark current contributions.

   d. Dark Conductivity—Mobilities. Recall that σ_0 in Eq. (16) is a function of the mobility μ:

$$\sigma_0 = e\mu N_0, \tag{59}$$

where N_0 is the effective density of states of conductivity levels in the material.

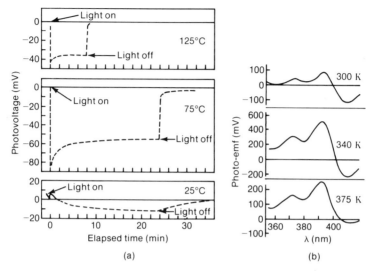

Fig. 35. (a) Temperature-dependent behavior of a Ag/tetracene (single-crystal)/Ag photo-voltaic cell indicative of thermal detrapping, illuminated at 400 nm [153]; (b) reversal of photovoltage response with temperature in anthracene [146].

The mobility consists of both diffusion (thermally activated) and drift (field-enhanced) terms and can be a function of temperature depending on the transport mechanism [1, 42b,c, 75a, 249–251], lattice characteristics, trapping distributions, type of scattering mechanism, purity, etc. For transport mechanisms involving pure *tunneling*, σ_0 exhibits little tempera-ture dependence until very high temperatures are reached, at which time significant thermal vibrations of the lattice atoms lead to a scattering-limited situation and a decrease in μ with increasing temperature [1, 31, 42c, 75b, 82, 219, 246]. Figure 33 showed an example for which $\mu \propto T^{-3/2}$. For scattering from *charged imperfections,* however, the mobility generally increases with increasing temperature: $\mu \propto T^{3/2}$. Often multiple scattering effects are present. See Ref. [31, Chapter 8]. Generally, mobilities ≤ 1 cm/V sec that increase with increasing temperature are indicative of *hopping* mechanisms, although there are cases where μ is either temperature-independent or decreases with increasing T. For example, for phonon-assisted hopping or charge transport by polarons [221] μ de-creases with increasing temperature [75b, 82]. *Band conduction* mecha-nisms [1, 42b, 75–81, 83–85, 222] generally show a complex temperature dependence based on trapping distributions and the extent of lattice in-teractions, etc. An increase in μ with increasing T can also be indicative of multiple trapping effects (and a frequency-dependent conductivity) [42c, 247]. See the general discussions in Refs. [1, 31] and references contained therein. In polymers the mobility is dominated by traps and the disordered nature of the polymer (Chapter 1).

e. Photocurrents. Recall that the observed output photocurrent I_{ph}, as given by Eq. (15), is the difference between the pure photogenerated current I_L and the dark current I_d. Again, care must be taken to exclude temperature effects on dark currents from any discussion of the photocurrents. Recalling Eq. (17),

$$I_{ph} = \Phi\mu ev\tau E/d = geN,$$

we see that the photocurrent can be temperature-dependent if Φ, μ, or τ is temperature-dependent (as is true for many organic materials). For systems with a uniform trap distribution one expects relative temperature independence; however, for an exponential trap distribution I_{ph} increases with increasing T; see Ref. [31, p. 73]. In addition, the conduction and transport mechanisms for *all* photogenerated species must be considered. For example, exciton dissociation to give free carriers can be thermally enhanced [1, 56, 136, 218, 231, 252–256], although the generation of excitons may not be. Release of previously trapped carriers (from dislocations, impurity sites, etc.) can occur with either thermal (infrared light, heat) or optical excitation if the trap depths are $\approx kT$ or $\approx h\nu$, respectively. Complex conduction mechanisms may be involved.

f. Structural Changes. Finally, increased temperatures can lead to changes in the dark currents or photocurrents as a result of the annealing out of potential trapping sites, actual changes in crystal packing (phase changes), melting, or diffusion of material into or out of an electrode or at an interface. The elimination of trapping sites (defects, grain boundaries, etc.) should decrease recombination or other deactivation modes, whereas phase changes can lead to significant changes in the charge-generation and conduction mechanisms involved, including changes in the bandgap energy E_g or creation of or a change in conduction anisotropy. See for example Refs. [1, p. 371, 31, 233] and further discussion in Section III.M.

2. Photovoltages

As mentioned previously, if the effective Fermi level changes as a function of T, so does the contact potential (and as a result, the barrier potential). Any decrease (usual case) in V_b should lead to a decrease in the maximum possible photovoltage, although additional complications can arise because of changes in surface states, annealing out of defects, thermal diffusion processes at electrode surfaces, etc. Because of high bulk resistivities, as I_{ph} increases with T, space-charge effects work to decrease the effective potential barrier across the device; see Ref. [1, Table 8.16].

In Eq. (54), $I \propto V^n$; n is a function of T and can be written as

$$n = [(T_c/T) + 1], \qquad (60)$$

where T_c is a temperature characteristic of the trap distribution [1]. For $T_c \ll T$, $n \to 1$ and ohmic behavior is observed.

3. Fill Factors

Since FF is a function of the series resistance, any change in R_s with temperature will also be reflected in the fill factor; the higher R_s, the lower FF. Degradation of FF with temperature could also indicate physical damage (or other change) to the junction or barrier interface due, e.g., to the migration of ionic species.

L. Time Dependence of Photovoltaic Parameters

The two major causes of time-dependent photovoltaic behavior can be classified as space-charge effects and stability effects. Several aspects of space-charge effects in photovoltaic devices have already been discussed in Sections II.A, II.B, III.E, and III.H.3.

1. Space-Charge Effects

a. Electrode Effects. We briefly discussed the detrimental effects of poor electrode contacts and the resulting time-dependent responses in Section III.F (Fig. 21). Often very erratic (and erroneous) measurements are obtained when the contact resistance R_c in the R_s term [see Eqs. (28) and (29) and Fig. 15b] is so large that $R_s \gtrsim R_{sh}$. If thick (tunneling no longer occurs) insulating layers occur on electrodes (particularly likely with electrodes such as Al) erratic time-dependent behavior can be obtained. See for example Ref. [22] and Fig. 36.

b. High Bulk Resistivities. The high bulk resistivities of organic materials and their numerous trapping levels (sometimes quite deep) can cause very large RC time constant values. In such cases, an electrical equilibrium is not established within the time frame of an average measurement.

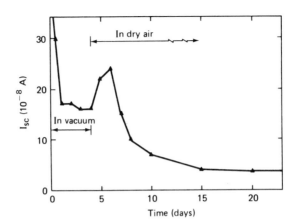

Fig. 36. Time-dependent photocurrent response of an Al/H₂Ph/Au photovoltaic cell indicating behavior both in a vacuum and after introduction of dry air. (From Ref. [22].)

In fact, this often occurs in cases involving ionic conduction processes.
Recall that under space-charge-limited conditions the carrier lifetime is
limited by the dielectric relaxation time.

Rise and decay times can be quite slow in organic materials because of
their high resistivities; however, this time dependence can be used to gain
information about conduction mechanisms, kinetics, and trap distributions. For example, hyperbolic decay curves are indicative of lifetimes
limited by bimolecular recombination. There have been numerous citings
of transient photocurrents and emfs in organic insulators. Many of these
have been due to electrode injection effects or to thermal or optical detrapping of carriers, or possible electrical polarization within the samples.
See Refs. [1, 4, 17, 54, 61, 124, 181, 257–263] and Figs. 37 and 38.

   c. Double Barriers. One of the major causes of time-dependent
photovoltaic behavior in semiconducting organic systems such as
chlorophyll and squarylium dyes is the presence of two blocking electrodes in the device. Figures 25 and 39 show time-dependent behavior for
single- and double-barrier chlorophyll cells [16]. Figure 40 and Table VI
illustrate time-dependent behavior of double-barrier (Ga/OHSq/In-O)
cells versus their single-barrier counterparts (Ga/OHSq/Pt) [24b]. Note
that the slow decrease with time (initial to steady-state values) of the
conversion efficiency for the double-barrier cell shows an approximately
22% drop compared to a decrease of only about 2% for the single-barrier
case.

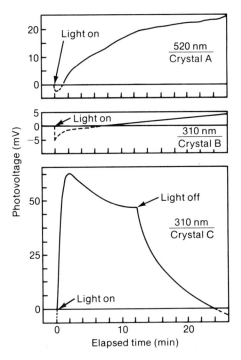

Fig. 37. Time dependence of observed photocurrents in an Ag/tetracene (single-crystal)/Ag sample. (From Ref. [153].)

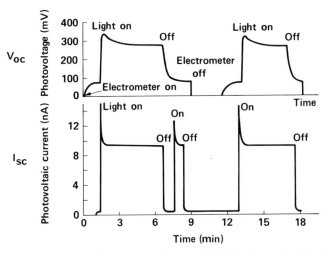

Fig. 38. Transient current spikes observed in an Al/Chl/Hg photovoltaic device. These were attributed to possible electrical polarization or inhomogeneous charge transport (grain-boundary conditions). (From Ref. [16].)

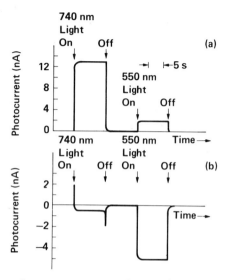

Fig. 39. Time-dependent photocurrent responses for (a) a single-barrier Cr/Chl-a/Hg cell and (b) a double-barrier Cr/Chl-a/Hg-In photovoltaic cell. (From Ref. [16].)

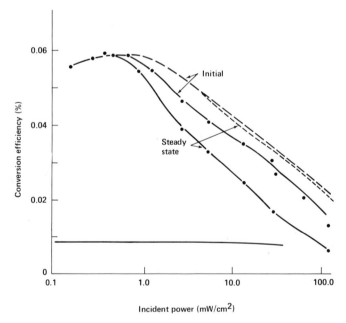

Fig. 40. Time-dependent decrease in conversion efficiency for a double-barrier Schottky barrier cell of Ga/OHSq/In-O. Note the increased drop in steady-state efficiency at higher intensities. The dashed line indicates the values for a single-barrier Ga/OHSq/Pt cell (AM0 light) where there was a drop of only about 2–5% in the steady-state values even at high intensities. (From Ref. [24b].)

2. Stability Effects

Many organic materials suffer from an inability to remain electrically, chemically, or structurally stable for long periods of time under constant use in the photovoltaic mode. Some systems may degrade within hours or days. The primary reasons for such degradation can be attributed to either chemical or structural changes. For example, under the high fields experienced by most thin-film photovoltaic devices (10^3–10^5 V/cm), irreversible electrochemical reactions or actual diffusion of materials can occur at electrode surfaces or junction interfaces. If small pinholes exist, repeated passage of current under high-field conditions can cause electrical shorting at these spots. Depending on the complexity and nature of the electrode materials (alloys, for example) and the porosity of the organic films, degradation of photovoltaic parameters may be due to the diffusion of particular species into the organic voids; see for example Ref. [210].

Obviously, if an organic material is susceptible to oxidation or reduction under atmospheric conditions, special precautions must be taken to protect the device, especially where thin-film devices are involved. A controlled-environment measuring cell was reported by Mehl and Wolff [264]. Potential dopants (oxygen, water, charge-transfer agents, etc.) need to be considered as either detrimental or beneficial. Of course, the whole field of photogalvanic effects is based on controlling reversible (optimum case) electrochemical reactions between organic materials (solids, liquids, solutions) and electrodes or electrolytes.

M. Structural Effects

The most obvious effects of chemical structure on photovoltaic parameters relate to the absorption properties and the transport or conduction mechanism(s) involved.

As discussed briefly in Section II.B, the breadth and intensity of absorption is influenced by structural aspects such as crystallinity (polycrystalline, single-crystalline, amorphous), the presence of dimer, trimer, or other aggregate species, actual crystal structure (phase), etc. Such absorption changes obviously affect both dark currents and photocurrents, and thus photovoltaic properties. For example, see Fig. 41 [209].

Charge transport (electronic or ionic) can be considerably impeded by recombination at grain boundaries or other trapping effects in polycrystalline materials. For example, studies comparing the photovoltaic properties of amorphous versus polycrystalline Ga/OHSq/Pt devices showed an order-of-magnitude decrease in the short-circuit photocurrents and conversion efficiencies (Table XII). In addition, there have already been ex-

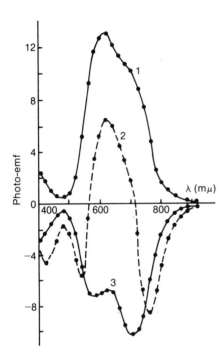

Fig. 41. Effect of structure or orientation on photovoltaic properties. The photo-emf of H_2Ph in relation to the form of aggregation of the pigments. Curve 1, crystalline α form; curve 2, mixture of crystalline α and β forms; curve 3, crystalline β form. (From Ref. [209].)

TABLE XII *Comparison of Photovoltaic Parameters for Amorphous and Polycrystalline Ga/230-nm OHSq/In-O Cells*

Film	V_{oc} (V)	J_{sc} ($\mu A/cm^2$)	FF	η (%)
Amorphous	0.52	20	0.27	1.3×10^{-3}
Polycrystalline	0.53	1.2	0.25	7.5×10^{-5}

tensive discussions of the effect of structure on conduction mechanisms; see Refs. [1, 2, 37, 39–42].

IV. SUMMARY OF DESIRABLE PHOTOVOLTAIC MATERIALS PROPERTIES

The previous discussions point out the numerous and complex factors involved in the generation of electrical current from organic photovoltaic devices. Table XIII is an attempt to summarize some of the more important properties to look for in potentially efficient organic photovoltaic materials.

TABLE XIII   *Desirable Properties of an Efficient Organic Photovoltaic Material[a]*

High quantum efficiency for charge generation (in particular, by visible light)
High absorption coefficient over as broad a region of photon energies as possible (match the solar spectral radiance if possible)
Ready availability from cheap, nontoxic materials
Good electronic transport properties such as high mobility and low bulk resistivity
Low dark conductivity in device configuration
Ready film-forming capability and easy device fabrication
Good performance at high intensities (100–130 mW/cm^2)
Long-term thermal and chemical stability

---

[a] Solar cell applications are highlighted by parenthetical statements.

V. NOTES ON POLYMER SYSTEMS

Although fairly high efficiencies were predicted [23] for photovoltaic energy conversion for organic charge-transfer polymer systems (PVK–TNF) some time ago, only recently have there been reports of promising experimental work on photovoltaic devices made of such "semiconductive" organic polymer systems as doped polyacetylenes [132, 133, 210], polypyrroles [200], polytetrathiafulvalenes, etc. Understanding the basic principles of energy or charge transport in these systems is even more complicated than for the organic (monomeric) materials discussed thus far; however, they remain highly promising materials for the future. A recent review by Reucroft and Ullal [25b] discusses the photovoltaic properties of polymer films.

REFERENCES

1. F. Gutmann and L. E. Lyons, "Organic Semiconductors." Wiley, New York, 1967.
2. H. Meier and W. Albrecht, *Ber. Bunsenges. Phys. Chem.* **68**, 64 (1964).
3. A. M. Hermann and A. Rembaum, *Polym. Lett.* **5**, 445 (1967).
4. N. N. Usov and V. A. Benderski, *Sov. Phys.—Semicond.* **2**, 580 (1968).
5. P. J. Reucroft and W. H. Simpson, *Photochem. Photobiol.* **10**, 79 (1969).
6. M. I. Fedorov and V. A. Benderski, *Sov. Phys.—Semicond.* **4**, 1198, 1720 (1971).
7. L. E. Lyons and O. M. G. Newman, *Aust. J. Chem.* **24**, 13 (1971).
8a. N. A. Dimond and T. K. Mukherjee, *Discuss. Faraday Soc.* **51**, 102 (1971).
8b. P. H. Fang, A. Golubovic, and N. A. Dimond, *Jpn. J. Appl. Phys.* **11**, 1298 (1972).
9. J. P. Dodelet, J. LeBrech, and R. M. Leblanc, *Photochem. Photobiol.* **29**, 1135 (1979).
10. Y. I. Vertsimakha and M. V. Kurik, *Microelectronica* **1**, 275 (1972).
11. P. J. Reucroft, A. K. Ghosh, and K. Keever, *J. Polym. Sci. Polym Phys. Ed.* **10**, 2305 (1972).
12. A. K. Ghosh and T. Feng, *J. Appl. Phys.* **44**, 2781 (1973).
13. P. J. Reucroft and A. K. Ghosh, *Phys. Rev. B* **8**, 803 (1973).
14. P. H. Fang, *J. Appl. Phys.* **45**, 4672 (1974).
15. K. Okumura, *J. Appl. Phys.* **45**, 5317 (1974).

16a. C. W. Tang and A. C. Albrecht, *Mol. Cryst. Liq. Cryst.* **25**, 53 (1974).

16b. C. W. Tang and A. C. Albrecht, *J. Chem. Phys.* **62**, 2139 (1975); **63**, 953 (1975); *Nature (London)* **254**, 507 (1975).

17. A. K. Ghosh, D. L. Morel, T. Feng, R. S. Shaw, and C. A. Rowe, Jr., *J. Appl. Phys.* **45**, 230 (1974).

18. Yu. S. Shumov and L. N. Strekova, *Zh. Fiz. Khim.* **54**, 197 (1980); Yu. S. Shumov, G. G. Komissarov, and M. F. Chaplii, *ibid.* **53**, 2838 (1979).

19. M. Matsumara, H. Uohashi, M. Furusawa, N. Yamamoto, and H. Tsubomura, *Bull. Chem. Soc. Jpn.* **48**, 1965 (1975).

20. F. J. Kampas and M. Gouterman, *J. Phys. Chem.* **81**, 690 (1977).

21. H. Meier, W. Albrecht, U. Tschirwitz, E. Zimmerhackl, and N. Geheeb, *Ber. Bunsenges. Phys. Chem.* **81**, 592 (1977).

22. F.-R. Fan and L. R. Faulkner, *J. Chem. Phys.* **69**, 3334, 3341 (1978).

23. P. J. Reucroft, K. Takahashi, and H. Ullal, *Appl. Phys. Lett.* **25**, 664 (1974); *J. Appl. Phys.* **46**, 5218 (1975).

24a. V. Y. Merritt and H. J. Hovel, *Appl. Phys. Lett.* **29**, 414 (1976).

24b. V. Y. Merritt, *IBM J. Res. Dev.* **22**, 253 (1978).

25a. E. R. Menzel and R. O. Loutfy, *Chem. Phys. Lett.* **72**, 522 (1980).

25b. P. J. Reucroft and H. Ullal, *Solar Energy Mater.* **2**, 217 (1979–1980).

25c. F. J. Kampas, K. Yamashita, and J. Fajer, *Nature (London)* **284**, 40 (1980).

26. D. L. Morel, A. K. Ghosh, T. Feng, E. L. Stogryn, P. E. Purwin, and C. Fishman, *Appl. Phys. Lett.* **32**, 495 (1978).

27. A. K. Ghosh and T. Feng. *J. Appl. Phys.* **49**, 5982 (1978).

28. J. M. Woodall and H. J. Hovel, *Appl. Phys. Lett.* **30**, 492 (1977); H. J. Hovel, *IBM J. Res. Dev.* **22**, 112 (1978).

29. For a recent review of inorganic photovoltaic materials and their best attained device conversion efficiencies, see E. A. Perez-Albuerne and Y.-S. Tyan, *Science* **208**, 902 (1980).

30a. For recent reviews of photogalvanic devices see K. Honda and K. Yoshi, *Shinpojuma "Kagaku no Mirai"* 24 (1979); Zheng-Hui Liu, Zhi-Chu Bi, Pe-Shu Cheng, Yan-Ning Zhu, Yu-Shu Li, and T. H. Tien, *K'o Hsueh T'ung Pao* **24**, 1027 (1979).

30b. See also Takashi Katsu, Keietsu Tamagake, and Yuzaburo Fujita, *Kokaga-ku Toronkai Koen Yoshishu* **76** (1979) (*Chem. Soc. Jpn.*); T. I. Quickenden and G. K. Yim, *J. Phys. Chem.* **84**, 670 (1980); E. Yu. Katz, Yu. N. Kozlov, and B. A. Kiselev, *Energy Convers.* **19**, 73 (1979).

31a. H. Dember, *Phys. Z.* **32**, 554, 856 (1931); **33**, 207 (1932).

31b. J. Frenkel, *Phys. Z. Sowjetunion* **8**, 185 (1935).

32. L. Pensak, *Phys. Rev.* **109**, 601 (1958); B. Goldstein and L. Pensak, *J. Appl. Phys.* **30**, 155 (1959).

33. G. Cheroff and S. P. Keller, *Phys. Rev.* **111**, 98 (1958).

34. J. T. Wallmark, *Proc. IRE* **45**, 474 (1957).

35. R. K. Mueller and R. L. Jacobson, *J. Appl. Phys.* **30**, 121 (1959).

36. T. S. Moss, L. Pincherle, and W. Woodward, *Proc. Phys. Soc.* **66B**, 743 (1953).

37. R. H. Bube, "Photoconductivity of Solids," Wiley, New York, 1960; *Phys. Rev.* **83**, 393 (1961).

38. H. J. Hovel, *Semicond. Semimet.* **11** (1975).

39. J. P. McKelvey, *in* "Solid State and Semiconductor Physics" (F. Seitz, ed.). Harper, New York, 1966.

40. S. M. Sze, *in* "Physics of Semiconductor Devices," Chapter 8. Wiley (Interscience), New York, 1960.

41. C. Kittel, "Introduction to Solid State Physics," 4th ed., Chapter 9, p. 293, Chapter 11, p. 359. Wiley, New York, 1971.

42a. H. Meier, "Die Photochemie der organischen Farbstoffe." Springer-Verlag, Berlin, 1963.

42b. H. Meier, "Organic Semiconductors: Dark- and Photoconductivity of Organic Solids." Verlag Chemie, Weinheim, 1974.

42c. H. Meier, *Top. Curr. Chem.* **61**, 85 (1976).

43. B. Levy and M. Lindsey, *Photogr. Sci. Eng.* **16**, 389 (1972); B. Levy, M. Lindsey, and C. R. Dickson, *ibid.* **17**, 115 (1973); B. Levy, *ibid.* **18**, 347 (1974).

44. T. S. Moss, *Proc. Phys. Soc. London* **66B**, 993 (1953); *Physica* **20**, 989 (1954).

45. R. N. Zitter, *Phys. Rev.* **112**, 852 (1958).

46. J. Tauc, *Czech. J. Phys.* **6**, 421 (1956).

47. S. W. Kurnick and R. N. Zitter, *J. Appl. Phys.* **27**, 278 (1956).

48. H. Bulliard, *Ann. Phys.* **9**, 52 (1954).

49. W. Gartner, *Phys. Rev.* **105**, 823 (1957).

50. O. Garreta and J. Grosvalet, *Prog. Semicond.* **1**, 167 (1956).

51a. H. Inokuchi, Y. Maruyama, and H. Akamatu, *Bull. Chem. Soc. Jpn.* **34**, 1093 (1961).

51b. Y. Maruyama, H. Inokuchi, and Y. Harada, *Bull. Chem. Soc. Jpn.* **36**, 1193 (1963).

52. H. T. Fan, *Phys. Rev.* **75**, 1631 (1949).

53. A. F. Ioffe, "Physics of Semiconductors," pp. 126, 193, 194. (Engl. transl.). Infosearch Ltd., London, 1960; *J. Phys. USSR* **10**, 49 (1946).

54. R. W. Smith, *Phys. Rev.* **97**, 1525 (1955).

55. J. Kommandeur, *J. Phys. Chem. Solids* **22**, 347 (1966).

56. L. E. Lyons, *J. Chem. Soc.* 5001 (1957).

57. P. A. Anderson, *Phys. Rev.* **47**, 958 (1935).

58. R. C. Nelson, *J. Opt. Soc. Am.* **46**, 1016 (1956).

59. A. M. Goodman, *J. Appl. Phys.* **36**, 1411 (1965).

60. A. Rembaum and J. Moacanin, "Polymeric Semiconductors." Jet Propulsion Lab, Pasadena, California, 1964.

61. H. Kallmann and M. Silver (eds.), "Electrical Conductivity in Organic Solids." Wiley (Interscience), New York, 1961.

62. W. Helfrich, "Physics and Chemistry of the Solid State," (D. Fox, M. Labes, and A. Weissberger, eds.), Vol. 2. Wiley (Interscience), New York, 1965; *ibid,* Vol. 3, Chapter 1, 1967.

63. P. T. Landsberg, *Solid State Electron.* **18**, 1043 (1975).

64. T. Nishino, F. Yamano, and Y. Hamakawa, *Jpn. Appl. Phys.* **14**, 1885 (1975).

65. A. Many, J. Levinan, and I. Teuchev, *Mol. Cryst.* **5**, 273 (1969).

66. G. Björklund and H. G. Grimmeiss, *Ark. Fys.* **40**, 183 (1969).

67. G. T. Noel, U.S. National Technical Inform. Service, AD Rep. No. 777139/IGA (1973).

68. W. Hwang, *Diss. Abstr. Int. B* **35**, 2189 (1974).

69. M. C. Teich and G. J. Wolga, *J. Opt. Soc. Am.* **57**, 542 (1967).

70. M. Ideda, H. Sata, K. Morimoto, and Y. Murakami, *Photogr. Sci. Eng.* **19**, 60 (1975).

71. W. L. McCubbin, *J. Polym. Sci.* **C30**, 181 (1970).

72. A. V. Vannikov, *Kokl. Akad. Nauk. SSSR [Engl. transl.:* **152**, 874 (1963)].

73. T. S. Moss, G. J. Burrell, and B. Ellis, "Semiconductor Opto-Electronics," Chapters 5 and 6. Wiley, New York, 1973.

74. K. Masuda and M. Silver (eds.), "Energy and Charge Transfer in Organic Semiconductors." Plenum Press, New York, 1974.

75a. H. Meier, *Chimia* **27**, 263 (1973).

75b. H. Meier and W. Albrecht, *Z. Naturforsch.* **24a**, 257 (1969).

75c. H. Meier, "Spectral Sensitization." Focal Press, London, 1968.
75d. H. Meier and W. Albrecht, *Ber. Bunsenges. Phys. Chem.* **69**, 160 (1965).
76. W. G. Williams, *Disc. Faraday Soc.* **51**, 61 (1971).
77. G. A. Cox and P. C. Knight, *J. Phys. C: Solid State Phys.* **7**, 146 (1974).
78. C. Hamann and M. Starke, *Phys. Status Solidi* **4**, 509 (1964).
79. G. H. Heilmeier and S. E. Harrison, *Phys. Rev.* **132**, 2010 (1963).
80. D. D. Eley and R. Pethig, *Proc. Int. Conf. Conduct. Low-Mobil. Mat., 2nd, Eilat, Israel* p. 397 (1971); R. Pethig and K. Morgan, *Phys. Status Solidi b* **43**, K119 (1971).
81. J. Dresner, *J. Chem. Phys.* **52**, 6343 (1970).
82. R. G. Kepler, *Phys. Rev.* **119**, 1226 (1960).
83. W. Mey and A. M. Hermann, *Phys. Rev. B* **7**, 1652 (1973).
84. O. H. LeBlanc, *J. Chem. Phys.* **39**, 2395 (1963).
85. A. Schmillen and W. W. Falter, *Z. Phys.* **218**, 401 (1969).
86. A. C. Albrecht, Photovoltaic effects in layered organic dyes, presented at the *State Univ. N.Y. Summer Inst. Sci. Technol. Conf. Elect. Photoconduct. Properties of Polym., New Paltz, New York* (May 18, 1978).
87. G. Sadasiv, "Photoelectronic Devices," Vol. 1, p. 111. Plenum Press, New York, 1971.
88a. H. Meier, W. Albrecht, and U. Tschirwitz, *Photogr. Sci. Eng.* **18**, 276 (1974).
88b. H. Meier, *Photochem. Photobiol.* **16**, 219 (1972).
89a. A. Rose, "Concepts in Photoconductivity and Allied Problems." Wiley (Interscience), New York, 1963.
89b. A. Rose, *RCA Rev.* **12**, 362 (1951).
90. See also a recent review of conductive polymers by J. Mort, *Science* **208**, 819 (1980).
91. L. E. Lyons and G. C. Morris, *J. Chem. Soc.* 5192 (1960); M. Batley and L. E. Lyons, *Mol. Cryst.* **3**, 357 (1968).
92. F. I. Vilessov and A. N. Terenin, *Dokl. Akad. Nauk. SSSR* **133**, 1060 (1960).
93. M. Pope, *J. Chem. Phys.* **36**, 2810 (1962).
94. M. Kotani and M. Akamatu, *Disc. Faraday Soc.* **51**, 94 (1971).
95a. G. A. Somorjai, Principles of surface chemistry, *in* "Fundamental Topics in Physical Chemistry." Prentice-Hall, Englewood Cliffs, New Jersey, 1972.
95b. M. E. Musser and S. C. Dahlberg, *Surf. Sci.* **91**, L23 (1980).
96. L. E. Lyons, *Proc. R. Aust. Inst.* **37**, 329 (1970).
97. H. Inokuchi, *Bull. Chem. Soc. Jpn.* **24**, 222 (1951); **29**, 131 (1956).
98. D. C. Northrop and O. Simpson, *Proc. R. Soc. London Ser. A* **234**, 124 (1956).
99. M. Sano and H. Akamatu, *Bull. Chem. Soc. Jpn.* **34**, 1569 (1961).
100. A. Terenin and I. Akimov, *Z. Phys. Chem. (Leipzig)* **217**, 307 (1961).
101a. Y. Harada and H. Inokuchi, *Bull. Chem. Soc. Jpn.* **39**, 1443 (1966).
101b. M. Kochi, Y. Harada, and H. Inokuchi, *Bull. Chem. Soc. Jpn.* **40**, 531 (1967).
102. H. J. Wintle, *Conf. Record 1976 IEEE Int. Symp. Elec. Insul., Montreal* p. 248 and Ref. 11 therein.
103. H. B. Michaelson, *IBM J. Res. Dev.* **22**, 72 (1978).
104. H. Meier, *Angew. Chem.* **77**, 633 (1965); *Angew. Chem. Int. Ed. Engl.* **4**, 619 (1965).
105. M. A. Lampert, *Rep. Prog. Phys.* **27**, 329 (1964).
106. M. Pope and H. Kallmann, *Discuss. Faraday Soc.* **51**, 7 (1971).
107. J. Kostelec, U.S. Patent 3,009,006 (1961).
108. M. Yoshida, *Bull. Kobayashi Inst. Phys. Res.* **13**, 109 (1963).
109. M. Calvin and D. R. Kearns, U.S. Patent 3,037,947 (1962).
110. A. D. Adler, U.S. Patent 3,935,031 (1976).
111. H. Meier and A. Haus, *Angew. Chem.* **72**, 631 (1960).

112. W. Noddack, H. Meier, and A. Haus, Z. *Wiss. Photogr. Photophys. Photochem.* **55**, 7 (1961).
113. R. J. Soukup and L. A. Akers, J. *Appl. Phys.* **49**, 4031 (1978).
114a. J. Mort and D. M. Pai (eds.) "Photoconductivity and Related Phenomena." Elsevier, Amsterdam, 1976.
114b. J. S. Bonham, *Aust. J. Chem.* **29**, 2123 (1976).
115a. C. Hamann, *Phys. Status Solidi* **4**, K97 (1964).
115b. E. Krikorian, U.S. Clearinghouse Fed. Sci. Tech. Inform A.D., AD-679601 (1968).
116. S. C. Dahlberg and M. E. Musser, *Surf. Sci.* **88**, 667 (1979).
117. S. J. Fonash, J. *Appl. Phys.* **46**, 1286 (1975).
118. H. C. Card and E. S. Yang, *Appl. Phys. Lett.* **29**, 51 (1976).
119. H. Schewchun, R. Singh, and M. A. Green, J. *Appl. Phys.* **48**, 765 (1977).
120. A. K. Ghosh, C. Fishman, and T. Feng, J. *Appl. Phys.* **49**, 3490 (1978).
121. R. J. Stirn and Y. C. M. Yeh, presented at the *IEEE Photovoltaic Specialists Conf., 12th* (1976); R. B. Godfrey and M. A. Green, *Appl. Phys. Lett.* **33**, 637 (1978), **34**, 790 (1979); M. A. Green, F. D. King, and J. Shewchun, *Solid-State Electron.* **17**, 557 (1974); M. A. Green and R. B. Godfrey *Appl. Phys. Lett.* **29**, 610 (1976); N. G. Tarr and D. L. Pulfrey, *ibid.* **34**, 295 (1977).
122. H. Meier and W. Albrecht, Z. *Phys. Chem.* **39**, 249 (1963).
123. H. Meier, W. Albrecht, and U. Tschirwitz, *Ber. Bunsenges. Phys. Chem.* **73**, 795 (1969).
124. D. R. Kearns, G. Tollin, and M. Calvin, J. *Chem. Phys.* **32**, 1013 1020 (1960).
125. Y. Aoyagi, K. Masuda, and S. Namba, J. *Phys. Soc. Jpn.* **31**, 524 (1971).
126. H. Inokuchi, Y. Maruyama, and H. Akamatu, *in* "Electrical Conductivity in Organic Solids." Wiley, New York, 1961.
127. W. Noddack and H. Meier, Z. *Elektrochem.* **57**, 691 (1953).
128. H. Meier, Z. *Electrochem.* **59**, 1029 (1955).
129. H. Meier, Z. *Wiss. Photogr. Photophys. Photochem.* **50**, II, 301 (1955).
130. H. Hoegl, J. *Phys. Chem.* **69**, 755 (1965).
131. B. Löhr, R. Arneth, and D. Winkelmann, *in* "Current Problems in Electrophotography" (W. F. Berg and K. Hauffe, eds.), p. 219. de Gruyter, Berlin, 1972.
132. H. Shirakawa, E. J. Louis, A. G. MacDiarmid, C. K. Chiang, and A. J. Heeger, J. *Chem. Soc. Chem. Commun.* 578 (1977); C. K. Chiang, S. C. Gau, C. R. Fincher, Jr., Y. W. Park, A. G. MacDiarmid, and A. J. Heeger, *Appl. Phys. Lett.* **33**, 18 (1978); C. K. Chiang, M. A. Druy, S. C. Gau, A. J. Heeger, E. J. Louis, A. G. MacDiarmid, Y. W. Park, and S. Shirakawa, J. *Am. Chem. Soc.* **100**, 1013 (1978); C. K. Chiang, Y. W. Park, A. J. Heeger, H. Shirakawa, E. J. Louis, and A. G. MacDiarmid, J. *Chem. Phys.* **69**, 5098 (1978); C. K. Chiang, C. R. Fincher, Jr., Y. W. Park, A. J. Heeger, H. Shirakawa, E. J. Louis, S. C. Gau, and A. G. MacDiarmid, *Phys. Rev. Lett.* **39**, 1098 (1977).
133. T. C. Clarke and G. B. Street, *Solid State Commun.* **28**, 873 (1978); T. Tani, P. M. Grant, W. D. Gill, G. B. Street, and T. C. Clarke, *ibid.* **33**, 499 (1980).
134. W. G. Schneider and T. C. Waddington, J. *Chem. Phys.* **25**, 358 (1956).
135. D. M. J. Compton and T. C. Waddington, J. *Chem. Phys.* **25**, 1075 (1956).
136. A. G. Chynoweth, J. *Chem. Phys.* **22**, 1029 (1954).
137. A. Bree and L. E. Lyons, J. *Chem. Phys.* **25**, 1284 (1956).
138. A. Bree, D. J. Carswell, and L. E. Lyons, J. *Chem. Soc.* 1728 (1955).
139. L. Landau, quoted by H. Kallmann and M. Pope, J. *Chem. Phys.* **30**, 585 (1959).
140. L. Landau, *Rev. Sci. Instrum.* **30**, 44 (1959).

141. H. Killesreiter and H. Baessler, *Chem. Phys. Lett.* **11**, 411 (1971).
142. P. J. Reucroft, *J. Chem. Phys.* **36**, 1114 (1962).
143. M. Pope and H. Kallmann, *J. Chem. Phys.* **30**, 585 (1959).
144. M. Pope and H. Kallmann, *in* "Electrical Conductivity in Organic Solids" (H. Kallmann and M. Silver, eds.), p. 83. Wiley (Interscience), New York, 1961.
145. M. Pope, H. Kallmann, A. Chen, and P. Gordon, *J. Chem. Phys.* **36**, 2486 (1962).
146. V. V. Vladimirov, M. V. Kurik, and Y. P. Piryantinskii, *Dokl. Akad. Nauk. SSSR* **181**, 1365 (1968) [*English trans.: Sov. Phys. Dokl.* **13**, 789 (1969)].
147. A. Marchetti and D. R. Kearns, *J. Chem. Phys.* **44**, 1301 (1966).
148. H. Kuroda and E. A. Flood, *Can. J. Chem.* **39**, 1981 (1961).
149. E. Silins *et al., Phys. Status Solidi a* **25**, 339 (1974).
150. E. Silins *et al., Pr. Nauk. Inst. Chem. Org., Fiz. Politech Wroclaw.* **7**, 72 (1974).
151. M. Calvin and D. Kearns, *J. Chem. Phys.* **29**, 950 (1958).
152. P. H. Fang, *Jpn. J. Appl. Phys.* **13**, 1232 (1974).
153. P. J. Reucroft, P. L. Kronick, and E. E. Hillman, *Mol. Cryst. Liq. Cryst.* **6**, 247 (1969); Final rep. to U.S. Air Force Cambridge Research Labs., Contract No. AF19(628)-5511 Project No. 8659 (1968).
154. E. L. Frankevich and E. I. Balabanov, *Phys. Status Solidi* **14**, 523 (1966).
155. A. Golubovic, *Proc. OAR Res. Appl. Conf., Washington, D.C.* p. 211 (1967).
156. T. K. Mukherjee, *Record Photovoltaics Specialists Conf., 6th, Cocoa Beach, Florida* **1**, 7 (1967).
157. H. Baessler, G. Hermann, N. Riehl, and G. Vaubel, *J. Phys. Chem. Solids* **30**, 1579 (1969).
158. P. H. Fang, U.S.N.T.I.S. AD Rep. 1974/IGA from Govt. Rep. Announce. (U.S.), **74**, 84 (1974).
159. P. H. Fang, M. Hirata, and M. Hirata, Air Force Cambridge Research Lab, Final Report 1970 AFCRL Contract No. F19628-70-C-0263.
160. E. Krikorian and R. J. Sneed, *J. Appl. Phys.* **40**, 2306 (1969).
161. H. Kuroda and E. A. Flood, *J. Chem. Phys.* **33**, 952 (1960).
162. H. Kuroda and E. A. Flood, *Can. J. Chem.* **39**, 1475 (1961).
163. T. K. Mukherjee, *J. Phys. Chem.* **74**, 3006 (1970).
164. R. M. Hochstrasser and M. Ritchie, *Trans. Faraday Soc.* **52**, 1363 (1956).
165. H. Akamatu, H. Inokuchi, and M. Matsunga, *Bull. Chem. Soc. Jpn.* **29**, 213 (1956).
166. H. Akamatu and H. Inokuchi, *Proc. Conf. Carbon, 3rd* p. 51. Pergamon, New York, 1959.
167. A. D. Tavares, *J. Chem. Phys.* **53**, 2520 (1970).
168. D. N. Bailey, D. M. Hercules, and D. K. Roe, *J. Electrochem. Soc.* **116**, 190 (1969).
169. H. Meier, *Z. Physik. Chem. (Leipzig)* **208**, 325 (1958); E. Rexer, "Organische Halbleiter." Akademie Verlag, Berlin, 1966.
170. H. Bauser and H. H. Ruf, *Phys. Status Solidi* **32**, 135 (1969); A. E. Binks, A. G. Campbell, and A. Sharples, *J. Polym. Sci. Part A-2* **8**, 529 (1970).
171. J. G. Simmons, *in* "Handbook of Thin Film Technology" (L. I. Maissel and R. Glang, eds.). McGraw-Hill, New York, 1970.
172. R. G. Kepler, *Phys. Rev. Lett.* **18**, 951 (1967).
173. E. Courtens, A. Bergman, and J. Jortner, *Phys. Rev.* **156**, 948 (1967).
174. C. L. Braun, *Phys. Rev. Lett.* **21**, 215 (1968).
175. G. R. Johnston and L. E. Lyons, *Chem. Phys. Lett.* **2**, 489 (1968).
176. G. Castro, *IBM J. Res. Dev.* **15**, 27 (1971).
177. J. Fourny, G. Delacote, and M. Schott, *Phys. Rev. Lett.* **21**, 1085 (1968).
178. P. Holtzman, R. Morris, R. C. Jarnagin, and M. Silver, *Phys. Rev. Lett.* **19**, 506 (1967).

179. F. C. Strome, Jr., *Phys. Rev. Lett.* **20**, 3 (1968).
180. A. Bergman and J. Jortner, *Phys. Rev. B* **9**, 4560 (1974).
181. G. Castro and J. F. Hornig, *J. Chem. Phys.* **42**, 1459 (1965).
182. N. Geacintov, M. Pope, and H. Kallmann, *J. Chem. Phys.* **45**, 2639 (1966).
183. R. F. Charken and D. R. Kearns, *J. Chem. Phys.* **49**, 2846 (1968).
184. R. G. Williams and B. A. Lowry, *J. Chem. Phys.* **56**, 5736 (1972).
185. J. Jortner, *Phys. Rev. Lett.* **20**, 244 (1968).
186. J. P. Hernandez and S. I. Choi, *J. Chem. Phys.* **50**, 1524 (1969).
187. N. E. Geacintov and M. Pope, *J. Chem. Phys.* **47**, 1194 (1967).
188. M. Pope and J. Burgos, *Mol. Cryst.* **3**, 215 (1967).
189. N. E. Geacintov and M. Pope, *J. Chem. Phys.* **50**, 814 (1969).
190. H. Killestreiter and H. Baessler, *Phys. Status Solidi b* **53**, 193 (1972); H. Killestreiter, *ibid*, **51**, 657 (1972).
191. J. Singh and H. Baessler, *Phys. Status Solidi b* **62**, 147 (1974).
192. A. Coret, S. Nikitine, J. P. Zielinger, and M. Zouaghi, *Proc. Int. Conf. Photoconduct., 3rd* (E. M. Pell, ed.), p. 81. Pergamon, New York, 1969.
193. L. E. Lyons, *Search* **7**, 339 (1976).
194. G. Chanussat and A. M. Glass, *Phys. Lett* **59A**, 405 (1976).
195. R. H. Batt, C. L. Braun, and J. F. Hornig, *J. Chem. Phys.* **49**, 1967 (1968); *Appl. Opt. Suppl. 3, Electrophotogr.* 20 (1969).
196. D. L. Stockmann, *in* "Current Problems in Electrophotography" (W. F. Berg and K. Hauffe, eds.), p. 202. de Gruyter, Berlin, 1972.
197. A. C. Albrecht and A. J. Twarowski, Dept. of Chemistry, Cornell Univ., Ithaca, New York, 14853, private communication.
198. Z. D. Popovic, *Appl. Phys. Lett.* **34**, 694 (1979).
199. B. S. Barkhalov and Y. A. Vidadi, *Thin Solid Films* **40**, L5 (1977).
200. K. K. Kanazawa, A. F. Diaz, R. H. Geiss, W. D. Gill, J. F. Kwak, J. A. Logan, J. F. Rabolt, and G. B. Street, *J. Chem. Soc. Chem. Commun.* 854 (1979).
201a. D. M. Ivory, G. G. Miller, J. M. Sowa, L. W. Schacklette, R. R. Chance, and R. H. Baughman, *J. Chem. Phys.* **71**, 1506 (1979).
201b. S. D. Phadke, *Indian J. Pure Appl. Phys.* **17**, 261 (1979).
202. H. J. Queisser, *Solid State Electron.* **5**, 1 (1962); R. L. Cummerow, *Phys. Rev.* **95**, 16 (1954); E. S. Rittner, *Photoconduct. Conf., Atlantic City, New Jersey* (R. G. Breckenridge, ed.), pp. 215–268. Wiley, New York, 1956; E. S. Rittner, *Phys. Rev.* **96**, 708 (1954).
203. M. A. Green, *Appl. Phys. Lett.* **27**, 287 (1975).
204. W. Schockley and H. J. Queisser, *J. Appl. Phys.* **32**, 510 (1961).
205. W. A. Anderson and A. E. Delahoy, *Proc. IEEE* **60**, 1457 (1972); D. L. Pulfrey and R. F. McQuat, *Appl. Phys. Lett.* **24**, 167 (1974); W. A. Anderson, A. E. Delahoy, and R. A. Milano, *J. Appl. Phys.* **45**, 3913 (1974).
206. A. T. Vartanian and I. A. Karpovich, *Dokl. Akad. Nauk. SSSR* **111**, 675 (1956).
207. A. E. Ennos, *Br. J. Appl. Phys.* **8**, 113 (1957); E. M. Da Silva and F. Kaufman, IBM Thomas J. Watson Research Center, Yorktown Heights, New York, unpublished results.
208. S. M. Lindsay, *Phys. Status Solidi a* **53**, 311 (1979).
209. E. K. Putseiko, *in* "Elementary Photoprocesses in Molecules" (B. S. Neporent, ed.). Consultants Bureau, New York, 1968; see also E. K. Putzeiko and A. Terenin, *Zh. Fiz. Khim.* **30**, 1019 (1956); E. Putzeiko, *Dokl. Akad. Nauk. SSSR* **124**, 796 (1959).
210. K. Seeger and W. D. Gill, IBM Research Rep. RJ-2568, June 19, 1979, IBM Research Laboratory, San Jose, California 95193.

211a. A. Terenin, *in* "Electrical Conductivity in Organic Solids" (H. Kallmann and M. Silver, eds.), p. 39. Wiley (Interscience), New York, 1961.

211b. A. T. Vartanian and I. A. Karpovitch, *Zh. Fiz. Khim.* **32**, 178, 274, 543 (1958).

212. K. Hasegawa and W. G. Schneider, *J. Chem. Phys.* **40**, 2533 (1964).

213. G. H. Heilmeier and S. E. Harrison, *J. Appl. Phys.* **34**, 2732 (1963).

214. R. Raman, L. Azarraga, and S. P. Glynn, *J. Chem. Phys.* **41**, 2516 (1964).

215. M. Sano, *Bull. Chem. Soc. Jpn.* **34**, 1668 (1961).

216. T. S. Moss, "Optical Properties of Semiconductors." Butterworths, London, 1959.

217. F. N. Hooge and D. Pokler, *Phys. Chem. Solids* **25**, 977 (1964).

218. H. Meier, *Z. Phys. Chem. Leipzig* **208**, 340 (1958).

219. H. Hänsel, *Ann. Phys.* **24**, 147 (1970).

220. L. I. Boguslavskii and A. V. Vannikov, "Organic Semiconductors and Biopolymers." Plenum Press, New York, 1970 (translated from Russian by B. J. Hazzard).

221. T. Holstein, *Ann. Phys.* **8**, 343 (1959).

222. M. Sukigara and R. C. Nelson, *Mol. Phys.* **17**, 387 (1969).

223. H. B. DeVore, *Phys. Rev.* **102**, 86 (1956).

224. A. M. Goodman, *J. Appl. Phys.* **30**, 144 (1959).

225. V. M. Agranovich and Yu. V. Konobeev, *Opt. Spektrosc.* **11**, 269 (1961).

226. L. E. Lyons and J. C. Mackie, *J. Chem. Soc.* 5186 (1960).

227. (a) B. Rosenberg, *in* "Electrical Conductivity in Organic Solids" (H. Kallmann and M. Silver, eds.), p. 291. Wiley (Interscience), New York, 1961; *J. Chem. Phys.* **31**, 238 (1959).

228. A. Bree, P. J. Reucroft, and W. G. Schneider, *in* "Electrical Conductivity in Organic Solids" (H. Kallmann and M. Silver, eds.), p. 113. Wiley (Interscience), New York, 1961.

229. H. Kokado and W. G. Schneider, *J. Chem. Phys.* **40**, 2937 (1964).

230. J. W. Eastman, G. M. Androes, and M. Calvin, *J. Chem. Phys.* **36**, 1197 (1962).

231. I. Nakada and Y. Ishihara, *J. Phys. Soc. Jpn.* **19**, 695 (1964).

232. D. C. Hoesterey and G. M. Letson, *Phys. Chem. Solids* **24**, 1609 (1963).

233. S. Z. Weisz, R. C. Jarnagen, M. Silver, M. Semhong, and J. Balberg, *J. Chem. Phys.* **40**, 3365 (1964).

234. S. E. Harrison and J. M. Assour, *J. Chem. Phys.* **40**, 365 (1964); L. E. Lyons and J. C. Mackie, *J. Chem. Soc.* 5186 (1960); H. Kokado and W. G. Schneider, *J. Chem. Phys.* **40**, 2937 (1964); J. W. Eastman, G. M. Androes, and M. Calvin, *ibid.* **36**, 1197 (1962).

235. W. Liptay, W. Eberlein, H. Weidenberg, and O. Elfleen, *Ber. Bunsenges Phys. Chem.* **71**, 548 (1967).

236. H. Labhart, *Tetrahedron Suppl. 2* **19**, 223 (1963).

237. V. K. Jain and S. C. Jain, *Phys. Status Solidi a* **30**, K69 (1975).

238. See for example, G. D. Pettit, J. J. Cuomo, T. H. DiStefano, and J. M. Woodall, *IBM J. Res. Dev.* **22**, 372 (1978); J. J. Cuomo, J. F. Ziegler, and J. M. Woodall, *Appl. Phys. Lett.* **26**, 557 (1975).

239. E. Bock, J. Ferguson, and W. G. Schneider, *Can. J. Chem.* **36**, 507 (1958).

240. I. C. Smith and E. Bock, *Can. J. Chem.* **40**, 1216 (1962).

241. G. Drefahl and H. J. Henkel, *Z. Phys. Chem.* **206**, 93 (1956).

242. A. T. Vartanyan, *Izv. Akad. Nauk. SSSR* **16**, 160 (1952).

243. H. Fröhlich and G. L. Sewell, *Proc. Phys. Soc. (London)* **74**, 643 (1959).

244. A. Many, E. Harnik, and D. Gerlich, *J. Chem. Phys.* **23**, 1733 (1955).

245. H. Kallmann, *Discuss. Faraday Soc.* **27**, 240 (1958).

246. J. Bardeen and W. Schockley, *Phys. Rev.* **80**, 72 (1950); W. Schockley, "Electrons and Holes in Semiconductors." Van Nostrand-Reinhold, Princeton, New Jersey, 1950.

247. H. A. Pohl, *J. Polym. Sci. Part C* **17**, 23 (1967).
248. J. Kommandeur, G. L. Korinek, and W. G. Schneider, *Can. J. Chem.* **35**, 998 (1957).
249. C. G. B. Garrett, *in* "Semiconductors" (N. H. Hannay, ed.), Von Nostrand-Reinhold, Princeton, New Jersey, 1959.
250. S. Aftergut and G. P. Brown, *in* "Organic Semiconductors" (J. J. Brophy and J. W. Buttrey, eds.), p. 87. Armour Research Foundation of Illinois, Institute of Technology, Macmillan, New York, 1962.
251. H. A. Pohl and D. A. Opp, *J. Phys. Chem.* **66**, 2121 (1962).
252. G. M. Delacote, *J. Chem. Phys.* **40**, 4315 (1965).
253. R. G. Kepler, *in* "Organic Semiconductors" (J. J. Brophy and J. W. Buttrey, eds.), p. 1. Armour Research Foundation of Illinois, Institute of Technology, Macmillan, New York, 1962.
254. A. G. Chynoweth and W. G. Schneider, *J. Chem. Phys.* **22**, 1021 (1954).
255. A. T. Vartanyan and I. A. Karpovitch, *Zh. Fiz. Khim.* **32**, 543 (1958).
256. I. Nakada, *Phys. Chem. Organ. Solid State* **1**, 745 (1963).
257. P. Mark and W. Helfrich, *J. Appl. Phys.* **33**, 205 (1962).
258. M. Silver *et al., J. Appl. Phys.* **33**, 1988 (1962).
259. A. Many, M. Simhony, S. Z. Weisz, and J. Levinson, *Phys. Chem. Solids* **22**, 285 (1961).
260. R. W. Smith and A. Rose, *Phys. Rev.* **97**, 1531 (1955).
261. W. Helfrich and P. Mark, *Z. Phys.* **166**, 370 (1962).
262. I. Nakada, K. Ariga, and A. Ichimaya, *J. Phys. Soc. Jpn.* **19**, 1587 (1964).
263. R. Raman, L. Azarraga, and S. P. McGlynn, *J. Chem. Phys.* **41**, 2516 (1964).
264. W. Mehl and N. E. Wolff, *J. Phys. Chem. Solids* **25**, 1221 (1964).

BIBLIOGRAPHY OF PAPERS ON PHOTOVOLTAIC EFFECTS OBSERVED IN ORGANIC SYSTEMS

(Also included are references to articles on such subjects as photoconductivity, photoredox properties, etc., of those organic systems which have exhibited photovoltaic behavior. Generally excluded are photogalvanic (photoelectrochemical) effects, work on lipid or bilayer membranes, and dye sensitization of inorganic semiconductors.)

A. Aromatic Hydrocarbons

(1) Anthracenes

1. W. G. Schneider and T. C. Waddington, *J. Chem. Phys.* **25**, 358 (1956).
2. D. M. J. Compton and T. C. Waddington, *J. Chem. Phys.* **25**, 1075 (1956).
3. A. G. Chynoweth, *J. Chem. Phys.* **22**, 1029 (1954).
4. A. Bree and L. E. Lyons, *J. Chem. Phys.* **25**, 1284 (1956).
5. A. Bree, D. J. Carswell, and L. E. Lyons, *J. Chem. Soc.* 1728 (1955).
6. L. Landau, quoted by H. Kallmann and M. Pope, *J. Chem. Phys.* **30**, 585 (1959).
7. L. Landau, *Rev. Sci. Instrum.* **30**, 44 (1959).
8. H. Killestreiter and H. Baessler, *Chem. Phys. Lett.* **11**, 411 (1971).
9. P. J. Reucroft, *J. Chem. Phys.* **36**, 1114 (1962).
10. M. Pope and H. Kallmann, *J. Chem. Phys.* **30**, 585 (1959).
11. M. Pope and H. Kallmann, *in* "Electrical Conductivity in Organic Solids" (H. Kallmann and M. Silver, eds.), p. 83. Wiley (Interscience), New York, 1961.
12. M. Pope, H. Kallmann, A. Chen, and P. Gordon, *J. Chem. Phys.* **36**, 2486 (1962).

13. O. C. Northrop and D. Simpson, *Proc. R. Soc. London Ser. A* **244**, 377 (1958).
14. V. V. Vladimirov, M. V. Kurik, and Y. P. Piryantinskii, *Sov. Phys.—Dokl* **13**, 789 (1969) [*English transl.: Dokl. Akad. Nauk. SSSR* **181**, 1365 (1968)].
15. A. Marchetti and D. R. Kearns, *J. Chem. Phys.* **44**, 1301 (1966).
16. K. Kato and C. L. Braun, *J. Chem. Phys.* **72**, 172 (1980).
17. S. C. Dahlberg and M. E. Musser, *J. Chem. Phys.* **71**, 2806 (1979).
18. H. Mitsudo, *Jpn. J. Appl. Phys.* **18**, 1853 (1979).
19. S. Kittaka and Y. Murata, *Jpn. J. Appl. Phys.* **18**, 295 (1979).

(2) Pentacenes

1. H. Kuroda and E. A. Flood, *Can. J. Chem.* **39**, 1981 (1961).
2. E. Silins *et al., Phys. Status Solidi a* **25**, 339 (1974).
3. E. Silins *et al., Pr. Nauk. Inst. Chem. Org., Fiz. Politech Wroclaw* **7**, 72 (1974).
4. See Ref. [17] under anthracenes.

(3) Perylenes

1. M. Calvin and D. Kearns, *J. Chem. Phys.* **29**, 950 (1958).
2. D. Kearns, G. Tollin, and M. Calvin, *J. Chem. Phys.* **32**, 1020 (1960).
3. D. Kearns and M. Calvin, U.S. Patent 3,037,947 (1962).
4. See Ref. [15] under anthracenes.
5. H. Inokuchi, Y. Maruyama, and H. Akamatu, *Bull. Chem. Soc. Jpn.* **34**, 1093 (1961).

(4) Naphthacenes

1. P. H. Fang, *Jpn. J. Appl. Phys.* **13**, 1232 (1974).
2. M. Matsumura, H. Uohashi, M. Furusawa, N. Yamamoto, and H. Tsubomura, *Bull. Chem. Soc. Jpn.* **48**, 1965 (1975).
3. A. K. Ghosh and T. Feng, *J. Appl. Phys.* **44**, 2781 (1973).

(5) Tetracenes (including tetrathiatetracenes)

1. P. J. Reucroft, P. L. Kronick, and E. E. Hillman, *Mol. Cryst. Liq. Cryst.* **6**, 247 (1969); Final report to U.S. Air Force Cambridge Research Labs, contract AF19(628)-5511, project 8659 (1968).
2. See Ref. [3] under naphthacenes.
3. See Ref. [2] under napthacenes.
4. E. L. Frankevich and E. I. Balabanov, *Phys. Status Solidi* **14**, 523 (1966).
5. A. Golubovic, *Proc. OAR Res. Appl. Conf., Washington D.C.* p. 211 (1967).
6. T. K. Mukherjee, *Record Photovoltaics Specialists Conf., 6th, Cocoa Beach, Florida* **1**, 7 (1967).
7. H. Baessler, G. Hermann, N. Riehl, and G. Vaubel, *J. Phys. Chem. Solids* **30**, 1579 (1969).
8. L. E. Lyons and O. M. G. Newman, *Aust. J. Chem.* **24**, 13 (1971).
9. P. H. Fang, A. Golubovic, and N. A. Dimond, *Jpn. J. Appl. Phys.* **11**, 1298 (1972).
10. P. H. Fang, *U.S.N.T.I.S. AD Rep. 1974/IGA,* from *Govt. Rep. Announce (U.S.)* **74**, 84 (1974).
11. P. H. Fang, M. Hirata, and M. Hirata, *Air Force Cambridge Research Lab., Final Rep. 1970,* AFCRL contract F19628-70-C-0263.
12. E. Krikorian and R. J. Sneed, *J. Appl. Phys.* **40**, 2306 (1969).
13. D. D. Kolendritskii, M. V. Kurik, and Y. P. Piryatinskii, *Ukr. Fiz. Zh.* **24**, 1662 (1979).
14. See Ref. [14] under anthracenes.
15. S. Arnold, M. Pope, T. K. T. Hsieh, *Phys. Status solidi b* **94**, 263 (1979).

(6) Miscellaneous aromatic systems

1. H. Kuroda and E. A. Flood, *J. Chem. Phys.* **33**, 952 (1960).
2. H. Kuroda and E. A. Flood, *Can. J. Chem.* **39**, 1475 (1961).
3. T. K. Mukherjee, *J. Phys. Chem.* **74**, 3006 (1970).
4. Y. Maruyama, H. Inokuchi, and Y. Harada, *Bull. Chem. Soc. Jpn.* **36**, 1193 (1963).
5. R. M. Hochstrasser and M. Ritchie, *Trans. Faraday Soc.* **52**, 1363 (1956).
6. H. Akamatu, H. Inokuchi, and M. Matsunga, *Bull. Chem. Soc. Jpn.* **29**, 213 (1956).
7. H. Akamatu and H. Inokuchi, *Proc. Conf. Carbon, 3rd* p. 51. Pergamon, New York, 1959.
8. H. Inokuchi, Y. Maruyama, and H. Akamatu, *Bull. Chem. Soc. Jpn.* **34**, 1093 (1961).
9. M. Kotani and H. Akamatu, *Discuss. Faraday Soc.* **51**, 94 (1971).
10. A. D. Tavares, *J. Chem. Phys.* **53**, 2520 (1970).
11. D. N. Bailey, D. M. Hercules, and D. K. Roe, *J. Electrochem. Soc.* **116**, 190 (1969).

B. Carotenes

1. V. I. Veselovskii and V. I. Ginzburg, *Zh. Fiz. Khim.* **24**, 366 (1950).
2. G. G. Komissarov and Y. S. Shumov, *Dokl. Akad. Nauk. SSSR* **171**, 1205 (1966).
3. G. G. Komissarov and Y. S. Shumov, *Biofizika* **13**, 421 (1968) [*English transl.: Biophysics* **13**, 503 (1968)].
4. B. Rosenberg, *J. Chem. Phys.* **31**, 238 (1959).
5. B. Rosenberg, *J. Chem. Phys.* **34**, 812 (1961).
6. B. Rosenberg, *J. Opt. Soc. Am.* **51**, 238 (1961).
7. B. Rosenberg, *Symp. Elec. Conduct. Organ. Solids* (H. Kallmann and M. Silver, eds.), p. 291. Wiley (Interscience), New York, 1961.
8. W. Arnold and H. K. Maclay, *Brookhaven Nat. Lab. Symp.* **11**, 1 (1958).
9. See Ref. [136] under References.
10. B. Rosenberg, R. J. Heck, and K. Aziz, *J. Opt. Soc. Am.* **54**, 1018 (1964).
11. H. T. Tien and N. Kobamoto, *Nature (London)* **224**, 1107 (1969).

C. Chlorophylls and Related Pigments (excluding Phthalocyanines)

(1) Chlorophyll

1. N. I. Barboi and I. I. Dilung, *Biofizika* **15**, 608 (1970) [*English transl.: Biophysics* **15**, 635 (1970)].
2. V. B. Yestigneyes, I. G. Savkina, and V. A. Gavrilova, *Biofizika* **7**, 298 (1962).
3. P. J. Reucroft and W. H. Simpson, *Discuss. Faraday Soc.* **51**, 202 (1971).
4. See Ref. [5] under References.
5. W. H. Simpson, F. A. Freeman, and P. J. Reucroft, *Photochem. Photobiol.* **11**, 319 (1970).
6. V. B. Yevstigneyev and A. N. Terenin, *Dokl. Akad. Nauk. SSSR* **81**, 223 (1951).
7. See Ref. [8] under carotenes.
8. K. J. McCree, *Biochem. Biophys. Acta* **102**, 90 (1965).
9. A. Terenin, E. Putzeiko, and I. Akimov, *Discuss. Faraday Soc.* **27**, 83 (1959).
10. D. R. Kearns and M. Calvin, *J. Chem. Phys.* **29**, 950 (1958).
11. D. R. Kearns and M. Calvin, *J. Am. Chem. Soc.* **83**, 2110 (1961).
12. See Ref. [2] under perylenes.
13. A. Bromberg, C. W. Tang, and A. C. Albrecht, *J. Chem. Phys.* **60**, 4058 (1974).
14. R. C. Nelson, *J. Chem. Phys.* **27**, 864 (1957).

15. E. Katz, *in* "Photosynthesis in Plants" (J. Franck and W. E. Loomis, eds.). Iowa State College Press, Ames, Iowa, 1949.
16. D. F. Bradley and M. Calvin, *Proc. Nat. Acad. Sci. U.S.A.* **43**, 563 (1955).
17. A. T. Vartanyan, *Izv. Akad. Nauk. SSSR Ser. Fiz.* **20**, 1541 (1956).
18. W. Arnold and H. K. Sherwood, *Proc. Nat. Acad. Sci. U.S.A.* **43**, 105 (1957).
19. B. Rosenberg, *J. Chem. Phys.* **28**, 1108 (1958).
20. B. Rosenberg and E. Rabinowitch, *Discuss. Faraday Soc.* **27**, 254 (1959).
21. V. B. Yevstigneyev and I. G. Savkina, *Biofizika* **8**, 181 (1963).
22. A. N. Terenin and E. Putseiko, *Tr. V. Mezhdnarad. Biokhim. Kongr. Akad. Nauk. SSSR* (1961).
23. E. K. Putseiko, *Dokl. Akad. Nauk. SSSR* **150**, 343 (1963).
24. E. K. Putseiko, *in* "Elementary Photoprocesses in Molecules" (B. S. Neporent, ed.), p. 281. Consultants Bureau, New York, 1968.
25. B. Rosenberg and J. F. Camiscoli, *J. Chem. Phys.* **35**, 982 (1961).
26. G. G. Komissarov, *Abstr. Int. Biophys. Congr., 2nd, Vienna* (1966).
27. G. G. Komissarov, *Biofizika* **12**, 592 (1967).
28. Y. S. Shumov and G. G. Komissarov, *Zh. Fiz. Khim.* **42**, 539 (1968).
29. Y. S. Shumov and G. G. Komissarov, *Biofizika* **13**, 984 (1968).
30. C. W. Tang and A. C. Albrecht, *Mol. Cryst. Liq. Cryst.* **25**, 53 (1974).
31. C. W. Tang and A. C. Albrecht, *J. Chem. Phys.* **62**, 2139 (1975).
32. C. W. Tang and A. C. Albrecht, *J. Chem. Phys.* **63**, 953 (1975).
33. C. W. Tang and A. C. Albrecht, *Nature (London)* **254**, 507 (1975).
34. E. K. Putseiko, *Dokl. Akad. Nauk. SSSR* **124**, 796 (1959).
35. F. Douglas and A. C. Albrecht, *Chem. Phys. Lett.* **14**, 150 (1972).
36. I. S. Meilanov, V. A. Benderskii, and L. A. Blyumeufeld, *Biofizika* **15**, 959 (1970) [*English transl.: Biophysics* **15**, 822 (1970)].
37. V. B. Yevstigneyev, A. A. Kazakova, and B. A. Kislev, *Biofizika* **18**, 53 (1973).
38. Y. A. Shkuropatov, V. I. Mel'nikova, and Y. M. Stolovitskii, *Biol. Nauch-Tekhn-Prog.* 42 (1974), from *Zh. Biolkhim.* (1974), Abstract 15F875 (in Russian).
39. V. B. Yestigneyev, Y. A. Shkuropatov, and Y. M. Stolovitskii, *Stud. Biophys.* **49**, 27 (1975).
40. J. P. Dodelet, J. LeBrech, and R. M. Leblanc, *Photochem. Photobiol.* **29**, 1135 (1979).
41. A. F. Janzen and J. R. Bolton, *J. Am. Chem. Soc.* **101**, 6342 (1979).
42. E. L. Frankevich, M. M. Tribel, J. A. Sokolik, and L. J. Kolesnikova, *Pr. Nauk. Inst. Chem. Org. Fiz. Politech. Wroclaw* **16**, 109 (1978).
43. G. A. Corker and I. Lundstrom, *J. Appl. Phys.* **49**, 686 (1978); *Photochem. Photobiol.* **26**, 139 (1977).

(1) Related pigments (porphines, pheophytins)

1. L. A. Drachev, A. A. Kondrashin, V. D. Samuilvov, and V. P. Skulachev, *FEBS Lett.* **50**, 219 (1975).
2. S. Saphon, J. B. Jackson, and H. T. Witt, *Biochim. Biophys. Acta* **408**, 67 (1975).
3. V. B. Yestigneyev and O. D. Bekasova, *Biofizika* **15**, 807 (1970) [*English transl.: Biophysics* **15**, 836 (1970)].
4. Rueppel and Hagins, *Biochem. Physiol. Visual Pigments Symp.* p. 257. Springer, New York, 1973.
5. See Ref. [34] under chlorophyll.
6. See Ref. [1] under chlorophyll.
7. A. Szent-Gyorgyi, *Bioenerget. Fiz. (Moscow)* (1960).
8. See Ref. [36] under chlorophyll.

9. M. E. Musser and S. C. Dahlberg, *Thin Solid Films* **66**, 261 (1980).
10. H. T. Tien, J. Higgins, and J. Mountz, *Sol. Energy: Chem. Convers. Storage Symp.*, p. 203 (1979).
11. T. Katsu, K. Tamagake, and Y. Fujita, *Kokagaku Toronkai Koen Yoshishu* **76** (1979).
12. D. C. Brune, J. Fajer, and S. P. Van, *Nat. Bur. Std. Spec. Publ. (U.S.)* **526**, 204 (1978).
13. F. J. Kampas and M. Gouterman, *J. Phys. Chem.* **81**, 690 (1977).
14. Z. H. Liu, Z. C. Bi, D. S. Cheng, Y. N. Zhu, Y. S. Li, and T. H. Tien, *K'o Hsueh T'ung Pao* **24**, 1027 (1979).

D. *Phthalocyanines, Merocyanines, Cyanines*

1. See Ref. [209] under References.
2. Y. S. Shumov and G. G. Komissarov, *Biofizika* **19**, 830 (1974).
3. G. G. Komissarov, Y. S. Shumov, and O. L. Morovova, *Biofizika* **15**, 1120 (1970) [*English transl.: Biophysics* **15**, 1162 (1970)].
4. G. G. Komissarov *et al.*, *Dokl. Akad. Nauk. SSSR* **187**, 3 (1969).
5. See Ref. [10] under chlorophyll.
6. See Ref. [109] under References.
7. D. Kearns, G. Tollin, and M. Calvin, *J. Chem. Phys.* **32**, 1020 (1960).
8. G. H. Heilmeier, *Org. Cryst. Symp., 3rd, Chicago, Illinois* paper 28 (1965).
9. G. H. Heilmeier and A. Zononi, *Phys. Chem. Solids* **25**, 603 (1964).
10. See Ref. [201] under References.
11. See Ref. [256] under References.
12. See Ref. [206] under References.
13. See Ref. [209] under References.
14. E. K. Putseiko, *Dokl. Akad. Nauk. SSSR* **124**, 796 (1959).
15. A. Terenin, *Proc. Chem. Soc.* 321 (1961).
16. A. Terenin, *Symp. Elec. Conduct. Organ. Solids* (H. Kallmann and M. Silver, eds.), p. 47. Wiley (Interscience), New York, 1961.
17. V. E. Kojevin and V. E. Lashkarev, *Radiotekn. Elektron.* **2**, 260 (1957).
18. I. A. Akimov, *Dokl. Akad. Nauk. SSSR* **128**, 691 (1959).
19. V. I. Sevastyanov, G. A. Alferov, A. N. Asanov, and G. G. Komissarov, *Biofizika* **20**, 1004 (1975).
20. See Ref. [9] under chlorophyll.
21. See Ref. [6] under References.
22. M. I. Fedorov, V. A. Benderskii, and N. N. Usov, *Dokl. Akad. Nauk. SSSR* **183**, 1117 (1968) [*English transl.: Proc. Acad. Sci. USSR* **183**, 915 (1968)].
23. See Ref. [7] under References.
24. See Ref. [17] under References.
25. Y. A. Vidadi, K. S. Kocharli, B. S. Barkhalov, and S. A. Sadreddinov, *Phys. Status Solidi a* **34**, 1677 (1976).
26. V. A. Ilatovskii and G. G. Komissarov, *Zh. Fiz. Khim.* **49**, 1352, 1353 (1975).
27. A. Y. Shkuropatov and M. M. Vankevich, *Biol. Nauch-Tekhn. Prog.* 45 (1974); from *Zh. Biol. Khim.* Abstr. 15F889 (1974) (in Russian).
28. V. B. Yevstigneev, A. Y. Shkuropatov, and Y. M. Stolovitskii, *Stud. Biophys.* **49**, 27 (1975).
29. See Ref. [2] under hydroxy squarylium.
30. D. L. Morel, A. K. Ghosh, T. Feng, E. L. Stogryn, P. E. Purwin, R. F. Shaw, and C. Fishman, *Appl. Phys. Lett.* **32**, 495 (1978).
31. A. K. Ghosh and T. Feng, *J. Appl. Phys.* **49**, 5982 (1978).

32. V. A. Benderskii, M. I. Alyanov, M. I. Fedorov, and L. M. Fedorov, *Dokl. Akad. Nauk. SSSR* **239**, 856 (1979).
33. B. S. Barkhalor and Y. A. Vidadi, *Thin Solid Films* **40**, L5 (1977).
34. M. I. Fedorov, E. E. P. Zinov, V. N. Shashaurov, and L. I. Nutrikhina, *Izv. Vyssh. Uchebn. Zaved. Fiz.* **20**, 157 (1977).
35. V. A. Ilatovskii, I. B. Dmitriev, A. K. Podchufarov, and G. G. Komissarov, *Org. Poluprovodn.* 55 (1976).
36. V. A. Ilatovskii, J. B. Dmitriev, and G. G. Komissarov, *Zh. Fiz. Khim.* **52**, 121 (1978).
37. G. A. Stepanova, L. S. Volkova, M. A. Gainullina, and F. F. Yumakulova, *Zh. Fiz. Khim.* **51**, 1771 (1977).
38. Y. S. Shumov, G. G. Komissarov, and M. F. Chaplii, *Zh. Fiz. Khim.* **53**, 2838 (1979).
39. Y. S. Shumov and L. N. Strekova, *Zh. Fiz. Khim.* **54**, 197 (1980).
40. S. C. Dahlberg and M. E. Musser, *Surf. Sci.* **90**, 1 (1979); *J. Chem. Phys.* **70**, 5021 (1979).
41. G. A. Chamberlain and P. J. Cooney, *Chem. Phys. Lett.* **66**, 88 (1979).
42. R. O. Loutfy and J. H. Sharp, *J. Chem. Phys.* **71**, 1211 (1979).
43. F. R. F. Fan, *Diss. Abstr. Int. B* **39**, 5875 (1979).
44. D. L. Morel, *Mol. Cryst. Liq. Cryst.* **50**, 127 (1979).
45. F. R. Fan and L. R. Faulkner, *J. Chem. Phys.* **69**, 3334, 3341 (1978).
46. M. L. Petrova, M. I. Rudenok, and F. T. Novik, *Vestn. Leningr. Univ. Fiz. Khim.* **22**, 61 (1976).
47. K. J. Hall, J. S. Bonham, and L. E. Lyons, *Aust. J. Chem.* **31**, 1661 (1978).
48. Y. S. Shumov, V. I. Mityaev, S. S. Chakhmakhchyan, and G. G. Komissarov, *Zh. Fiz. Khim.* **52**, 1807 (1978).
49. V. A. Ilatovskii, J. B. Dmitriev, and G. G. Komissarov, *Zh. Fiz. Khim.* **52**, 1000 (1978).
50. See Refs. [25a,c] under References.

E. Other Dyes

(1) Hydroxy squarylium

1. V. Y. Merritt and H. J. Hovel, *Appl. Phys. Lett.* **29**, 414 (1976).
2. V. Y. Merritt, *IBM J. Res. Dev.* **22**, 353 (1978).

(1) Miscellaneous dyes

1. H. Meier and A. Haus, *Z. Elektrochem.* **64**, 1105 (1960).
2. S. Mizushima and T. Komatsu, *Jpn. J. Appl. Phys.* **7**, 550 (1968).
3. H. Baba and K. Nitta, *Semicond. Abstr.* **4**, 1137 (1957).
4. H. Baba, K. Chitoku, and K. Nitta, *Nature (London)* **177**, 672 (1956).
5. G. Tomita, *Z. Naturforsch. b* **24**, 520 (1969).
6. R. E. Kay and E. R. Walwick, U.S. Patent 3,900,945 (1975).
7. Y. S. Lebedev and G. A. Korsunovskii, *Zh. Fiz. Khim.* **49**, 900 (1975).
8. U. Schoeler, K. H. Tews, and H. Kuhn, *J. Chem. Phys.* **61**, 5009 (1974).
9. J. S. Huebner, *Biochim. Biophys. Acta* **406**, 178 (1975).
10. M. Kryszewski, S. Sapieha, J. Tyczkowski, and M. Zielinski, *Pr. Nauk. Inst. Chem. Org. Fiz. Politech. Wroclaw* **7**, 390 (1974).
11. See Ref. [2] under hydroxy squarylium.
12. M. E. Musser and S. C. Dahlberg, *Surf. Sci.* **91**, L23 (1980).
13. T. I. Quickenden and G. K. Yim, *J. Phys. Chem.* **84**, 670 (1980).
14. D. R. Rosseinsky, R. E. Malpas, and T. E. Booty, *Transition Met. Chem. N.Y.* **3**, 254 (1978).

15. M. Kaneko, S. Sato, and A. Yamada, *Makromol. Chem.* **179**, 1277 (1978).
16. E. W. Williams, *IEEE J. Solid State Electron. Dev.* **1**, 185 (1977).

F. Polymers

1. See Ref. [102] under References.
2. A. M. Hermann and A. Rembaum, *Polym. Lett.* **5**, 445 (1967).
3. A. M. Hermann and A. Rembaum, *in* "Electrical Conduction Properties of Polymers" (A. Rembaum and R. F. Landel, eds.), p. 107. Wiley (Interscience), New York, 1967; *J. Polyn. Sci. Part C* **17**.
4. See Ref. [1] under References.
5. See Ref. [23] under References.
6. G. Oster, G. K. Oster, and M. Kryszewski, *Nature (London)* **191**, 164 (1961).
7. V. S. Myl'nikov, *in* "Elementary Photoprocesses in Molecules" (B. S. Neporent, ed.), p. 315. Consultants Bureau, New York, 1968.
8. V. S. Myl'nikov, *Zh. Fiz. Khim.* **42**, 1150 (1968).
9. K. Okumura, *J. Appl. Phys.* **45**, 5317 (1974).
10. See Ref. [15] under miscellaneous dyes.
11. M. J. Cohen and J. S. Harris, Jr., *Appl. Phys. Lett.* **33**, 812 (1978).
12. T. Tani, P. M. Grant, G. B. Street, and T. C. Clarke, *Solid State Commun.* **33**, 499 (1980).
13. S. Tazuke, *Kagaku (Kyoto)* **33**, 696 (1978).
14. S. D. Phadke, *Indian J. Pure Appl. Phys.* **17**, 261 (1979).
15. See Ref. [132] under References.
16. See Ref. [133] under References.
17. See Ref. [200] under References.
18. See Ref. [210] under References.
19. See Ref. [25b] under References.

G. Miscellaneous Organic Systems and Charge-Transfer or Metal Complexes

1. N. A. Dimond and T. K. Mukherjee, *Trans. Faraday Soc.* **51**, 102 (1971)—(arylidene-1,3-indanones).
2. P. J. Reucroft, P. L. Kronick, and E. E. Hillman, *J. Electrochem. Soc.* **114** 1054 (1967)-(TCNE-THF CT complexes).
3. O. V. Kolniniv, Z. V. Zvonkova, and V. P. Glushkova, *Zh. Fiz. Khim.* **43**, 1498 (1969) [*English transl.: Russ. J. Phys. Chem.* **43**, 832 (1969)]—(TCNE-THF).
4. M. Andraud, J.-P. Baratange, A. Helene, and F.-J. Taboury, *C. R. Acad. Sci. Paris Ser. C* **266**, 1200 (1968).
5. See Ref. [10] under other dyes—(spiropyrans).
6. G. Tomita, *Z. Naturforsc. b* **24**, 520 (1969)—(acriflavine).
7. O. V. Kolninov, V. M. Vozzhennikov, Z. V. Zvonkova, E. G. Rukhadze, V. P. Glushkova, and A. P. Terent'ev, *Dokl. Akad. Nauk. SSSR* **181**, 1420 (1968) [*English transl.: Proc. Acad. Sci. USSR* **181**, 628 (1968)] and references therein—(xanthogenates).
8. O. V. Kolninov, V. M. Vozzhennikov, Z. V. Zvonkova, E. G. Rukhadze, V. P. Glushkova, and A. P. Terent'ev, *Dokl. Akad. Nauk. SSSR* **168**, 1327 (1966) [*English transl.: Proc. Acad. Sci. USSR* **175**, 639 (1967)—(dithiocarbamates)].
9. M. R. Padhye, T. S. Varadarajan, and L. N. Chaturvedi, *Proc. Nucl. Phys. Solid State Phys. Symp.* **21C**, 379 (1978).

Chapter 5

Thermally Stimulated Discharge Current Analysis of Polymers

Stephen H. Carr
DEPARTMENT OF MATERIALS SCIENCE AND ENGINEERING
NORTHWESTERN UNIVERSITY
EVANSTON, ILLINOIS

I. THERMALLY STIMULATED DISCHARGE CURRENT TECHNIQUE (TSDC)

TSDC analysis of polymer solids is rapidly becoming recognized as a very rich source of information from polymeric materials. The kinds of insight being gained from TSDC analysis include quantitative measurement of the following: impurity concentrations, the wide varieties of possible molecular motions, characterization of the states of macromolecules and their local environments, chemical effects (state of cure, degradation chemistry), and anisotropy in microstructure.

The TSDC analysis involves a thermoelectric schedule as follows. The common situation is to start with any dielectric (this includes virtually all polymers) in a parallel plate capacitor configuration, elevate its temperature, apply an electrical field across the electrodes, cool the material to some reduced value, and then remove the external electrical field. The specimen now possesses an electrical polarization that will persist for a very long time. This polarized state induces in adjacent electrodes charges equal in magnitude and opposite in sign to that at the specimen surface; these are called image charges. These external electrodes are then short-circuited to each other through a device that can measure electrical currents with great sensitivity. The TSDC phase of this schedule then commences with the establishment of a uniform, slow (about 1°C/min) heating

215

rate. As the temperature rises, the persistent polarization induced during the initial phase begins to decay, and whenever a temperature range is encountered over which the decay rate matches the time scale of the experiment, there follows a release of some of these image charges. As a result, one then sees some current flowing in the external circuit through the current-measuring device. A plot of this current as a function of temperature is a thermally stimulated discharge current thermogram. Figure 1 shows this schematically.

Samples used during TSDC experiments are commonly in the form of films. This permits the externally applied electrical potentials (volts) to be reasonably small, while still giving strong electrical fields (volts per centimeter). Samples may have acquired a polarization prior to the TSDC experiment, in which case the starting materials are considered *electrets* themselves. However, one may still expect to see TSDC currents generated even from nonpolarized polymers, at least in some cases. Thus, the TSDC method may also serve to evaluate the magnitude of the electrical polarization on which such characteristics as piezoelectricity and pyroelectricity are based. Commonly, the electrodes used are actually very thin (approximately 100 Å) layers of metal deposited either by sputtering or vacuum evaporation. This configuration prevents the existence of an air gap between the dielectric material to be tested and the electrodes used to do the testing. As will be detailed later in this chapter, variations on this physical configuration may be justified.

The reason that TSDC is a thermal analysis technique being used increasingly relates to two factors: high sensitivity and high resolution. The sensitivity arises because electrical techniques can yield signals derived from exceedingly subtle effects, and the resolution effect arises from the fact that relaxation half-widths are intrinsically foreshortened in the case

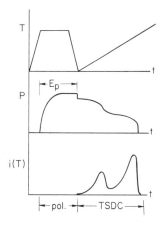

Fig. 1. Sequence of events during a TSDC experiment. The temperature schedule followed is shown at the top. The first stage (pol.) involves application of external voltage E_p across the specimen at an elevated temperature, and the second stage, TSDC, involves measurement of the current flowing between the short-circuited electrodes as the specimen is heated at a constant rate. In the example depicted here, the polarization P developed during the polarizing stage is shown to decay arbitrarily in two stages and, as a direct consequence, a peak in discharge current $i(T)$ is seen coincident with each decay process.

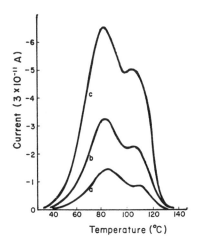

Fig. 2. TSDC thermograms from poly-(ethylene terephthalate) film polarized at 120°C under fields E_p of (a) −44.4 kV/cm, (b) −83.3 kV/cm, and (c) −166.7 kV/cm. [Adapted from Asano and Suzuki (1972, Fig. 2), reprinted with permission.]

of very slow test frequencies ω. In fact, the relationship that determines the effective test frequency of a TSDC analysis test is shown by

$$\omega = A/bkT_m^2. \tag{1}$$

Here A is the apparent activation energy, b is the reciprocal of the heating rate, k is Boltzmann's constant, and T_m is the temperature at which TSDC reaches its maximum. As the test frequency diminishes, the characteristic half-widths of any relaxation peak likewise decline. Similarly, the shifting of a peak to lower temperatures as a result of the measurements being made at lower frequencies tends naturally to increase the separation of individual peaks along the temperature axis. Thus, closely spaced discharge peaks will become more distinct in the thermogram as the test frequency goes down. Commonly, relaxation processes, which occur on the same time scale as a TSDC experiment in the cryogenic temperature ranges, correspond to frequencies of 10^{-5} Hz or lower, while higher-temperature peaks (with their concomitantly higher activation energies) may have corresponding test frequencies of 10^{-3} Hz.

A typical TSDC thermogram is shown in Fig. 2. Here it can be seen that increasing the polarizing voltage likewise increases the magnitude of the currents released. These currents arise from charges in the electrodes themselves, which are images of the polarization charge possessed by the dielectric (the test material itself). Thus, the surface of the dielectric, which has a positive charge, will induce in its contacting electrode a negative charge of equal magnitude; the other side of the dielectric will have a negative charge and will, in turn, induce an equal charge of opposite sign in its contacting electrode. When the dielectric is heated into temperature ranges where the origins of the polarization relax and thereby disappear, the image charges are released from their electrodes and flow toward each other via the external circuit. It is the current flowing in this external circuit that provides the output signal for the TSDC thermogram. Unfortunately, in the case where some of the polarization in the dielectric arises from charged species that possess some mobility, it is possible for these charges to diffuse during the TSDC experiment, either toward each other through the dielectric itself or alternatively away from the dielectric into the electrode materials, thereby annihilating their charging effects. It will be discussed subsequently how these effects can be identified and characterized.

Persistent electrical polarization, from which the discharging currents are usually a reflection, is commonly thought to arise from a variety of contributions. The first distinction among these contributions is divided between heterocharge and homocharge (Adams, 1927; Gemant, 1935; Gross and Denad, 1945; Perlman, 1971; Natarajan, 1972). A heterocharge is a polarization (coulombs per square centimeter) at a surface whose sign is opposite that of the polarizing electrode adjacent to it. A homocharge, conversely, is a polarization whose sign on a given surface is the same as that of the polarizing electrode next to it. One may think of a heterocharge as being the result of the dielectric reaction of a material to an impressed external electrical field, while a homocharge can be envisioned as resulting from charges being transferred directly from a polarizing electrode to the surface of the dielectric adjacent to it. Another way of viewing total polarization is to apportion it between contributions arising from dipolar moieties in the solid having a preferred orientation and real charged molecular species that have become displaced such that the centroid of positive charges does not coincide along the thickness direction with that of the negative charges. The physical situation is summarized in Fig. 3.

Most polymeric materials contain chains along which are spaced groups of atoms possessing a permanent dipole moment. Examples are ester

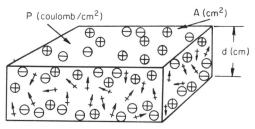

Fig. 3. Schematic representation of a polarized slab of a dielectric of thickness d and surface area A. Real charges are represented by circles containing either a plus or a minus sign; dipolar moieties are represented by arrows. The net polarization P on the surface comes from the sum of internal electric fields arising from preferential orientation of the dipoles and net displacement of real charges.

linkages, amide linkages, carbonyl groups, and nitrile groups. If any fraction of these dipolar moieties has a physical orientation in which as many groups, on average, point up as point down, then no polarization will result from them. However, if these dipoles are, on bulk average, disposed in space such that a few more of them point in one direction than in the opposite direction, then one will expect to obtain a finite amount of permanent polarization P_{dip}. This relationship is shown quantitatively by (Hill *et al.*, 1969)

$$P_{dip} = (\kappa_\infty + 2)N\mu_0\langle\cos\theta\rangle/3, \qquad (2)$$

where κ_∞ is the permittivity of the medium measured at infinite frequency, N is the number of dipoles having permanent dipole moment μ_0 per unit volume, and $\langle\cos\theta\rangle$ results from bulk-averaging over all orientation angles θ between the dipolar axes and the electrical polarization direction. Thus, it is seen that, as the number of dipoles per unit volume increases or as the average degree to which the dipoles are aligned parallel with the field direction increases, the depolarization resulting from dipoles will correspondingly increase.

Polymer electrets commonly have this dipolar polarization generated by heating to elevated temperatures and then applying a strong external electric field E. Although the electrical field inside the dielectric I may not be exactly the same as E, it nonetheless has a finite value which in turn imposes a torque Θ on the dipoles and tends to orient them in a parallel alignment:

$$\Theta = [3\kappa_0/(2\kappa_0 + 1)]\mu_0 E \sin\theta, \qquad (3)$$

where κ_0 is the permittivity of the medium at zero frequency. After the electret has been formed and the temperature has been dropped, this orientation of the dipoles is largely retained because thermally activated motions necessary to permit a return of the orientation distribution in these dipoles toward random become exceedingly sluggish. In fact, it may require years or even centuries before an appreciable fraction of this orientation is lost. Only by returning the temperature once again toward that at which the dipolar orientations were created is it possible to foreshorten this depolarization time. In the case of polymers, this depolarization process occurs at a rate having a characteristic temperature-dependent time constant τ, and this fact becomes the very basis of the use of TSDC in analyzing polymeric materials. For example, it is now possible to observe what are called secondary and primary dispersion effects (Lacabanne and Chatain, 1973) in polymers. Secondary dispersion effects are commonly ascribed to molecular motions involving very small portions of a polymer chain, typically those involving a single chemical repeat unit or a side group.

Primary dispersions involve motions of whole segments of backbones, possibly represented by tens of repeat units. Thus, the motional freedom imparted to a given dipole once secondary motions become possible is sufficient to permit the onset of some randomization motions, but the complete disorientation process will require the mobility of whole polymer chain segments, which becomes possible only when primary dispersion effects are possible. Thus, TSDC measurements on polarized polymer solids should be expected to reveal the temperature ranges over which each of these individual relaxation processes occurs.

The other source of polarization, that arising from real charges, is responsible for heterocharging and/or homocharging. Examples of real charged species are ions injected from the exterior of the sample, electrons implanted by irradiation, ionic species indigenous to the material before any thermoelectric treatments were imposed, and counterions bound to ionic sites along the polymer chains. Such charges may be concentrated with equal quantities of opposite signs on opposite sides of the film, or they may be of unequal concentration, giving rise to an asymmetric electret condition. The extreme example of this latter situation is what is called a monopole electret, and it is characterized by simply having an excess of charges of one sign located asymmetrically across the thickness of the electret. A monopole electret is opposed to a dipole electret, which is the case where the dielectric has the same number of oppositely charged species on opposite sides of the film. Injected ions or electrons can be created by exposure of the sample to ion or electron accelerators prior to insertion in TSDC measuring devices (Sessler and West, 1975; Legrand *et al.*, 1977; Hasegawa and Morimoto, 1974; Marconi Co., 1974; Gross *et al.*, 1973). Ions can also be injected by electromigration (Osaki and Ishida, 1973) or by photoinjection (Sapieha and Wintle, 1977; Takai *et al.*, 1976; Pillai *et al.*, 1977). Likewise, exposure of electrets to a corona (as may be created, for example, by radio-frequency excitation of a gas at reduced pressure) will probably have a combination of effects on an electret, including ion injection and chemical reaction of the surfaces leading to some development of ionic species in the electret itself (Creswell and Perlman, 1970a; Moreno and Gross, 1976; Jordan, 1975; Fukunaga and Yamamoto, 1973; Kodera, 1975). Subsequently, during TSDC experiments, the redistribution of these ions is usually observed in a rather discrete temperature range, usually 20–40°C *above* the primary dispersion temperature (T_g). The discharge current peak corresponding to the onset of this redistribution depends on actual electrical conductivity in the dielectric, and it is, therefore, a measure of the bulk redistributing as a result of transport of charged ionic species themselves (Seytre *et al.*, 1973; Borisova *et al.*, 1975; Mehendru *et al.*, 1976a). Such discharge peaks are often labeled ρ

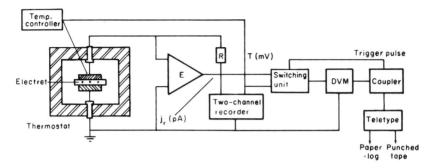

Fig. 4. Circuit and components schematic diagram for a typical fully automated TSDC apparatus. The input to the electrometer E is a small current in the picoampere range. [From van Turnhout (1975, Fig. 9-1), published with permission.]

peaks and are, therefore, indicative of the amount of polarization possessed by a polymer electret that was not due to dipolar orientation. One needs to bear in mind that the discharging of polarization resulting from ionic species can occur by drifting of these charges either into electrodes (therefore, out of the sample) or toward each other within the sample under the influence of the internal electrical field I.

Equipment for conducting TSDC experiments is historically tailor-made for each individual investigator's purposes (van Turnhout, 1975, Chapter 9). However, at least one commercial apparatus is currently available through the Toyo-Seiki Company.* Other apparatus has been described by Yalof and Hedvig (1975). Figure 4 is a common configuration for such pieces of equipment. Voltages commonly supplied are in the range 0–10 kV, and the currents often measured lie in the range 10^{-14}–10^{-5} A. Temperature stability and control is *highly* important, the most critical environmental control factor being the linearity of the heating rate, b^{-1}. This rate needs to be held within 0.1% of the selected value; otherwise, noticeable artifacts will be developed in the discharge current thermogram. The atmosphere in these instruments is best maintained at a reduced pressure, although a high vacuum is unnecessary and may actually interfere with the best thermal control. Likewise, very low pressures may induce a corona to form in the gas space of the sample chamber. Continuously flowing purges are often undesirable, as they may introduce static charging on various dielectric components of the sample-holding

---

* This equipment is fashioned after the design developed by E. Fukada of the Japanese Institute of Physical and Chemical Research. Currently, the Toyo-Seiki electret thermal analyzer is distributed in the United States by Atlas Electric Devices Company, Chicago, Illinois.

fixture itself. TSDC measurements at elevated hydrostatic pressures are being made in the author's laboratory and elsewhere (Ai *et al.*, 1979).

II. CHARACTERIZATION OF MOLECULAR RELAXATION PROCESSES

Imagining for a moment that polymers are homogeneous solids throughout which are dispersed dipolar moieties, one can calculate [Eq. (2)] a polarization P_{dip} knowing the entire distribution of orientations possessed by these dipoles. Note that $\langle \cos \theta \rangle$ is a function that ranges from $+1$ to -1 and, therefore, can take into account orientations projecting by some amount parallel with the imposed electrical field or by amounts projecting in opposition to the impressed electrical field. In the case where these dipoles do not interact with each other, the value of $\langle \cos \theta \rangle$ is found to be a function of temperature T and polarizing voltage, as given by

$$\langle \cos \theta \rangle = \varepsilon_0 \mu_0 E / 3kT. \tag{4}$$

In such dielectrics, the rate at which polarization grows or decays is given by a Debye equation:

$$dP(t)/dt = -\alpha P(t) + \varepsilon_0(\kappa_0 - \kappa_\infty)\alpha E, \tag{5}$$

where $P(t)$ is the time polarization of the dielectric as a function of time and $\alpha = 1/\tau(T)$. As mentioned in the previous section, the heating of electrets will eventually bring the temperature to levels at which α becomes large (τ, the relaxation time, becomes very small) and, consequently, the derivative becomes large. As polarization is lost from electrets in a TSDC apparatus, the corresponding image charges in the electrodes are released and permitted to flow through the external circuit to cancel each other. Thus, the time derivative becomes exactly equal to the discharge current i. In TSDC experiments, E is typically 0, so one may then write from Eq. (5) an expression for the discharge current:

$$i(t) = -dP(t)/dt = \alpha P(t). \tag{6}$$

This expression permits calculation of $P(t)$ as a function of time:

$$P(t) = P_0 \exp\left(-\int_0^t \alpha \, dt\right), \tag{7}$$

where P_0 is the polarization present at time 0. Combining Eqs. (6) and (7), and using the chain rule to change the integration from time to temperature, permits one to express the discharge current as a function of temperature $i(T)$:

$$i(T) = -\alpha P_0 \exp\left(-b \int_0^T \alpha \, dT\right), \tag{8}$$

where the inverse heating rate $b = dt/dT$. Thus, the discharge current $i(T)$ depends on both the heating rate b^{-1} and the relaxation frequency α.

In the case of polymers in which some dipolar motions relate simply to isolated dipoles, it is found that the Arrhenius relationship can be used to express the temperature dependence of α:

$$\alpha(T)_{\text{Arr}} = \alpha_0 \exp(-A/kT), \tag{9}$$

where α_0 is the natural relaxation frequency of the dipole in question. However, certain relaxation processes are observed to be dependent not upon absolute temperature T but rather the difference in temperature between some nonzero value T' and the prevailing temperature T. In these cases, an Eyring-type process matches more closely that which is observed:

$$\alpha(T)_{\text{Eyr}} = \alpha_0 \exp[-A(T)/k(T - T')]. \tag{10}$$

For long-chain macromolecular materials, some relaxation processes are dependent upon the cooperative effects necessarily involved in chain segment mobility. In these cases, the empirical relationship commonly designated WLF seems most appropriate:

$$\alpha(T)_{\text{WLF}} = \alpha_g \exp[C_1(T - T_g)/(C_2 + T - T_g)], \tag{11}$$

where α_g is the relaxation frequency in the glassy state, T_g is the liquid–glass transition temperature, and C_1 and C_2 are constants.

From the above, one has the basis for calculating an analytical expression for the discharge current obtained during a continuous heating experiment, specifically a TSDC experiment. Insertion of Eq. (9) or (10) or (11) into Eq. (8) should give the desired result. However, this is a mathematically difficult task, and resort is made to approximations. The primary problem is that the integration leads to a convergent infinite series, and so an approximation needs to be invoked in order to obtain a tractable result (Cowell and Woods, 1967; van Turnhout, 1971):

$$i(T) = C_3 \exp\left[-\frac{A}{kT} - \frac{C_4(kT)^2}{A^2}\exp\left(-\frac{A}{kT}\right)\right]. \tag{12}$$

This relationship involves two adjustable parameters, C_3 and C_4, but it does faithfully reproduce experimentally obtained TSDC peaks as well as peaks obtained from such different experiments as thermoluminescence. An example of such curve fitting to actual TSDC measurements is shown in Fig. 5. Here, one sees two distinctly different discharging processes each of which exhibits the characteristic skewed shape predicted by Eq. (12). Thus, the thermally stimulated discharge of polarization occurs by a process that starts out slowly, rises exponentially as the temperature con-

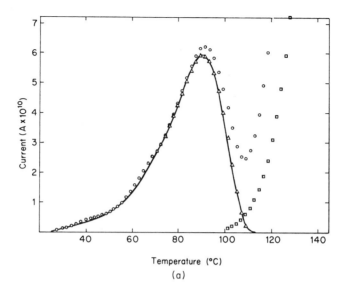

(a)

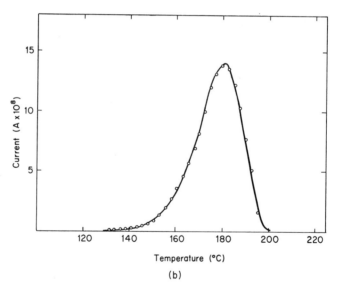

(b)

Fig. 5. TSDC thermogram (circles) obtained from polyacrylonitrile polarized at 130°C under a field of 56 kV/cm. Solid lines and triangles represent a fitting of Eq. (12) to peaks centered at (a) 90°C and (b) 180°C. The squares in Fig. 5a show the low-temperature toe of the 180°C peak by subtracting the fitted curve from the actual data. See Comstock *et al.* (1977).

tinues to increase, reaches a maximum, and then drops sharply as the randomization of preferentially oriented dipoles becomes complete. One notes from Eq. (12) that activation energy A for the relaxation process appears in the equation. It is not possible to obtain a good fit between theoretical prediction and experimental data unless the proper value of A has been, in fact, selected. Thus, curve fitting can be regarded as a means for experimental determination of the activation energy. Furthermore, Eq. (12) is intended to apply only to cases where the depolarization process occurs uniformly across the thickness dimension of a specimen.

Inspection of Eq. (12) also reveals that there should be a linear relationship between $\ln i(T)$ and $1/T$, at least in the early stages of depolarization. The slope of such plots, at temperatures considerably below the location of the peak maximum T_{max} should be approximately given by

$$\frac{d \ln i(T)}{d1/T}\bigg|_{T \ll T_{max}} \cong -\frac{A}{k}. \tag{13}$$

Plots such as are suggested by this treatment are termed the *initial rise method* (van Turnhout, 1971; Nicholas and Woods, 1964). Data from Fig. 5 are plotted thusly in Fig. 6, where it is seen that, for the first approximately 30% of the charge lost, this relationship holds fairly well (Sessler and West, 1976). Appropriate agreement between activation energies obtained by the initial rise method and by curve-fitting techniques, as mentioned above, is good.

Differentiating Eq. (8) and solving for the condition at the peak maximum results in the relationship given by

$$\frac{d1/\alpha}{dT}\bigg|_{T=T_{max}} = -b. \tag{14}$$

What this relationship says is that when the change in relaxation time of the dipoles with respect to temperature becomes identical with the reciprocal of the heating rate, then the rate at which the charge remaining in the sample is being lost likewise reaches a maximum. Taking Eq. (9) (the Arrhenius relationship) as one plausible expression for α, one can rewrite Eq. (14) as

$$\alpha(T_{max})bkT_{max}^2/A = 1. \tag{15}$$

Alternatively, as one recalls from treatments of dielectric dispersion data, there are observed maxima at various frequencies ω_{max} in the imaginary component κ'' of the complex permittivity κ^*, which occur when the condition

$$\omega_{max}\tau(T) = 1 \tag{16}$$

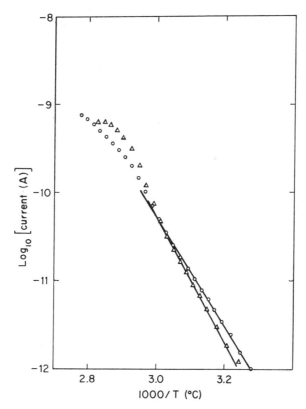

Fig. 6. TSDC data plotted according to Eq. (13). Circles and triangles are data from two different preparations of polyacrylonitrile films. Straight lines yield slopes from which activation energies of 27.7 and 31.7 kcal/mole, respectively, were obtained. [From M.S. thesis of R. J. Comstock, Northwestern University, Evanston, IL. (1974).]

is met. By inspection, one can observe that combining Eqs. (15) and (16) explains the origin of Eq. (1). Reconciliation between dielectric dispersion data and measured TSDC thermograms has been accomplished with good success (Sessler and West, 1976).

It follows from what is contained in the preceding paragraph that a complex TSDC thermogram, comprised of many overlapping peaks, can be analyzed incrementally to obtain activation energies for the discharging process at successively higher temperatures. An example of the decomposing of a TSDC maximum is found in Berticat *et al.* (1978). Plots of the initial rise of each successively higher discharge increment on a log current versus reciprocal temperature yields the curve in Fig. 7. From the slopes of each line, it is possible to determine activation energies that

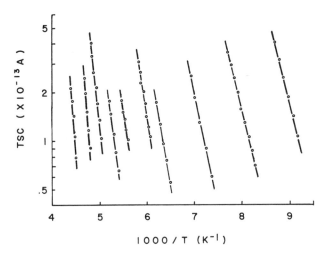

Fig. 7. TSDC data plotted according to Eq. (13) from a partial heating experiment on poly(bisphenol A carbonate). [From Aoki and Brittain (1976, Fig. 6), reprinted with permission.]

prevail for the discharge in each small temperature range. Figure 8 shows a plot of these activation energies as a function of the temperature at which each datum was determined, and such a relationship essentially represents the activation energy spectrum throughout a specified range of temperatures. It is seen that the activation energy is larger for higher temperatures, as one would expect. However, there are certain temperature ranges over which a common value of activation energy prevails, and so it becomes plausible to suppose that a common molecular event is responsible for the discharge effects throughout any particular temperature interval. Thus, these temperature intervals provide evidence for the existence of multiple molecular events associated with a broad TSDC peak (Gobrecht and Hofman, 1966; Creswell and Perlman, 1970b; Vanderschueren, 1972; Aoki and Brittain, 1976; Suzuoki, 1978). The interesting thing to note here is that correlations are being discovered between these individual values of activation energies and such mechanical characteristics as yield stress (Aoki and Brittain, 1977) and creep (Berticat *et al.*, 1978). TSDC analyses of many other commercial polymers have been reported and give support to the notions that a significant contribution to polymerization arises from the relaxation of dipoles that had acquired a preferential orientation during the electrical polarization step. Examples of these studies are contained in Table I.

An alternative way to make a direct determination of activation energy for the discharge processes occurring at different temperatures is called

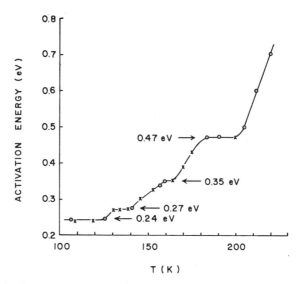

Fig. 8. Activation energy distribution through the cryogenic temperature range for poly(bisphenol A carbonate). The two runs correspond to two successive partial heating experiments on the same specimen. ⊙, first run; ×, second run. [From Aoki and Brittain (1976, Fig. 8), reprinted with permission.]

fractional polarization. It is a method in which the specimen is polarized intermittently as it is being cooled, rather than having polarization start at some high temperature and be maintained continuously while being cooled to the beginning of a TSDC experiment. What fractional polarization means is that only portions of the total dielectric reaction to impressed electrical fields are developed. A subsequent TSDC determination will sequentially release these increments in polarization without polarization being released in a given temperature range having a tail that superposes on polarization that will release at some higher temperature range. Since the discharge thermogram will be composed of a series of individual peaks, it is then possible to plot these peaks as shown in Fig. 7. Consequently, one then obtains activation energies throughout that same temperature range. Unfortunately, experience (Vanderschueren and Linkens, 1977) with this method has revealed that it shows rather less resolution of individual processes and, further, that the values of activation energy obtained from it lie below those commonly obtained by TSDC analysis or by dielectric dispersion analyses.

Persistent electrical polarization can be developed in polymer solids by techniques other than the thermal polarization method referred to above. These include mechanical means, exposure to electrical coronas, direct

TABLE I *Partial Bibliography of TSDC Analyses of Common Polymers*

Polymer	References[a]
Poly(vinyl chloride)	1
Poly(vinyl acetate)	2
Polystyrene	3
Polypropylene	4
Poly(ethylene terephthalate)	5
Poly(vinyl alcohol)	6
Poly(methyl methacylate)	7
Poly(vinylidene fluoride)	8
Polyethylene	9
Poly(vinyl fluoride)	10
Poly(bisphenol A carbonate)	11
Polycaproamide (nylon 6)	12
Poly-p-chlorostyrene	13
Poly-4-vinylpyridine	14
Cellulose (regenerated)	15
Polyacrylonitrile	16
Polyoxymethylene	17
Polyvinylbutyral	18
Polytetrafluoroethylene	19
Thiourea formaldehyde resins	20
Phenolic resins	21
Polyester urethanes	22
Epoxy resins	23
Poly(2,6-dimethyl-1,4-phenylene oxide)	24
Copolymer of ethylene and vinyl acetate	25
Copolymer of styrene and p-chlorostyrene	26
Copolymer of vinyl chloride and vinyl acetate	27
Copolymer of chlorinated ethylene and acrylonitrile–styrene	28
Blends of polyethylene and polyacrylonitrile	29
Blend of poly(2,6-dimethyl-1,4-phenylene oxide) either polystyrene, poly-p-chlorostyrene, or poly(p-chlorostyrene-co-styrene)	30
Hemoglobin	31
Poly-L-proline	32

[a] References:

(1) Pillai *et al.* (1969, 1972a,b, 1973); Talwar and Sharma (1978).

(2) Pillai *et al.* (1972b); Mehendru *et al.* (1975, 1977).

(3) Marchal *et al.* (1978); Bui *et al.* (1974); Alexandrovich *et al.* (1976); Bhargava and Srivastava (1979); Draconu and Dumitrescu (1978).

(4) Matsui and Murasaki (1973); Takamatsu and Fukada (1972).

(5) Asano and Suzuki (1972); Marchal *et al.* (1978); Takai *et al.* (1976); Lushchelkin and Voiteshanak (1975); Kojima *et al.* (1976); Borisova *et al.* (1975); Vanderschueren and Linkens (1978b).

(6) Jain *et al.* (1975); Sharma *et al.* (1980).

(7) Creswell and Perlman (1970b); Solunov and Vasilev (1974); Gubkin and Ogloblin (1972); van Turnhout (1977);

(continued)

TABLE I *(Continued)*

Vanderschueren (1974); Vanderschueren and Linkens (1977); Lamarre *et
al.* (1980).

(8) Murayama and Hashizume (1976);
Tamura *et al.* (1977); Sharp and Garn
(1976).

(9) Takamatsu and Fukada (1972); Fischer
and Roehl (1974); Hashimoto *et al.*
(1975, 1978); Ieda *et al.* (1979); Perret
and Fournie (1975).

(10) Reardon and Waters (1976).

(11) Aoki and Brittain (1976, 1977); Kryszewski and Ulanski (1976); Wissler and
Crist (1980).

(12) Ikeda and Matsuda (1976).

(13) Marconi Co. (1974).

(14) Gable *et al.* (1973).

(15) Baum (1973); Pillai and Mollah (1980).

(16) Comstock *et al.* (1977); Stupp and Carr
(1975).

(17) Goel and Pillai (1979).

(18) Jain *et al.* (1979).

(19) Gross *et al.* (1976).

(20) Nalwa *et al.* (1979).

(21) Goel *et al.* (1978).

(22) Baturin *et al.* (1976).

(23) Tanaka *et al.* (1977); Woodard (1977);
Pillai and Goel (1973); Su *et al.* (1980).

(24) Alexandrovich *et al.* (1976).

(25) Linkens *et al.* (1976).

(26) Alexandrovich *et al.* (1976).

(27) Gupta *et al.* (1977, 1979).

(28) Takamatsu and Nakajima (1974).

(29) Kartalov (1973).

(30) Alexandrovich *et al.* (1976).

(31) Rechle *et al.* (1970).

(32) Guillet *et al.* (1977).

injection of electrons, and implantation and/or incorporation of ionic species. Studies by Sacher (1973) have revealed ways in which molecular orientation induced during the manufacture of poly(ethylene terephthalate) film can produce an electrical anisotropy detectable by TSDC analysis. Similar studies have been reported elsewhere (Tsygel'nyi, 1975). Exposure of films to excited gas plasmas, especially those produced in the electrical breakdown of air, are widely exploited in the preparation of permanently polarized polymers. In some cases, large electrical fields are established between the plasma and the grounded side of the polymer film in question (McKinney and Davis, 1977). Examples of other studies on corona-charged polymers are found in Creswell and Perlman (1970), Fukunaga and Yamamoto (1973), Jordan (1975), Creswell *et al.* (1972), and Kodera (1975).

The problem with corona charging of polymers from a scientific standpoint is that a collection of individual events may occur for any given combination of corona-charging conditions and the polymeric material in question (Moreno and Gross, 1976). For example, ionized gas molecules accelerate in the corona and embed themselves with various binding energies in the polymeric material. Likewise, electrons that are released in the corona and can travel into the specimen (provided the polarity of the charging field is suitable) can develop the desired polarized state. In some cases, it is plausible to imagine these trapping events involving a chemical reaction with the polymer molecules themselves. This may be due to the ion itself bonding to the chain. It may alternatively be that the energy of the ion can, in turn, be transferred to parts of polymer chains, and these ionized species then may proceed to react elsewhere in the polymer solid itself. In cases where the polymer chains become so chemically modified, it is even possible that ionizable groups, such as carboxylic acid groups, can be formed if oxygen is present during the corona-poling step. Similarly, injection of charge directly by exposure of polymeric films to electron beams (Marconi Co., 1974; Hasegawa and Morimoto, 1974; Legrand *et al.*, 1977; Sessler and West, 1975; Gross *et al.*, 1973; Borisova *et al.*, 1975) and x rays (Suzuoki, 1978) has been performed. TSDC analysis of polymers, which have been charged by exposure to electron beams or to coronas, can permit evaluation of both dipolar effects and charge transport effects.

Because of the electrical gradients and local electrical fields created when these polarization methods are used, dipoles may adopt some preferred orientation. Thus, a subsequent TSDC thermogram often reveals discharge maxima coincident with temperature ranges over which dipolar relaxation effects are expected. Other discharge current maxima are also observed, and these are usually attributed to the motion of real charges implanted during the polarization steps (Creswell and Perlman, 1970; Moreno and Gross, 1976). What can be determined here is the magnitude of charge released as a result of implanted and mobile ionic species. Unfortunately, some of the charged species are generated internally to the film after it has been irradiated. Furthermore, any induced dipolar orientation is of a heterocharge nature, while implanted charge is usually of a homocharge nature. Thus, overlapping peaks of opposite polarity may be observed and thus imperfectly determined because of the difficulty in decomposing such a thermogram. Furthermore, the stability of charges trapped inside such polymer solids is often dependent upon the physical or chemical nature of the trapping site and, as a result, the release of such charges occurs in a stagewise manner. Finally, real charged sites generated during the initial polarization irradiation can also be bound to the chain as pendant and ionizable side groups. These charged species cannot drift at all, and as a result the polarization they represent cannot be measured by TSDC analysis. Recent work by Collins (1975) and others (De-Reggie *et al.*, 1978) has described ways of making direct measurement of the distribution (Natarajan, 1975) of charges across the thickness of a polarized polymer film. Discrimination between electronic and ionic conductivity in the discharge of implanted species has been studied (Seanor, 1968).

Ionic species can also be incorporated into polymers either at the time the solid is itself being prepared or while it is being polarized. It is possible to prepare "doped" polymers (Jain *et al.*, 1975; Wissbrun and Hannon,

1975; Mehendru *et al.*, 1976b; Latour and Donnet, 1976) by dissolving ionic species directly into polymer solutions from which subsequently dried films are cast. These materials usually exhibit spontaneous polarization due simply to the anisotropic distribution of charges that result during the final stages of evaporation (Stupp and Carr, 1975; Mehendru *et al.*, 1976b). Furthermore, the injection of ions from electrodes is also observed to occur (Kojima *et al.*, 1976; Osaki and Ishida, 1973), depending upon the metal ions that can be liberated from them. It has become common practice for gold to be the material of choice for electrodes, since it is the least frequently implicated as being a source of ionic impurities implanted in dielectric solids. Alternatively, ions can be deliberately introduced from electrodes if the electrodes are, in fact, electrolyte solutions (Turyshev *et al.*, 1977). The use of solutions as conducting electrodes has employed water, dioxane, methanol, and amyl acetate as solvents. Although these polymers are capable of storing a considerable amount of charge by simple displacement of the ionic species they contain, films prepared from polyelectrolytes store an even greater charge because of the counterions they naturally contain (Seytre *et al.*, 1973; Bornzin and Miller, 1978). A quantitative treatment of the transport such ionic species represents has been reported by Saito *et al.* (1974). Activation energies for charge transport were in the range of 2 eV, strongly suggesting that a cooperative motion of chain segments was required for the passage of charge from point to point. Presumably, activation energies that require such motions involve ionic species that exist as individual, and therefore dissolved, entities inside the polymer solid phase itself (Linder and Miller, 1973).

The amount by which ionizing radiation can cause discharging (or charging, for that matter) of polymer films can also be studied by TSDC analysis. In a sense, electrets can serve as very useful dosimeters for ionizing radiation, although in these cases the measurement of surface charge alone can suffice to detect the irreversible damage done during exposure to some form or forms of ionizing radiation. Research related to this subject is found in Pineri *et al.* (1976), Bowlt (1976), Perret and Fournie (1975), and Vanderschueren and Linkens (1978a). What is observed in many cases is a very pronounced increase in the magnitude of the ρ peak (see below), indicating the creation of charged species, at least some fraction of which are mobile at a suitably high temperature (specifically, the one at which space charges are released in a large quantity).

One of the current challenges in TSDC analysis is establishing exactly where in a thermogram the role of ionic species is manifested. In many cases, one is able to correlate discharge peaks with dielectric loss maxima observed dielectrically, and in this way it is common for the assignment of dipolar motions to be extended from dielectric studies to the TSDC analysis. Such a procedure, however, becomes more tenuous when the relaxation process in question coincides with the major dispersion, the glass transition. Observations of TSDC maxima that coincide with T_g lend confidence to the proposition that it is the dipolar nature of the side groups on the chains or portions of the main chain themselves that acquire mobility. However, activation energies for the glass transition are often calculated to be in the range 1–3 eV, and, as such, it is also possible that the dielectric effect might actually be due to the transport of real charge. The ambiguity in this assignment still needs to be resolved. In only one case (Stupp and Carr, 1978) has direct measurement of dipolar orientation been followed through the simultaneous depolarization of the polymer. In this work, the slight preference in orientation of nitrile side groups in polyacrylonitrile was observed to decay over exactly the same temperature range as the major relaxation process.

TSDC analysis of a large number of polymers reveals, however, that at a temperature about 30–50°C above T_g a second discharge peak is observed. This is normally called the ρ peak, so designated because it implies resistivity and is related to the transport of charge (van Turnhout, 1975; Vanderschueren, 1972). An example of just how common the ρ peak is can be seen in Fig. 9, which shows TSDC thermograms for a homologous series of acrylate polymers (Vanderschueren and Linkens, 1977). The thermal activation of the ρ peak conforms most closely to the WLF relationship, while, as implied by the previous paragraph, the main discharge

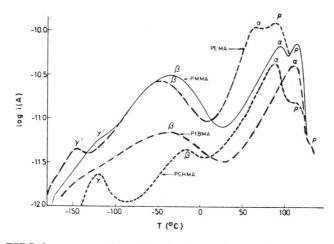

Fig. 9. TSDC thermograms of a series of acrylate polymers showing the ρ peak at a temperature about 30°C above the primary dispersion (α peak). [From Vanderschueren and Linkens (1977, Fig. 3b), reprinted with permission.]

peak assigned to the glass transition in fact obeys the Arrhenius relationship (Bui *et al.*, 1974). In an earlier study by Stupp and Carr (1975), it was
shown that the dipolar peaks decayed after a single temperature scan, but
that the ρ peak recurred in slightly diminished form upon a repeated
temperature scan. This is characteristic of a peak arising from space
charges and is presumably due to the fact that ions that had been displaced macroscopic distances during the initial polarization detrap and
drift at retarded rates, even though the specimen is in temperature range
where their discharge can occur (Pillai *et al.*, 1975). Furthermore, ρ peaks
are not seen in dielectric dispersion measurements, but they are seen in
TSDC thermograms (van Turnhout, 1977). Even if the concentration of
mobile anions and cations is unequal, ρ peaks will be observed. Furthermore, ρ peaks will be seen if one type or another of the anions or cations is
not mobile, as would be the case for ionic sites bound to the polymer
chains themselves.

The coincidence of the observation of a major relaxation, the ρ peak, at
temperatures 30–50°C above T_g and the proposition of a liquid–liquid
transition (T_{ll}) (Gillham and Boyer, 1977) must be noted (Lacabanne *et
al.*, 1980). The exact molecular nature of the T_{ll} transition is still the
subject of some controversy (Neumann and MacKnight, 1981), but it is
attributed, at least in part, to a distinct change in the nature of molecular
mobility. Further work needs to be done to establish just how plausible
the proposition is that ρ peaks in TSDC thermograms are, in fact, manifes-
tations of the same effect reported from studies that observe a T_{ll} transition. Whatever the situation actually is, at least, can be regarded as involving either a new mode of chain segmental motions that permit rapid
transport of ions or motions that abolish trapping sites at which ions were
localized in their initially polarized condition.

The primary variant of the TSDC method of polymer analysis is the
thermally stimulated polarizing current (TSPC) technique. This technique
has been reported in some detail by von Turnhout (1978) and others
(Wieder and Kaufman, 1953; Vanderschueren and Linkens, 1978b). One
of the primary advantages of the TSPC method is that one can scan the
relaxation spectrum without using an experiment that requires previous
heating of the specimen to an elevated temperature. This technique has
been exploited effectively by Vandershueren and Linkens (1978b). In their
study, it was possible to distinguish clearly between polarization contributions arising from space-charge drifting and from dipolar orientation.
Other studies have emphasized how this method can give information on
the relaxation times of polarizable chain segments (Vanderschueren and
Linkens, 1978b).

It is also possible to make instructive variations on the TSDC method itself. These have been summarized by van Turnhout (1978) and are shown in Fig. 10. Figure 10a illustrates the standard TSDC analysis configuration. The overall electrical field in this configuration is zero, since the internal polarization gives rise to an internal electrical field equal in magnitude but opposite in sign to that imposed externally by the image charges in the electrodes. Thus, integrating in the thickness direction x from the bottom surface to the top surface located at d yields a null value for the overall electrical potential of this three-layered structure. One measures, as has been stated previously in this article, the current that flows in this external circuit when the sample is heated, following the stagewise decay of polarization through the concomitant release of image charges from the contracting electrodes. An alternative configuration is shown in Fig. 10b, in which a finite air gap of thickness s exists between the top of the dielectric (nonmetallized surface) and the upper electrode. In this case, it is possible to observe the neutralization of space charges inside a polarized dielectric because of their subsequent drift upon heating. This occurs because the electrical potential across the dielectric is not zero. The fact that the upper electrode is placed some distance from the upper surface means that part of the image charge it should have is in the lower, contacting electrode. Thus, integrating the electrical fields due to both the internal polarization and the external charges contained in the electrodes results in a nonzero internal electrical field. It is this nonzero internal electrical field that causes space charges to drift when polariza-

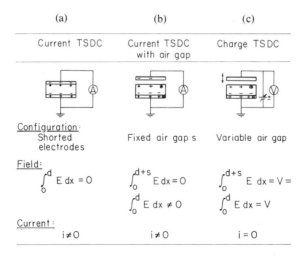

	(a)	(b)	(c)
	Current TSDC	Current TSDC with air gap	Charge TSDC

Fig. 10. Three test configurations for performing thermally stimulated measurements on polarized dielectrics. See text for detailed description. [From van Turnhout (1975, Fig. 9-1), reprinted with permission.]

tion occurs. The primary advantage is that a homocharge possessed by the film will usually decay at higher temperatures than trapped space charges. The third configuration, shown in Fig. 10c, illustrates a method essentially for measuring surface potential by a vibrating electrode method. In this kind of experiment, a backing voltage is applied to the upper electrode, as it is physically oscillated up and down in the space just above the electret. An electronic feedback system regulates the applied potential such that the overall bias on the configuration is always nullified. The backing voltage applied, then, is the experimental variable detected while the entire specimen is heated.

Although a very wide variety of physical aspects of polymers can be characterized via their relaxation characteristics and, therefore, via TSDC analysis, there are a few general areas that are also worthy of mention. The first involves the physical aging of polymers. In some cases, the TSDC method can simply be used as a screening method for determining the service life of an electret material (Sharp and Garn, 1976). Sometimes these aging effects are simply due to the uptake of water from the surrounding atmosphere, and if this is in fact the case, then they can also be characterized via TSDC analysis (Pillai *et al.*, 1972b; Vanderschueren, 1974). The primary area in which TSDC analysis provides information on aging relates to the role relaxation processes have in strength, stress relaxation, and creep. These effects involve various combinations of the various kinds of relaxation effects, including those of side groups and those involving the main polymer chain backbones themselves. Consistent with this picture is the fact that some thermally activated processes detected by TSDC analysis obey the Arrhenius rate law, while others obey the WLF kind of rate law. Direct correlation has been made recently (Pillai *et al.*, 1972b) between yield stress and activation energy of one part of the β-relaxation process in poly(bisphenol A carbonate). These data suggest that the rate-limiting step in the yielding event is one that can be detected by TSDC analysis. Other studies (Berticat *et al.*, 1978; Lamarre *et al.*, 1980) have made direct measurements of creep and correlated them directly with TSDC data. A detailed study of the long-term relaxation effects in poly(vinyl chloride) has emphasized how TSDC analysis can produce the information needed to predict effects related to long-chain rearrangements. This study spans the range of mechanical properties from bulk-level stress–strain relationships to microscopical strain such as that encountered in creep (van Turnhout *et al.*, 1977; Struik, 1978).

REFERENCES

Adams, E. P. (1927). *J. Franklin Inst.* **204,** 469.
Ai, B., Stoka, C. P., Giam, H. T., and Destruel, P. (1979). *Appl. Phys. Lett.* **34,** 821.
Alexandrovich, P., Karasz, F. E., and MacKnight, W. J. (1976). *J. Appl. Phys.* **47,** 425.
Aoki, Y., and Brittain, J. O. (1976). *J. Appl. Polym. Sci.* **20,** 2879.

Aoki, Y., and Brittain, J. O. (1977). *J. Polym. Sci. Polym. Phys. Ed.* **15**, 199.

Asano, Y., and Suzuki, T. (1972). *Jpn. J. Appl. Phys.* **11**, 1139.

Baturin, S. M., Manelis, G. B., Melentev, A. G., Nadgornyi, E. M., Ol'khov, Yu. A., and Shteinberg, V. G. (1976). *Vysokomol. Soedin. Ser. A* **18**, 2461.

Baum, G. A. (1973). *J. Appl. Polym. Sci.* **17**, 2855.

Berticat, P., Chatain, D., Monpagens, J. G., and Lacabanne, C. (1978). *J. Macromol. Sci.-Phys.* **B15**, 549.

Bhargava, B., and Srivastava, A. P. (1979). *Indian J. Phys.* **53A**, 47.

Borisova, M. E., Koikov, S. N., Paribok, V. A., and Fomin, V. A. (1975). *Vysokomol. Soedin. Ser. B* **17**, 488.

Bornzin, G. A., and Miller, I. F. (1978). *J. Electrochem. Soc.* **125**, 409.

Bowlt, C. (1976). *Contemp. Phys.* **17**, 461.

Bui, A., Carchano, H., Gustavino, J., Chatain, D., Gautier, P., and Lacabanne, C. (1974). *Thin Solid Films* **21**, 313.

Collins, R. E. (1975). *Appl. Phys. Lett.* **26**, 675.

Comstock, R. J., Stupp, S. I., and Carr, S. H. (1977). *J. Macromol. Sci. Phys.* **13**, 101.

Cowell, T. A. T., and Woods, J. (1967). *Br. J. Appl. Phys.* **18**, 1045.

Creswell, R. A., and Perlman, M. M. (1970a). *J. Appl. Phys.* **41**, 2367.

Creswell, R. A., and Perlman, M. M. (1970b). *J. Appl. Phys.* **41**, 2365.

Creswell, R. A., Perlman, M. M., and Kabayama, M. A. (1972). *Proc. Symp. Dielectr. Properties Polym.* (F. E. Karasz, ed.), pp. 295–312. Plenum Press, New York.

DeReggie, A. S., Guttman, C. M., Mopsik, F. I., Davis, G. T., and Broadhurst, M. G. (1978). *Phys. Rev. Lett.* **40**, 413.

Draconu, I., and Dumitrescu, S. (1978). *Eur. Polym. J.* **14**, 971.

Fischer, P., and Roehl, P. (1974). *Annu. Rep. Conf. Electr. Insul. Dielectr. Phenomena* pp. 359–371. The Electrochemical Society, Princeton, New Jersey.

Fukunaga, K., and Yamamoto, K. (1973). Japanese Kokai No. 73 43,195, June 22.

Gable, R. J., Vijayraghavan, N. V., and Wallace, R. A. (1973). *J. Polym. Sci. Polym. Chem. Ed.* **11**, 2387.

Gemant, A. (1935). *Philos. Mag. Suppl.* **20**, 929.

Gillham, J. K., and Boyer, R. F. (1977). *J. Macromol. Sci.-Phys.* **B13**, 497.

Gobrecht, H., and Hofman, D. (1966). *J. Phys. Chem. Solids* **27**, 509.

Goel, M., Viswanathan, P. S., and Vasudevan, P. (1978). *Polymer* **19**, 905.

Goel, M., and Pillai, P. K. C. (1979). *J. Macromol. Sci. Phys.* **B16**, 397.

Gross, B., and Denard, L. F. (1945). *Phys. Rev.* **67**, 253.

Gross, B., Sessler, G. M., and West, J. E. (1973). *Appl. Phys. Lett.* **22**, 315.

Gross, B., Sessler, G. M., and West, J. E. (1976). *J. Appl. Phys.* **47**, 968.

Gubkin, A. N., and Ogloblin, V. A. (1972). *Vysokomol. Soedin. Ser. B* **14**, 420.

Guillet, J., Seytre, G., Chatain, D., Lacabanne, C., and Monpagens, J. C. (1977). *J. Polym. Sci. Polym. Phys. Ed.* **15**, 541.

Gupta, N. P., Jain, K., and Mehendru, P. C. (1977). *Proc. Nucl. Phys. Solid State Symp.* **20C**, 136.

Gupta, N. P., Jain, K., and Mehendru, P. C. (1979). *Thin Solid Films* **61**, 297.

Hasegawa, Y., and Morimoto, K. (1974). Japanese Kokai No. 74 125,898, December 2.

Hashimoto, T., Shiraki, M., and Sakai, T., (1975). *J. Polym. Sci. Polym. Phys. Ed.* **13**, 2401.

Hashimoto, T., Sakai, T., and Miyata, S. (1978). *J. Polym. Sci. Polym. Phys. Ed.* **16**, 1965.

Hill, N. E., Vaughan, W. E., Price, A. H., and Davies, M. (1969). "Dielectric Properties and Molecular Behaviour," Chapter 1. Van Nostrand-Reinhold, New York.

Ikeda, S., and Matsuda, K. (1976). *Jpn. J. Appl. Phys.* **15**, 963.

Ieda, M., Mizutani, T., and Mizuno, H. (1979). *IEE Conf. Publ. No. 177* 266.

Jain, V. K., Gupta, C. L., Jain, R. K., Agarwal, S. K., and Tyagi, R. C. (1975). *Thin Solid Films* **30**, 245.

Jain, K., Kumar, N., and Mehendru, P. C. (1979). *J. Electrochem. Soc.* **126**, 1958.
Jordan, I. B. (1975). *J. Electrochem. Soc.* **122**, 290.
Kartalov, P. (1973). *Nature (London)* **6**, 53.
Kodera, Y. (1975). German Offen. 2,432,377, January 23.
Kojima, K., Maeda, A., and Ieda, M. (1976). *Jpn. J. Appl. Phys.* **15**, 2457.
Kryszewski, M., and Ulanski, J. (1976). *Rocs. Chem.* **50**, 1441.
Kulshrestha, Y. K., and Srivastra, A. P. (1979). *Polym. J.* **11**, 515.
Lacabanne, C., and Chatain, D. (1973). *J. Polym. Sci. Polym. Phys. Ed.* **11**, 2315.
Lacabanne, C., Goyand, P., and Boyer, R. F. (1980). *J. Polym. Sci. Polym. Phys. Ed.* **18**, 277.
Lamarre, L., Schreiber, H. P., Wertheimer, M. R., Chatain, D., and Lacabanne, C. (1980). *J. Macromol. Sci. Phys.* **B18**, 195.
Latour, M., and Donnet, G. (1976). *J. Phys. Lett.* **37**, 145.
Legrand, M., Dreyfus, G., and Lewiner, J. (1977). *J. Phys. Lett.* **38**, 439.
Linder, C., and Miller, I. F. (1973). *J. Polym. Sci. Polym. Chem. Ed.* **11**, 1119.
Linkens, A., Vanderschueren, J., Choi, S.-H., and Gasoit, J. (1976). *Eur. Polym. J.* **12**, 137.
Lushcheikin, G. A., and Voiteshonok, L. I. (1975). *Vysokomol. Soedin. Ser. A* **17**, 429.
Marconi Co., Ltd. (1974). British Patent 1,368,454, September 25.
Marchal, E., Benoit, H., and Vogl, O. (1978). *J. Polym. Sci. Polym. Phys. Ed.* **16**, 949.
Matsui, M., and Murasaki, N. (1973). *In* "Electrets, Charge Storage and Transport in Dielectrics" (M. M. Perlman, ed.), pp. 141–154. Electrochemical Society, Princeton, New Jersey.
McKinney, J. E., and Davis, G. T. (1977). U.S. NTIS AD Rep. AD-A048903.
Mehendru, P. C., Jain, K., Chapra, V. K., and Mehendru, P. (1975). *J. Phys. D Appl. Phys.* **8**, 305.
Mehendru, P. C., Pathak, N. L., Singh, S., and Mehendru, P. (1976a). *Phys. Status Solidi A* **38**, 355.
Mehendru, P. C., Jain, K., and Mehendru, P. (1976b). *J. Phys. D* **9**, 83.
Mehendru, P. C., Jain, K., and Mehendru, P. (1977). *J. Phys. D Appl. Phys.* **10**, 729.
Moreno, R. A., and Gross, B. (1976). *J. Appl. Phys.* **47**, 3397.
Murayama, N., and Hashizume, H. (1976). *J. Polym. Sci. Polym. Phys. Ed.* **14**, 989.
Nalwa, H. S., Viswanathan, P. S., and Vasudevan, P. (1979). *Angew. Makromol. Chem.* **82**, 39.
Natarajan, S. (1972). *J. Appl. Phys.* **43**, 4535.
Natarajan, S. (1975). *Jpn. J. Appl. Phys.* **14**, 1145.
Neumann, R. M., and MacKnight, W. J. (1981). *J. Polym. Sci. Polym. Phys. Ed.* **19**, 369.
Nicholas, K. H., and Woods, J. (1964). *Br. J. Appl. Phys.* **15**, 783.
Osaki, S., and Ishida, Y. (1973). *J. Polym. Sci. Polym. Phys. Ed.* **11**, 801.
Perlman, M. M. (1971). *J. Appl. Phys.* **42**, 2645.
Perret, J., and Fournie, R. (1975). IEEE Conf. Publication No. 129, p. 83.
Pillai, P. K. C., and Goel, M. (1973). *J. Electrochem. Soc.* **120**, 395.
Pillai, P. K. C., and Mollah, M. (1980). *J. Macromol. Sci. Phys.* **B17**, 69.
Pillai, P. K. C., Jain, V. K., and Vij, G. K. (1969). *J. Electrochem. Soc.* **116**, 836.
Pillai, P. K. C., Jain, K., and Jain, V. K. (1972a). *Phys. Lett. A* **39**, 216.
Pillai, P. K. C., Jain, K., and Jain, V. K. (1972b). *Nuovo Cimento Soc. Ital, Fis. B* **11**, 339.
Pillai, P. K. C., Jain, K., and Jain, V. K. (1973). *Ind. J. Pure Appl. Phys.* **11**, 597.
Pillai, P. K. C., Jain, K., and Jain, V. K. (1975). *Nuovo Cimento Soc. Italian Fis. B* **28B**, 152.
Pillai, P. K. C., Agarwal, S. K., and Nair, P. K. (1977). *J. Polym. Sci. Polym. Phys. Ed.* **15**, 379.

Pineri, M., Berticat, P., and Marchal, E. (1976). *J. Polym. Sci. Polym. Phys. Ed.* **14**, 1325.

Reardon, J. P., and Waters, P. F. (1976). *Proc. Symp. Thermally Photostimul. Currents Insul., 1975* (D. M. Smyth, ed.), pp. 185–200. Electrochemical Society, Princeton, New Jersey.

Rechle, M., Nedetzka, T., Mayer, A., and Vogel, H. (1970). *J. Phys. Chem.* **74**, 2659.

Sacher, E. (1973). *J. Macromol. Sci. Phys.* **7**, 231.

Saito, S., Hirota, S., and Sasabe, H. (1974). *Rep. Prog. Polym. Phys. Jpn.* **17**, 399.

Sapieha, S., and Wintle, H. J. (1977). *Can. J. Phys.* **55**, 646.

Seanor, D. A. (1968). *J. Polym. Sci. Part A-2* **6**, 463.

Sessler, G. M., and West, J. E. (1975). *J. Electrostat.* **1**, 111.

Sessler, G. M., and West, J. E. (1976). *J. Appl. Phys.* **47**, 3480.

Seytre, G., Quang, T. P., and May, J. F. (1973). *C. R. Acad. Sci. Paris Ser. C* **277**, 755.

Sharma, R., Sud, L. V., and Pillai, P. K. C. (1980). *Polymer* **21**, 925.

Sharp, E. J., and Garn, L. E. (1976). *Appl. Phys. Lett.* **29**, 480.

Solunov, Khr., and Vasilev, T. (1974). *J. Polym. Sci. Polym. Phys. Ed.* **12**, 1273.

Struik, L. C. E. (1978). "Physical Aging in Amorphous Polymers and Other Materials." Elsevier, Amsterdam.

Stupp, S. I., and Carr, S. H. (1975). *J. Appl. Phys.* **46**, 4120.

Stupp, S. I., and Carr, S. H. (1978). *J. Polym. Sci. Polym. Phys. Ed.* **16**, 13.

Su, W.-F. A., Carr, S. H., and Brittain, J. O. (1980). *J. Appl. Polym. Sci.* **25**, 1355.

Suzuoki, Y. (1978). D. Eng. Dissertation, Nagoya Univ., Nagoya, Japan, p. 38ff.

Takai, Y., Osawa, T., Mizutani, T., Ieda, M., and Kojima, K. (1976). *Jpn. J. Appl. Phys.* **15**, 1597.

Takamatsu, T., and Fukada, E. (1972). *Kobimshi Kagaku* **29**, 505.

Takamatsu, T., and Nakajima, Y. (1974). *Rep. Prog. Polym. Phys. Jpn.* **17**, 391.

Talwar, I. M., and Sharma, D. L. (1978). *J. Electrochem. Soc.* **125**, 434.

Tanaka, T., Hayashi, S., and Shibayama, K. (1977). *J. Appl. Phys.* **48**, 3478.

Tamura, M., Hagiwara, S., Matsumoto, S., and Ono, N. (1977). *J. Appl. Phys.* **48**, 513.

Tsygel'nyi, I. M. (1975). *Fiz.-Khim. Melch. Mater.* **11**, 120.

Turyshev, B. I., Shuvaev, V. P., Sazhin, B. I., and Lobanov, A. M. (1977). *Plast. Massy* (10), 21.

Vanderschueren, J. (1972). *Polym. Lett.* **10**, 543.

Vanderschueren, J. (1974). *J. Polym. Sci. Polym. Phys. Ed.* **17**, 991.

Vanderschueren, J., and Linkens, A. (1977). *J. Electrostat.* **3**, 155.

Vanderschueren, J., and Linkens, A. (1978a). *J. Polym. Sci. Polym. Phys. Ed.* **16**, 223.

Vanderschueren, J., and Linkens, A. (1978b). *J. Appl. Phys.* **49**, 4195.

van Turnhout, J. (1971). *Polym. J.* **2**, 173.

van Turnhout, J. (1975). "Thermally Stimulated Discharge of Polymer Electrets." Elsevier, Amsterdam.

van Turnhout, J. (1977). *Proc. 1975 Int. Symp. Electrets and Dielectr.* (M. S. deCampos, ed.), p. 97. Academy of Brasileira Cienc, Rio de Janeiro, Brazil.

van Turnhout, J. (1978). *In* "Electrets" (G. M. Sessler, ed.), Chapter 2. Springer-Verlag, Berlin and New York.

van Turnhout, J., Klaase, P. Th. A., Ong, P.-H., and Struik, L. C. E. (1977). *J. Electrostat.* **3**, 171.

Yalof, S. A., and Hedvig, P. (1975). *Org. Coat. Plast. Preprints Am. Chem. Soc.* **35**(1), 417.

Wieder, H. H., and Kaufman, S. (1953). *J. Appl. Phys.* **24**, 156.

Wissbrun, K. F., and Hannon, M. J. (1975). *J. Polym. Sci. Polym. Phys. Ed.* **13**, 223.

Wissler, G. E., and Crist, B., Jr. (1980). *J. Polym. Sci. Polym. Phys. Ed.* **18**, 1257.

Woodard, J. B. (1977). *J. Electron. Mat.* **6**, 145.

Chapter 6

Polymeric Electrets

G. M. Sessler
TECHNICAL UNIVERSITY OF DARMSTADT
DARMSTADT, WEST GERMANY

I. INTRODUCTION

Electrets are quasipermanently charged dielectrics, i.e., dielectrics whose charge arrangement persists much longer than the time period over which it is studied. Electret charges may be real charges or polarization charges; the former consist of positive or negative carriers trapped at the surfaces or in the volume of the dielectric, while the latter are either aligned and frozen-in dipoles or real charges displaced within molecular or domain structures (such displaced charges are often referred to as *Maxwell–Wagner polarization*). Some of the charge arrangements of an electret are schematically shown in Fig. 1.

Electret properties were first mentioned by Gray in 1732 and by Faraday in 1839. The term "electret" was then introduced by Heaviside in 1892. It was not until the work of Eguchi in 1919 that systematic research into electret properties began. Since then work has progressed in many different veins. Examples are the study of electret-forming (charging or polarizing) techniques, research into charge-retention mechanisms and isothermal charge decay, the development of various thermally stimulated current analyses, and the application of electrets in a number of fields.

While early electret investigations were mostly performed with wax materials, work during the past two decades has been largely devoted to polymers. The recent attention to polymer electrets must be attributed to the excellent charge-storage capabilities of these materials and to the fact that they are available as thin, flexible films. There are basically two kinds of polymer materials that are of interest today: (1) highly insulating substances such as the Teflon materials polytetrafluoroethylene (PTFE) and its copolymer fluoroethylene–propylene (FEP), which are capable of extremely permanent real-charge storage, and (2) polar substances, such as poly(vinylidene fluoride) (PVDF), which exhibit permanent dipole alignments resulting in piezoelectric and pyroelectric activities. Both types of polymer electrets are under active investigation today and have already found many practical applications.

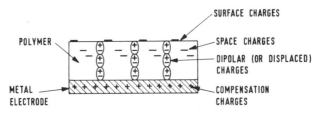

Fig. 1. Schematic cross section of a one-sided metallized electret having deposited surface charges, injected space charges, aligned dipolar charges (or microscopically displaced charges), and compensation charges. [From Sessler (1980).]

The present chapter is exclusively devoted to polymer electrets. Because of the broad scope of this book, the treatment of this specific topic has to be rather concise. For a more detailed review of polymeric as well as other electrets the reader is referred to the literature (see, e.g., Sessler, 1980).

II. FORMING METHODS

Electret forming, which means charging or polarizing of a dielectric, can be achieved by a variety of methods. These depend on air breakdown next to the surface to be charged, on injection of carriers by electrodes or particle beams, or on field-induced charge separation or dipole alignment within the dielectric. The forming process is frequently limited by internal or external breakdown phenomena which depend on the dielectric strength and the Paschen law, respectively; in other cases, forming is limited by saturation phenomena dependent on the number of traps or polarizable molecules in the material.

The most important forming techniques for polymers are the thermal, corona, liquid-contact, electron-beam, and photoelectric methods. These will be discussed in the following sections (see also reviews in Collins, 1973; Kiess, 1975; Euler, 1976; Sessler, 1980). Other processes depending on such phenomena as contact electrification, ionization by penetrating irradiation, etc., are of lesser importance for controlled electret forming but are often the cause of unwanted or hazardous charging effects. For a discussion of these phenomena, reference is made to the literature (Seanor, 1972a; Fuhrmann, 1978; Gross, 1980).

A. Thermal Methods

Thermal forming techniques consist of the application of an electric field to a polymer at an elevated temperature and subsequent cooling while the field is still applied. Such methods can be used on samples having electrodes intimately deposited on the dielectric or on samples separated from the electrodes by air gaps (see Fig. 2a).

Depending on the geometry, electrode configuration, field strength, temperature cycle, and material parameters, a number of different charging phenomena occur. In the presence of air gaps, breakdown processes can occur in the air, resulting in charge deposition on the polymer surfaces. When deposited electrodes are used, charges may be injected through the electrodes. Finally dipoles may be aligned or charges separated within the dielectric. Upon cooling, all charge motions are frozen in.

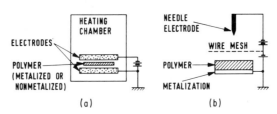

Fig. 2. Schematic representation of (a) the thermal charging method and (b) the corona-charging method.

If primarily surface charging is desired on samples not metallized on one face, the presence of an air gap is necessary. For the customary 6–25 μm polymer films, gaps with thicknesses of tens of micrometers to millimeters are used. With fields of 0.1–1 MV/cm, breakdown in the air gap occurs, resulting in charge deposition on the polymer surface. A maximum forming temperature somewhat above the glass transition temperature but well below the melting point (about 150–200°C for Teflon and 100–150°C for Mylar) is maintained for a few minutes to an hour (van Turnhout, 1975; Gubkin *et al.*, 1970). Use of the elevated temperature ensures that the deposited charge is deeply trapped and thus thermally stable (see Section V.C).

If volume charging, for example dipole alignment, is desired, samples with deposited electrodes are used to which high fields are applied. For example, polarization of PVDF is achieved at elevated temperatures (80–120°C) with fields of up to 2 MV/cm and at room temperature with fields of up to 4 MV/cm (Day *et al.*, 1974; Hayakawa and Wada, 1973).

The advantage of the thermal methods is the stability of the charge or polarization achieved. Drawbacks are the slight nonuniformities of the lateral charge distribution in the case of air-gap charging and the slowness of the process.

B.   Corona and Other Discharge Methods

The corona method depends on charge deposition on a polymer by a corona discharge (see Fig. 2b). The discharge is generated by application of a voltage of a few kilovolts between a needle or knife electrode and a back electrode located behind the sample to be charged. An additional wire mesh located between the needle electrode and the dielectric and biased to a potential of at least a few hundred volts is used to control the lateral current distribution and the total current to the sample (Ieda *et al.*, 1968). The current distribution is initially bell-shaped, with a superposed "modulation" by the "shadows" of the wires. However, the eventual distribution of the deposited charge is generally uniform if the charging is carried to the point where the entire sample surface assumes a saturation potential that equals the grid potential (Sessler, 1981b). For nonconductive materials, the current vanishes at this point. If the surface potential and the compensation current into the rear electrode are continuously monitored, the equivalent surface-charge density and the conduction current through the sample can be determined during the charging process (Moreno and Gross, 1976).

The corona-charging method has the advantage of being simple to implement and fast to operate. Since it is performed at room temperature, the thermal stability of corona-charged electrets is inferior to that of thermoelectrets.

Related breakdown-charging methods depend on the isothermal application of an electric field to a dielectric–air gap sandwich between parallel disk electrodes. If spark breakdowns occur in the gap, one obtains charge deposition on the dielectric. To avoid destructive arcing, another dielectric insert may be additionally placed between the electrodes, acting as a series resistor of high resistivity (Sessler and West, 1972; Ikezaki *et al.*, 1974).

If the ambient pressure is reduced to about 0.1 atm or less, the disruptive discharge converts to a uniform Townsend discharge which can also be used for the charging of polymers (Seiwatz and Brophy, 1966). Spark breakdown and Townsend breakdown methods are capable of depositing very large charge densities on polymers (up to 1.2×10^{-6} C/cm^2 on Mylar; see Sessler and West, 1972).

C. Liquid-Contact Method

This method depends on charge transfer from a soft wet electrode to a polymer surface it contacts (see Fig. 3a). The charge transfer is caused by a potential applied between this electrode and a metal layer on the oppo-

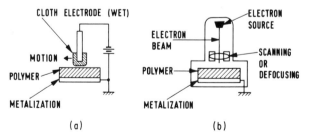

Fig. 3. Schematic representation of (a) the liquid-contact-charging method and (b) the electron-beam-charging method.

site surface of the dielectric. Many liquids, such as water and ethyl alcohol, can be used for the wetting. During the contact, the surface potential approaches (but does not reach) the applied potential (Chudleigh, 1972, 1976). The charge transfer occurs from a charge double layer located at the liquid–polymer interface.

Large areas of the surface may be charged by moving the electrode across the surface. To ensure charge retention on the dielectric, the electrode has to be withdrawn, or the liquid has to evaporate, before the voltage is removed. This method has also been used with nonwetting liquid–insulator contacts to record high-resolution charge patterns on polymers (Engelbrecht, 1974). Advantages of the liquid-contact method are simplicity, easy control of the deposited charge density, and uniformity of the lateral charge distribution.

D.   Electron-Beam Method

This method consists of the injection of electrons of proper energy into suitable polymers. If the range of the electrons is smaller than the thickness of the dielectric, the charging is directly caused by trapping of the injected carriers. The electron-beam method has been used in the past to charge thick dielectric plates (Gross, 1958) and has been adapted more recently to the charging of polymer films (Sessler and West, 1970), as will be discussed below. In addition, charging with electrons of range larger than the thickness of the dielectric has also been achieved. Such methods depend on secondary emission, backscattering, or injection from adjacent dielectrics. They are of lesser importance for polymers and will not be discussed here. Similarly, charging with ion beams is of little interest.

The charging of polymers with electrons of range smaller than the dielectric thickness is governed by the range energy relations. For example, the practical range of electrons of 10 and 50 keV in Teflon is about 1 and 20 μm, respectively (Matsukawa *et al.,* 1974). Thus, application of this method to polymer films up to 25 μm thick requires electron beams of less than 50 keV energy. Charging of such films is therefore conveniently achieved with small electron accelerators or electron microscopes. Such an implementation is shown in Fig. 3b. Beam scanning is often used in such setups to improve the lateral uniformity of the beam current.

The charging process for a sample metallized and grounded on one surface and irradiated by an electron beam through the other surface can be visualized as follows: When striking the surface, the electrons release some secondaries (emission yield at 10 keV $\approx$ 0.2), leaving a positively charged surface layer. The electrons then penetrate into the volume and generate secondary carrier pairs which are quickly trapped. The secondaries cause a radiation-induced conductivity (RIC) several orders of magnitude greater than the intrinsic conductivity. As a result of collisions, the primary electrons are eventually slowed down enough to be trapped also, forming initially a distribution of negative charge around the average range, which is about two-thirds of the practical range. The self-field of the charges, which is essentially directed toward the electrode on the rear of the sample, causes currents within the region of RIC that correspond to a further inward motion of electrons. After time periods on the order of minutes after termination of the irradiation, the measurable currents cease and a stable charge arrangement results (Gross *et al.*, 1974).

The advantages of this charging method are the lateral uniformity of the injected charge and control over charge depth and charge density. Electron-beam charging is widely used in research and industrial production.

E. Photoelectret Formation

This process is suitable for the charging of inorganic and organic photoconductors. The most important polymeric substance of this kind is polyvinylcarbazole (PVK)–trinitrofluorenone (TNF). If such materials are irradiated with ultraviolet or visible light under an applied field, a permanent "polarization" is generated. The effect is attributed to carrier generation by the light, displacement by the applied field, and eventual trapping. Depending on carrier mobilities, this results in a single charge cloud or in two separate charge clouds of opposite sign located either at the dielectric–electrode interfaces (barrier polarization) or in the volume (bulk polarization). Dipole polarization has been ruled out. The photoelectret effect has been reviewed extensively in the literature (see, e.g., Freeman *et al.*, 1961; Fridkin and Zheludev, 1961; Gill, 1976; Mort, 1980).

III. CHARGE-MEASURING METHODS

Experimental investigations of electret properties depend on a wide variety of methods based on the measurement of external potentials or currents under isothermal or nonisothermal conditions and on optical, x-ray, electron-beam, and other techniques (Sessler, 1980). Of greatest importance have been the methods utilizing electrical measurements. Of these we will review in the following methods for determining charge densities and distributions. Another important experimental technique depending on electrical measurements is the thermally stimulated current (TSC) method which will be described in Chapter 5.

Since the methods to be discussed depend on an evaluation of charges, potentials, or currents, some important relations concerning these quantities will first be reviewed.

We shall start with the important case of a one-sided metallized electret. Charge densities on such dielectrics are generally measured by means of induction methods utilizing the action of the external field of the electret on an adjacent electrode. A customary arrangement is depicted in Fig. 4, showing a charged dielectric sheet covered with an electrode on one side (lower electrode) and separated by an air gap from another (upper) electrode. The thickness dimensions s_1 and s_2 and the dielectric constants ε_1 and ε_2 are introduced in the figure.

The electret charge is given by a surface-charge density σ, located at $x = s_1$ and measured in charge per unit area, and a volume-charge density $\rho(x)$, measured in charge per unit volume. Generally, σ and ρ are composed of real-charge and polarization-charge contributions. If the real-charge densities at the surface and in the volume are denoted by σ_r and ρ_r, respectively, and the polarization by P, one has

$$\sigma(s_1) = \sigma_r(s_1) - P(s_1), \tag{1}$$
$$\rho(x) = \rho_r(x) - dP(x)/dx. \tag{2}$$

For the case of merely a surface-charge layer of density σ at $x = s_1$, Gauss's law and Kirchhoff's second law yield, respectively, for short-circuited electrodes:

$$\varepsilon_0(-\varepsilon_1 E_1 + \varepsilon_2 E_2) = \sigma, \tag{3}$$
$$s_1 E_1 + s_2 E_2 = 0, \tag{4}$$

where ε_0 is the permittivity of free space and E_1 and E_2 are the electric fields in the dielectric and air gap, respectively. Equations (3) and (4) yield for the fields E_1 and E_2

$$E_1 = -s_2 \sigma / \varepsilon_0 (\varepsilon_1 s_2 + \varepsilon_2 s_1), \tag{5}$$
$$E_2 = s_1 \sigma / \varepsilon_0 (\varepsilon_1 s_2 + \varepsilon_2 s_1). \tag{6}$$

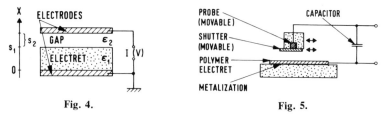

Fig. 4.   Fig. 5.

Fig. 4. Arrangement of electret with gap and electrodes.
Fig. 5. Schematic view of capacitive-probe setup.

If surface and volume charges are present, Eq. (5) is replaced by a more general relation which yields a field dependent on location (see, e.g., Sessler, 1980), while σ in Eq. (6) has to be replaced by the "effective" or "projected" surface-charge density $\hat{\sigma}$ given by

$$\hat{\sigma} = \sigma + (1/s_1)\int_0^{s_1} x\rho(x)\, dx. \tag{7}$$

Since the induction charge σ_i on the electrode at $x = s_1 + s_2$ is, under short-circuit conditions, related to the field E_2 by $\sigma_i = -\varepsilon_0\varepsilon_2 E_2$, the surface-charge density σ of an electret can be determined from σ_i according to Eq. (6) by

$$\sigma = -[(\varepsilon_1 s_2 + \varepsilon_2 s_1)/\varepsilon_2 s_1]\sigma_i. \tag{8}$$

For electrets with surface and volume charges, Eq. (8) yields the effective charge density $\hat{\sigma}$.

In another case of interest in the context of charge-density measurements, an external bias V_0 is applied to the upper electrode in Fig. 4 such that the external field E_1 of the electret is compensated. Replacing Eq. (4) by $s_1 E_1 = V_0$ one obtains together with Eq. (3) for the surface-charge density to be determined

$$\sigma = -\varepsilon_0\varepsilon_1 V_0/s_1, \tag{9}$$

which is independent of the air-gap thickness. For combined surface and volume charges, Eq. (9) yields $\hat{\sigma}$.

Another important case is that of a two-sided metallized electret of thickness $s_1 + s_2$ characterized by a charge layer of density σ at a certain depth in the material. With reference to Fig. 4, if $\varepsilon_1 = \varepsilon_2 = \varepsilon$ and assuming that the charge layer resides at $x = s_1$, one obtains from Eq. (8) a relation between σ and the induction charge σ_{i2} on the upper electrode:

$$\sigma_{i2} = -[s_1/(s_1 + s_2)]\sigma. \tag{10}$$

Similarly, the induction charge on the lower electrode is

$$\sigma_{i1} = -[s_2/(s_1 + s_2)]\sigma. \tag{11}$$

Currents may occur in an electret for a number of reasons, such as the application of heat, ionizing radiation, or illumination. If currents flow, the quantities E, P, and σ, so far assumed to be time-invariant, become functions of time. Quite generally, the total current density $i(t)$, which is the same everywhere in the dielectric, is given by

$$i(t) = \varepsilon_0\varepsilon\frac{\partial E(x, t)}{\partial t} + \frac{\partial P(x, t)}{\partial t} + i_c(x, t). \tag{12}$$

Here the terms on the right represent, respectively, the densities of the displacement current, the depolarization current, and the conduction current. The latter can be further resolved according to

$$i_c(x, t) = [g + \mu_+\rho_{r+}(x, t) + \mu_-\rho_{r-}(x, t)]E(x, t). \qquad (13)$$

In this equation, g is the ohmic conductivity, μ_+ and μ_- are the mobilities of positive and negative carriers, respectively, whose excess densities (over and above an intrinsic equilibrium density) are ρ_{r+} and ρ_{r-}, respectively.

A. Measurement of Effective Surface-Charge Density

According to Eq. (8), a measurement of σ_i on volume-charged electrets always yields the effective surface-charge density. To measure the total charge density,

$$\sigma_T = \sigma + \int\rho(x)\,dx, \qquad (14)$$

a second measurement is necessary, as will be shown below.

The effective surface-charge density on nonmetallized or one-sided metallized electrets may be determined with a *dissectible capacitor*. This instrument consists of two plate electrodes placed on the two sides of the electret and connected by a ballistic galvanometer or a capacitively shunted electrometer. If one of the electrodes is lifted from the dielectric, the induction charges of density σ_i will flow into the galvanometer and $\hat{\sigma}$ can be determined from $\hat{\sigma} = -\sigma_i$. The dissectible-capacitor method is the classical tool for measuring charge densities but has severe drawbacks related to contact electrification, breakdown effects, and spurious air gaps, which have largely eliminated it from modern electret research.

In wide use today is the *capacitive-probe method*. It avoids the drawbacks of the dissectible capacitor by using a well-defined, relatively large air gap between a probe and the electret (Krämer and Messner, 1964; Davies, 1967b; Wintle, 1970; van Turnhout, 1971). In one such setup shown schematically in Fig. 5 (Sessler and West, 1971), the probe is shunted by a large capacitor ($C \gg C_{\text{probe}}$) and can be shielded from the field of the electret by a shutter. If the probe is exposed to the field, a charge $a\sigma_i = -CV$ flows into it from the capacitor ($a = $ probe area), charging the latter to the voltage V. Since short-circuit conditions prevail, Eq. (8) holds and one obtains

$$\hat{\sigma} = [1 + (\varepsilon_1 s_2/\varepsilon_2 s_1)]CV/a, \qquad (15)$$

where all quantities on the right are accurately measurable.

Another modern tool for determining the charge density on one-sided metallized electrets is a *compensation method* utilizing a dynamic capacitor (Reedyk and Perlman, 1968; Sessler and West, 1968). If the electret or the opposing electrode is set into mechanical vibration (for example, by a loudspeaker), an ac voltage $V_\sim$ is generated by the setup, as shown in Fig. 6. Upon compensating the field in the air gap by application of a dc voltage V_0, the ac signal disappears. The charge density σ can then be determined from V_0 by means of Eq. (9). Thus, a measurement of the air-gap thickness is not necessary in this case. Local charge meters for measuring lateral charge distributions can also be designed (van Turnhout, 1971).

The total charge density σ_T can be determined by the *thermal pulse method* (Collins, 1975, 1976, 1977, 1980), which is also in extensive use for evaluating the charge centroid (see below). This procedure, schematically shown in Fig. 7, is based upon an evaluation of the potential change across the electret during diffusion of the thermal energy generated by a light pulse. The change ΔV in the voltage V is due to heat expansion of the material and variations in the dielectric constant. If the voltage changes $\Delta V(t_1)$ immediately following excitation and $\Delta V(t_2)$ after the establishment of thermal equilibrium are measured, one can determine σ_T from

$$\sigma_T = \hat{\sigma}\, \Delta V(t_1)/\Delta V(t_2), \qquad (16)$$

where $\hat{\sigma}$ is the effective charge density.

The methods discussed above relate to samples having at least one nonmetallized surface. On samples metallized on both surfaces charge densities may be determined in some cases by measuring *polarization and depolarization currents.* Examples are charge injection by electron-beam irradiation and depolarization of a dipole alignment. In both cases, every charge injected or disaligned will cause an induction charge to flow. For general and unknown charge distributions, however, depolarization tech-

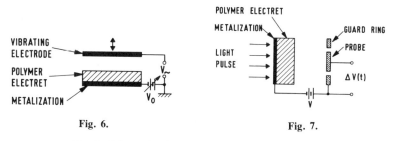

Fig. 6.   Fig. 7.

Fig. 6. Schematic view of dynamic capacitor.
Fig. 7. Schematic view of thermal pulse method.

niques generally do not yield an accurate measure of the total stored charge in two-sided metallized samples.

B. Measurement of Mean Charge Depth and Charge Distribution

Sectioning and planing techniques are well established for determining charge distributions on plate electrets. With resolutions of about 5 μm (Latour, 1980), these methods are primarily applicable to relatively thick electrets but are of little use for the investigation of thin-film samples. By their nature, these techniques are also destructive. For a detailed discussion of these methods we refer the reader to the literature (Gross, 1971; Walker and Jefimenko, 1973; van Turnhout, 1975).

For determining the mean charge depth (or centroid location) on precharged electrets the *thermal pulse technique* introduced above is most convenient. If the voltage changes $\Delta V(t_1)$ and $\Delta V(t_2)$ (see above) are measured, the centroid location $\bar{r}$, defined as $\bar{r}/s_1 = \dot{\sigma}/\sigma_T$, may be obtained by means of Eq. (16) from

$$\bar{r}/s_1 = \Delta V(t_2)/\Delta V(t_1). \tag{17}$$

Experiments of this kind are nondestructive, since the thermal pulse causes only a minor temperature increase in the sample.

The thermal pulse technique can in principle also be used to determine the actual distribution $\rho(x)$ of the charge. However, because of inaccuracies in the numerical deconvolution procedure necessary for such evaluations, a unique solution for $\rho(x)$ is generally not obtained (De Reggi *et al.*, 1978). If, in addition, one allows for measuring errors on the order of 0.1%, only a few spatial Fourier coefficients of the charge distribution can be found (von Seggern, 1978; Mopsik and De Reggi, 1980).

Related to the thermal pulse method is the *pressure-pulse technique* (Laurenceau *et al.*, 1977). Here, an acoustic pulse or step of short rise time is propagated through the sample. Again, currents or voltage changes are produced in an external circuit from which the charge distribution can be calculated (Gerhard, 1982b). Since the acoustic signal maintains its shape during propagation, deconvolution is not required in this case and $\rho(x)$ can be determined uniquely and accurately. Compared to the thermal pulse technique this method requires more complicated instrumentation capable of exciting the required acoustic signal. For this purpose, pulsed lasers (Rozno and Gromov, 1979; Alquie *et al.*, 1981; Sessler *et al.*, 1981, 1982a) or quartz crystals (Eisenmenger and Haardt, 1982) have been used. The resolution is determined by the length of the pressure pulse or by the risetime of the pressure step. Values of a few micrometers can be achieved presently (Sessler *et al.*, 1982a; Eisenmenger and Haardt, 1982).

Electron-beam sampling of electrets also yields the charge distribution (Sessler *et al.*, 1977, 1982b; Tong, 1980). If a two-sided metallized, charged sample is irradiated with an electron beam of slowly increasing energy, the conductive region in the sample will extend progressively deeper into the material. Thus all the charges residing originally in the dielectric will be gradually removed. The corresponding change in the induction charge q on the nonirradiated electrode yields $\rho(x)$ by means of

$$\rho(s) = -d^2(qs)/ds^2, \tag{18}$$

where s is the thickness of the nonconductive region. The resolution of the method is about 20% of the beam penetration depth.

A number of other methods are in use to determine charge centroids. One of these, applicable to precharged and one-sided metallized electrets, is based on a measurement of the effective surface charge and the *integrated depolarization current* of the electret (Sessler, 1972). Another method, designed to measure the charge centroid during electron-beam charging, depends on an evaluation of the induction charges on the two electrodes by means of a *split-Faraday cup* (Gross *et al.*, 1974; 1977).

IV. PERMANENT DIPOLE POLARIZATION

Certain polymers show, after proper poling, a permanent dipole polarization. The most important synthetic polymer exhibiting such a polarization is semicrystalline PVDF, which has the repeat unit $-CF_2-CH_2-$. Its importance lies in the fact that it is strongly piezoelectric and pyroelectric and thus useful in a wide variety of applications. In the following, we will concentrate on this material but also discuss other polar polymers.

A. Poling Process and Polarization

In polar polymers, a polarization P is obtained by alignment of the polar groups by an applied field E at elevated temperatures, where the molecular chains are sufficiently mobile (see Section II.A). The alignment is controlled by the Debye equation which, for a single dipole-relaxation frequency $\alpha(T)$, can be written (see, e.g., van Turnhout, 1980) as

$$dP(t)/dt + \alpha(T)P(t) = \varepsilon_0(\varepsilon_s - \varepsilon_\infty)\alpha(T)E, \tag{19}$$

where ε_s and ε_∞ are the static and optical dielectric constants, respectively. Under isothermal polarizing conditions, and if $P(0) = 0$ is assumed, the time dependence of the polarization follows from Eq. (19) as

$$P(t) = \varepsilon_0(\varepsilon_s - \varepsilon_\infty)E[1 - e^{-\alpha(T)t}], \tag{20}$$

which yields a saturation value of

$$P_s(\infty) = \varepsilon_0(\varepsilon_s - \varepsilon_\infty)E. \qquad (21)$$

The temperature dependence of the relaxation frequency often follows an Arrhenius law:

$$\alpha(T) = \alpha_r \exp(-U/kT), \qquad (22)$$

where α_r is the natural relaxation frequency and U the dipolar activation energy. Generally, polymers show a distribution of relaxation frequencies that may be discrete or continuous, the difference being that in the case of discrete distributions clearly separated groups of relaxation frequencies exist (separable, for instance, in TSC experiments), while in the case of continuous distributions such a separation is not possible. Multistage TSC experiments and fractional polarization data show that in polymers the distribution is generally due to a distribution of the activation energies U and not to a distribution of natural frequencies α_r (see van Turnhout, 1975, 1980; Vanderschueren and Gasiot, 1979).

In many polymers, three discrete relaxation frequencies (or groups of frequencies) exist, which are directly related to the γ-, β-, and α-relaxations. While the γ-relaxation is due to motions within side groups of the molecular chains, the β-relaxation originates from motions of the side groups themselves, and the α-relaxation is caused by joint motions of side groups and main chains. In polarized samples, the three relaxations manifest themselves as three distinct peaks in TSC measurements. A fourth TSC peak, usually at higher temperatures, is due to the motion of space charges (ρ-relaxation).

Typical TSC curves for a few polymers are shown in Fig. 8. The samples were thermally charged prior to the TSC experiments and show three or four peaks. The peak location and shape depend, among other factors, on the heating rate (see Chapter 5).

According to Eq. (21), the saturation polarization increases propor-

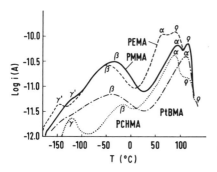

Fig. 8. TSC current spectra of thermally charged poly(methyl, ethyl, *tert*-butyl, and cyclohexyl) methacrylate electrets (Vanderschueren, 1974; see also van Turnhout, 1980).

tionally to E. Such an increase is also experimentally found in many polymers. As an example, results for polystyrene are depicted in Fig. 9, showing for lower fields the expected behavior of peak current and released charge. Similar results were found for other polymers (van Turnhout, 1975; Fischer and Röhl, 1976; Fischer, 1977/78; Vanderschueren *et al.*, 1978; Davis and Broadhurst, 1977).

The relation (21) between P_s and E has also been evaluated quantitatively for the γ-, β-, and α-peaks of a number of polymers using, instead of $\varepsilon_0 - \varepsilon_\infty$, the corresponding change $\Delta\varepsilon$ for each relaxation (van Turnhout, 1975). It was found that $P_s/\varepsilon_0 E$ agreed with $\Delta\varepsilon$ for most of the γ- and β-relaxations, while the results for the α-peak were affected by space-charge motions.

B. Effect of Material Parameters on Polarization

The storage of polarization charges in polymer electrets depends on a number of material parameters such as crystallinity, cross-linking, additives, irradiation history, stereoregularity, and water absorption. In the following, the effect of four of these parameters on the dipole polarization will be discussed.

Crystallinity has a rather severe effect on the dipolar TSC peaks. Two examples will demonstrate this. In polycarbonate, the β-peak is highest for an amorphous sample and decreases considerably with increasing crystallinity (Vanderschueren, 1974). In PVDF, on the other hand, the α-peak appears to be linked to the crystalline phases (Murayama and Hashizume, 1976).

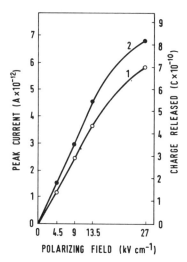

Fig. 9. Peak current (curve 1) and released charge (curve 2) of polystyrene electrets as a function of the polarizing field. [From Shrivastava *et al.* (1980).]

Another structural parameter that affects the dipolar alignment is the *steric arrangement* of the polar groups (Vanderschueren, 1974). In poly(methyl methacrylate) (PMMA), for example, the syndiotactic type (vinyl groups alternately on both sides of the main chain) exhibits a larger β-peak in TSC experiments than the isotactic type (vinyl groups on one side).

Some work has been devoted to the effect of *additives and dopants* on the polarization of polymers. It has been found that in polyethylene, ionic agents such as NaCl will shift the α-peak to higher temperatures (Lacabanne and Chatain, 1978). This is due to an increase in the potential barrier responsible for this relaxation. The lower-temperature β- and γ-peaks are not affected by ionic agents but by antioxidants. The latter increase the activation energy and decrease the preexponential factor. Doping with such an element as copper or iodine will also affect the dipolar and space-charge relaxations (Jain *et al.*, 1978).

A number of laboratories have studied the effect of *water or humidity* on the polarization of polymers (Vanderschueren and Linkens, 1978a, Weber, 1978). A general result of these studies is that, in a wide variety of polymers, the α- and β-relaxations are significantly affected by the presence of water in the sense that, with increasing water content, the corresponding TSC peaks shift to lower temperatures. This has been attributed to the formation of hydrogen bonds between the water molecules and the macromolecular units, resulting in a plasticizing effect which weakens intermolecular attraction and thus decreases the activation energy. In these cases, the peak amplitude can increase or decrease, depending on the nature of the polymers.

C. Decay of Polarization

The decay of the polarization of a dielectric with a single dipole relaxation frequency $\alpha(T)$ is governed by Eq. (19). If external fields or space-charge fields are absent ($E = 0$), and for isothermal conditions, this equation yields

$$P(t) = P(0) \exp(-\alpha t). \tag{23}$$

In this simple case the decay is thus exponential. More general formulas for distributed relaxation frequencies and nonisothermal conditions are given in the literature (van Turnhout, 1980).

Experimental decays are generally nonexponential with increasing time constants, indicating a distribution of relaxation frequencies [see, e.g., Tamura *et al.* (1974)]. The isothermal decay of the polarization of β-phase PVDF at elevated temperatures is depicted in Fig. 10, which shows the

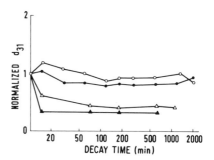

Fig. 10. Decay of normalized piezoelectric d_{31} constant of PVDF (polarized at 130°C and 800 kV/cm) at various temperatures. ○, 50°C; ●, 80°C; △, 100°C; ▲, 100°C. [From Murayama *et al.* (1976).]

decrease in the piezoelectric d_{31} constant with time (Murayama *et al.*, 1976). The decay is markedly nonexponential, and there appears to be eventual stabilization at a lower value of d_{31}. Other annealing experiments on corona-charged samples show a significant decay of the piezoelectric constant only at temperatures above 90°C; the decay is steep at 140°C (Das Gupta and Doughty, 1978). Results for biaxially oriented PVDF show decays of the d_{31} constant, that can be described by a single activation energy in the temperature range 60–88°C but are more complicated to interpret at higher decay temperatures (Blyler *et al.*, 1980). Other data for PVDF show a more pronounced decay for β-phase films than for α-phase films (Sussner, 1979). In PMMA, a nonexponential decrease in the polarization has been observed (Vanderschueren and Linkens, 1978a). The same is true for Mylar (PET) (van Turnhout, 1975).

D. Polarization of PVDF

Many studies indicate that the simple model of polarization discussed above has to be modified to explain the strong polarization of PVDF. The main aspects of our present understanding of the polarization of PVDF and its relation to the piezoelectric properties of this material will be briefly surveyed in the following (see also recent reviews by Broadhurst and Davis, 1980; Kepler and Anderson, 1980; Lovinger, 1981).

1. Effect of Crystalline Structure on Polarization

PVDF is a semicrystalline polymer. It is composed of lamellar crystals embedded in an amorphous background. The degree of crystallinity is 50–70%. It has been shown that the amorphous phase has the properties of a supercooled liquid, showing a liquid–glass transition at −50°C (Abkovitz and Pfister, 1975). The crystalline regions in PVDF can consist of four crystal phases termed α-, α_p-, β-, and γ-phases (Davies, 1981). Cooling of the molten polymer normally yields the α-phase; this phase is usually antipolar in the sense that the dipole moments of adjacent chains are pointing in opposite directions. Application of electric fields of about 1 MV/cm to the α-phase causes a change to the polar α_p-phase (McKinney and Davis, 1978). Higher fields or mechanical stretching of the films transforms the α-phase into the polar β-phase. The γ-phase, which is also polar, is obtained by crystallization from a number of solvents. The piezoelectric and pyroelectric properties are believed to be mainly due to the β-phase and the α_p-phase.

2. Role of Electrode Injection at Low and Moderate Fields; Uniformity

It has been known for some time that charge injection through the electrodes during formation and subsequent trapping contributes significantly to the polarization of PVDF if poling fields up to 1 MV/cm are used (Pfister *et al.*, 1973; Murayama *et al.*, 1975). In particular, poling of the material through blocking electrodes results in samples that exhibit only a very small pyroelectric effect, while poling through injecting electrodes yields strongly polarized samples. More recently, further experiments were performed with blocking and nonblocking electrodes to investigate the effect of charge injection (Sussner and Dransfeld, 1978). After poling between two blocking electrodes, the samples showed no piezoelectric activity; they showed small activity when poled between a positively biased blocking electrode and a negatively biased regular electrode and exhibited a 20 times greater effect when poled in reverse. Poling of a three-layer sandwich structure at a field of 5×10^5 V/cm according to Fig. 11a yields samples with very different polarization as indicated by the different piezoelectric responses (Fig. 11b). All these experiments show that charge injection through the positive electrode appears to be essential for good poling of PVDF at low and medium fields (see below).

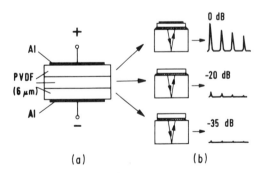

Fig. 11. (a) Poling of a stack of three PVDF films. (b) Piezoelectric response of the films. [From Sussner and Dransfeld (1978).]

Apart from this, the distribution of the polarization in PVDF can be very nonuniform. Clear evidence of this was found in measurements of the pyroelectric sensitivity: On samples poled at relatively low fields and temperatures, the pyroelectric sensitivity was greatest on the side of the sample facing the positively biased electrode during poling (Phelan *et al.*, 1974; Day *et al.*, 1974). These experiments showed that poling fields in excess of 1.7 MV/cm at temperatures above 70°C are necessary to produce uniform polarization. To further demonstrate the possible nonuniformities, previously polarized PVDF samples were excited by radio-frequency voltages which caused acoustic vibrations at frequencies where the wavelength of the piezoelectrically excited wave equals twice the film thickness and where it equals the film thickness (Sussner and Dransfeld, 1978). Since the second frequency is forbidden in a homogeneously polarized sample, the polarization must be asymmetric. In addition, flexural resonances of thin PVDF sheets were observed. Such modes are likewise impossible in a uniformly polarized film.

Measurements of the uniformity of the polarization in PVDF films and PVDF-PTFE copolymer were also carried out with the thermal pulse and pressure-pulse methods described in Section III.B. Experiments with thermally charged and corona-charged PVDF-TFE yield clear evidence of spatially nonuniform polarization centered near the electrode maintained at positive potential during charging. With increasing poling temperature and voltage, the polarization extends further into the material (De Reggi *et al.*, 1978). In PVDF samples thermally polarized at about 200 kV/cm, the polarization is also concentrated close to the positive electrode, while in samples corona-charged to high fields the polarization is in the center of the sample (Eisenmenger and Haardt, 1982).

3. *Dependence on Poling Conditions; The Molecular Picture*

The polarization achieved in PVDF depends on applied field, poling temperature, poling time, poling process, electrode conditions, and morphology of the particular sample. From the host of experimental data available, only a few important results will be discussed in the following.

The dependence of the polarization on the applied field in thermal poling is linear until saturation occurs (Murayama *et al.*, 1976). Results obtained for the nonpolar α-phase are depicted in Fig. 12; the 230- and 250-K peaks are due to dipolar polarization while the other peaks may relate to interfacial polarization (Mizutani *et al.*, 1981).

For long enough poling times, the response depends only on the poling field, and the poling temperature affects merely the rate at which the polarization builds up (Blevin, 1977). This is consistent with Eq. (20). For

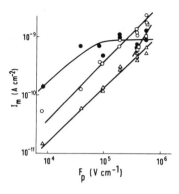

Fig. 12. Dependence on poling field of maximum currents I_m of four TSC peaks of PVDF. The peaks occur at approximately the following temperatures: ○, 230 K; △, 250 K; ●, 310 K; □, 350 K. [From Mizutani *et al.*, 1981).]

thermal charging, the temporal buildup of the polarization is controlled by at least two time constants, a fast one on the order of seconds and a slow one on the order of hours (Naegele and Yoon, 1978; Sussner and Dransfeld, 1978). For corona-charged samples, the polarization forms in about a second at room temperature, if fields of 2 MV/cm are applied, but takes longer to develop at lower temperatures or fields. The relation between polarization and field is characterized by a threshold value (Southgate, 1976; Das Gupta *et al.*, 1979).

For moderate-field poling (up to 1 MV/cm), electrode injection considerably affects the polarization in PVDF, as pointed out above.

Since the piezoelectric sensitivity is greater on the side of the material facing the positive electrode during charging, a negative space charge in the bulk of the polymer strengthening the field at the positive electrode has to be assumed (Broadhurst and Davis, 1980). In addition, space-charge dipoles might be formed by the deposition of charges of opposite polarity at the crystal boundaries, particularly if the poling is performed at high temperatures. Such dipoles weaken the piezoelectric and pyroelectric activities of the molecular dipoles.

Poling of PVDF at high fields (above 1 MV/cm) results in molecular changes that can be detected by optical and x-ray methods. The molecular rearrangement of the β-phase due to poling of this kind was detected by means of infrared transmission measurements by Southgate (1976) and was later confirmed by other authors. The molecular transition from the α-phase to the α_p-phase was demonstrated in x-ray scattering experiments (Das Gupta and Doughty, 1977; Naegele *et al.*, 1978). Such molecular changes are indicative of dipole reorientations and thus strengthen the argument for a predominantly dipolar origin of the observed polarization.

4. *Piezoelectricity and Pyroelectricity; Ferroelectricity*

The strong piezoelectric effect in PVDF was discovered in 1969 by Kawai. In 1971, the pyroelectric properties of this material were first described (Bergman *et al.*, 1971). Later it was found that PVDF samples subjected to fields cycled between large positive and negative values showed a hysteresis of the polarization and the piezoelectric response (Tamura *et al.*, 1974, 1977). Since these effects are strongly dependent upon β-crystal content and disappear below the glass-transition temperature they are attributed to cooperative phenomena of the β-crystals, influenced by the surrounding amorphous regions. The material thus exhibits a behavior resembling that of ferroelectrics. The ferroelectric polarization has been explained by a cooperative model based on the assumption that the molecular dipoles have two or more orientations available (Broadhurst and Davis, 1981).

Several models have been proposed to explain the piezoelectric and pyroelectric behavior of PVDF. Most of them are based on the existence of a dipole polarization in the crystalline regions embedded in amorphous material. These models, however, initially could not explain all observations. In particular, the relatively strong piezoelectric effect found in α-phase material after poling was irreconcilable with the antipolar behavior of this phase. This led to the development of other models based more on the effect of trapped charges. The situation became more transparent with the recent discovery of the polar α_p-phase and the α-to-α_p transition (see above). Presently, a consensus is developing that the electrical effects observed in PVDF are primarily due to the dipole polarization in the crystalline α_p-, β-, or γ-phases of this material.

The piezoelectric and pyroelectric activity is best explained (Broadhurst and Davis, 1980; Davies, 1981; Wada and Hayakawa, 1981; Tashiro *et al.*, 1981) by a model assuming the crystal to consist of lamellae dispersed in an amorphous liquid. The molecular charges are aligned so that their dipole moments are parallel, and a certain amount of real charge is trapped at the crystalline–amorphous interfaces normal to the crystal moment. If one further assumes different thermal expansion coefficients and compressibilities for the crystalline parts and for the sample as a whole, a change in the thickness dimension of the sample will obviously cause a change in the electrode charge or voltage, as shown in Fig. 13. This accounts for the piezoelectric activity. The case shown in the figure, namely, zero real charge at the crystalline–amorphous interfaces, gives excellent agreement between theory and experiment. Similarly, the pyroelectric effect can be explained. The similarity of the temperature

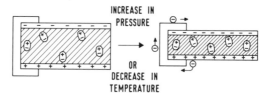

Fig. 13. Model of dipolar electret showing generation of the piezoelectric effect. [From Broadhurst and Davis (1980).]

dependence of the piezoelectric and pyroelectric activities also suggests a common origin of both effects (Das Gupta and Doughty, 1978).

The piezoelectric and pyroelectric properties of PVDF have been measured by a number of laboratories (Ohigashi, 1976; Tamura et al., 1977; Klaase and van Turnhout, 1979). Comparison of α_p-, β-, and γ-phase material shows the piezoelectric and pyroelectric constants of the β-phase material to be larger than those of the α_p-film by about a factor of 2 (Klaase and van Turnhout, 1979). This agrees well with the theoretical estimate derived by considering the polarizations of the two phases (Broadhurst et al., 1978).

Table I shows the piezoelectric and pyroelectric coefficients of PVDF and some other piezoelectric materials. It is evident from the table that the piezoelectric d constant and the pyroelectric coefficient of the polymer materials are considerably lower than the corresponding values of the ceramic PZT-5. However, because of the relatively large g constant of PVDF, the electromechanical coupling coefficient of this material, defined as $k = \sqrt{dg/c}$, where c is the elastic compliance, is smaller than that of PZT by only a factor of 2 to 3. In spite of the somewhat smaller response, PVDF finds frequent applications because of its excellent mechanical

TABLE I *Piezoelectric and Pyroelectric Constants of Some Materials*

| Material | Piezoelectric constants | | Pyroelectric coefficient, p (nC/cm² K) | Coupling coefficient, k_{31} |
	d_{31} (10^{-12} C/N)	g_{31} (10^{-3} V m/N)		
PVDF	20	200	4	0.12
PVF[a]	1	20	1.0	0.03
PZT-5[b]	171	11	6–50	0.34
Quartz	2 (d_{11})	50 (g_{11})	—	0.09 (k_{11})

[a] Poly(vinyl fluoride).
[b] Lead zirconate titanate.

properties, its availability as thin films, and its relatively low cost (Sessler, 1981a).

V. REAL-CHARGE STORAGE

A. Trapping Levels and Trap Densities

The quasipermanent retention of real charges in polymer electrets is due to the presence of trapping states capable of holding electrons or holes for long periods of time. Trapping in polymers is often interpreted in terms of a modified energy-band model (see Fig. 14a). According to this model, traps are localized states belonging to certain molecules or molecular groups. Since polymers are amorphous or polycrystalline, the energy levels, which are affected by their environment, are different in different molecular regions of the material. Thus, the trap depths are correspondingly distributed. Activation from such local states is by thermal energy.

Apart from these localized traps there are delocalized states, generally referred to as extended states, which are energetically located near the bottom of the conduction band and the top of the valence band (Fig. 14b). They are separated from the localized states by the so-called mobility edge at which the carrier mobility drops by several orders of magnitude (Mott and Davis, 1971; see also Seanor, 1972b). Carriers in such states move by quantum-mechanical hopping. Charge trapping in extended states is generally negligible in electret materials. The extended states, however, play a role in charge transport (see Chapter 1).

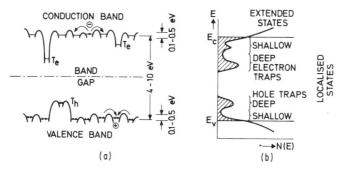

Fig. 14. (a) Energy diagram for a polymer. T_e, electron traps; T_h, hole traps. [From Bauser (1972).] (b) Density of states $N(E)$ for a polymer. Localized states (traps) are shaded; E_c and E_v are mobility edges. [From Sessler, 1980).]

Charge storage in polymers may occur in surface and volume traps. While it is readily possible to distinguish between these categories (see below), it is difficult to assess the molecular origin of the traps. Surface traps may be due to chemical impurities, specific surface defects caused by oxidation products, broken chains, adsorbed molecules, or differences in the short-range order of surface and bulk (Fuhrmann, 1978). For volume traps, investigations on substituted polyolefins have indicated the presence of three structural trapping levels, but no trapping due to impurity centers (Creswell *et al.*, 1972). While the primary levels are at atomic sites on the molecular chains, the secondary levels are between groups of atoms in neighboring molecules and the tertiary levels are in the crystalline regions or at crystalline–amorphous interfaces. The experimental evidence for or against most of these surface trap or volume trap categories is, however, presently weak.

For Teflon FEP, which is important because of its electret applications, the spatial location and thermal activation temperature of surface and volume traps have been determined recently by TSC measurements (von Seggern, 1979a). In these experiments, surface and volume traps for corona-charged electrets were separated by comparing short-circuit and open-circuit TSC data. Release temperatures of 155 and 200°C for the surface and volume traps, respectively, and 170°C for an intermediate trap were found. Also, a shallower trap level active under conditions where the deeper traps are filled (trap-filled limit) and discharging at 95°C was detected. Finally, information about the spatial distribution of the 155–200°C traps was derived from the open-circuit TSC data for electron-beam-charged samples depicted in Fig. 15. The results of this investigation, shown in Table II, suggest an extension of the surface layers to a depth of about 2 μm.

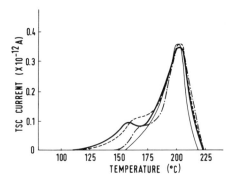

Fig. 15. Open-circuit TSC currents for 25-μm Teflon FEP samples charged with electron beams of the following energies: ——, 3 keV; ---, 5 keV; - · - ·, 7 keV; ——, 10 keV.

TABLE II *Distribution of Traps for Negative Charges in 25-μm Teflon FEP-A[a]*

Peak temperature (°C)	Location relative to charged surface	(μm) Kind of trap
95	0–25	Energetically shallower trap active under trap-filled limit conditions
155	0–0.5	Surface trap
170	0.5–1.8	Near-surface trap
200	1.8–25	Bulk trap

[a] From von Seggern (1979a).

Activation energies of trapping levels in a number of polymers have also been derived from TSC and other measurements. Typical values are 1.9 eV for Teflon (Chudleigh *et al.*, 1973), 1.8–2.2 eV for Mylar, 1.5 eV for polyethylene, and 1.2 eV for polyimide (van Turnhout, 1980). Traps of this depth hold charges for long periods of time; one obtains for the trap-release frequency

$$\nu = \nu_0 \exp(-U/kT) \qquad (24)$$

at room temperature values of 10^{-4}–10^{-22} sec^{-1}, assuming an escape frequency $\nu_0 = 10^{13}$ sec^{-1}.

Trap densities in polymers have been inferred from maximum stored-charge densities. Some experimental values are given in Table III. Since most of the data in the table are limited by breakdown and not trap saturation, the actual trap densities may be larger than those shown. In some polymer films, such as 25-μm Teflon, surface traps are capable of

TABLE III *Greatest Observed Charge Densities and Full-Trap Densities in Some Polymer Materials[a]*

Polymer and charging method	Thickness (μm)	Surface charge (S) or volume charge (V)	Projected charge density and sign (10^{-6} C/cm^2)	Full-trap density in volume and sign (10^{15} cm^{-3})
FEP, breakdown charged	12.5	Mostly S	0.5 (+, −)	—
FEP, electron-beam charged	25	V	—	0.14 (−)[b]
PET, breakdown charged	3.8	S, V	1.2 (+, −)	20 (+, −)
PC[c], breakdown charged	2.0	?	1.0 (+, −)	—

[a] Sessler and West (1972); Sessler, (1977).

[b] In deep traps with a TSC relaxation temperature of 200°C or higher.

[c] Polycarbonate.

storing more charge than volume traps and charging of such materials will generally produce a surface charge (an exception is electron-beam charging which results in a volume charge). In other polymers, such as Mylar, a specific surface trap has not been detected (von Seggern, 1981b; Broemme *et al.*, 1981).

B. Spatial Distribution

The location of the charge centroid on Teflon FEP films charged by the injection of electrons was determined by split-Faraday cup measurements (Gross *et al.*, 1977) and thermal pulse measurements (von Seggern, 1979a). Comparison of the initial, end-of-charging, and final charge depths indicates a significant charge drift during charging but less thereafter. The initial depth is somewhat smaller than the so-called practical range, while the final range, shown in Fig. 16, exceeds it substantially. The charge drift is due to the action of the self-field in the presence of a radiation-induced conductivity. Charge centroids in Teflon have also been calculated from the irradiation parameters (Berkley, 1979; Beers and Pine, 1980).

The charge centroid was also measured on one-sided metallized, positively and negatively liquid-charged Teflon electrets with the thermal pulse method (Collins, 1975). For both polarities, the charges reside originally on the surface of the samples. Injection into the volume occurs at 100 and 180°C for positive and negative carriers, respectively. The experiments further show that most of the positive charges, once released from the surface, move rapidly through the film, while the negative charges are

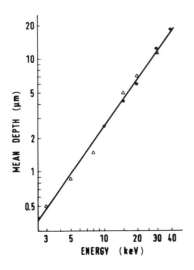

Fig. 16. Centroid location of charges injected by electron beams as a function of beam energy. ●, Split-Faraday cup measurement (Gross *et al.*, 1977); △, thermal pulse measurement. 25-μm Teflon FEP-A films. [From von Seggern (1979a).]

trapped in the volume. This indicates that for Teflon hole traps on the surface are deeper than volume traps, while the opposite is true for electron traps. The former result is substantiated by hole transit experiments (Gross *et al.*, 1979), and the latter by TSC data (von Seggern, 1979a).

Centroid data for the charge distribution were also obtained for corona-, Townsend-, and breakdown-charged electrets. For Teflon, negative corona charging does not yield measurable charge penetration (Moreno and Gross, 1976), while negative breakdown charging at high voltages yields charge depths of a few micrometers (van Turnhout, 1975; Sessler, 1972). On Mylar, even corona and Townsend charging leads to charge depths of 1–2 μm (van Turnhout, 1975; Seiwatz and Brophy, 1966).

Detailed charge distributions were measured by means of electron-beam sampling (Tong, 1980; Sessler and West, 1981). A typical result for an electron-beam-charged PET sample is shown in Fig. 17a. To avoid secondary emission, the sample was biased during charging. The charge distribution is characterized by low densities close to the irradiated electrode and a relatively narrow charge centered around the mean depth of 6 μm expected from Fig. 16. On samples where secondary emission during charging is not prevented, positive charge is found close to the irradiated electrode (Sessler *et al.*, 1982b).

Very recently, a pressure-pulse method utilizing excitation by a short laser pulse was used to determine directly and accurately the charge distribution in corona- and electron-beam charged PET and FEP films with micrometer resolution (Sessler *et al.*, 1982a). A typical result for electron-beam-charged FEP is shown in Fig. 17b. It is evident from the figure that

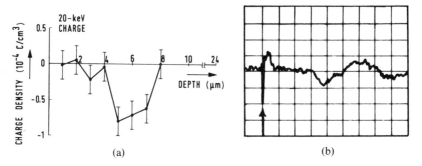

(a)   (b)

Fig. 17. (a) Charge distribution in 24-μm Mylar sample, as measured with electron-beam sampling. [From Sessler and West (1981).] (b) Charge distribution in 55-keV electron-beam-charged 75-μm Teflon FEP sample measured with pressure-pulse method (abscissa 10 nsec/div). Shown is the current response of the sample upon propagation of a pressure pulse in thickness direction. The pulse enters the sample at time indicated by arrow, transit time is 60 nsec. The current corresponds directly to charge density. [From Sessler *et al.* (1982a).]

the sample carries a significant positive charge layer between the injected and trapped electrons (broad negative peak) and the electrode of incidence (center of broad positive peak). The positive charges are due to injection of carriers caused by secondary emission during charging.

C. Charge Decay

The decay of the real charge on an electret can be due to external or internal causes. If the electret is stored under unshielded conditions, such that it is subject to air circulation, the charge decay is accelerated. The effect has been investigated for polypropylene and Teflon electrets with somewhat different results (van Turnhout et al., 1976; Anderson et al., 1973). While the polypropylene data show only a minor effect of airflow on charge decay, the Teflon data exhibit a pronounced decay. Generally, air currents are kept away from electrets by proper shielding. Then the external decay is diminished such that internal conduction determines the charge loss.

The charge decay (or, better, potential decay) of a number of one-sided metallized polymer electrets due to internal causes is shown in Fig. 18. The excellent charge stability of PTFE (negatively charged) and PC is evident from the figure. PTFE also shows superior charge-storage behavior in high-humidity atmospheres (van Turnhout, 1975). This is attributed to the hydrophobic behavior of this material. Positively charged Teflon shows a much more pronounced decay (von Seggern, 1981a).

Different charging methods lead to different initial charge distributions. This, in turn, affects the potential decay. A study of TSC currents of corona-charged, liquid-charged, and electron-beam-charged Teflon FEP electrets (von Seggern, private communication) shows corona-charged or liquid-charged electrets to decay from about 140°C on, as a result of the injection of carriers from surface to volume states. As opposed to this, the beam-charged samples exhibit initially a slight potential rise due to injection of holes left at the surface during charging as a result of secondary emission. Since the electrons are already in the volume, the decay starts

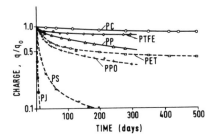

Fig. 18. Decay of electret charges in polycarbonate (PC), Teflon (PTFE), polypropylene (PP), Mylar (PET), polyphenylene oxide (PPO), polysulfone (PS), and polyimide (PI). [From van Turnhout (1971).]

only at 160°C. Thus, electron-beam-charged electrets have better stability than surface-charged samples (von Seggern, 1979a).

The charge stability of most electrets can be improved by the application of heat before, during, or after charging. For example, better stability has been noted for Teflon electrets annealed at 150°C or quenched from 250°C prior to charging (Ikezaki *et al.*, 1977; van Turnhout, 1980). The improvement has been attributed to the generation of deeper traps because of changes in crystalline grain size or crystallinity. The stability was also improved by application of heat up to 220°C to Teflon during charging (van Turnhout, 1975; Perlman and Unger, 1974; Ikezaki *et al.*, 1975). This is attributed to the retrapping, in deeper levels, of carriers from shallower levels. The transition can be from surface levels to volume levels or from relatively shallow volume levels to deep volume levels. The same phenomena are responsible for the increase in stability due to annealing after charging (van Turnhout, 1975). The charge loss due to such annealing may be made up by repeated charging.

The potential decay of electron-beam-charged Teflon at 150°C shows two exponential decay sections (Sessler and West, 1975). The two decay phenomena with very different time constants correspond to the decay regions at about 190 and 240°C, respectively, found in open-circuit TSC experiments on this material. These regions are due to two volume trap levels. A comparison of measured and predicted potential decays at 145°C over a more limited potential range (where the deeper volume trap does not yet release) is given in Section V.D.

At room temperature, the potential decay of Teflon electrets is also characterized by two or more decay constants. Since the number of deep volume traps (with TSC release temperatures of 200°C or more) is about 1.4×10^{14} cm^{-3} (see Table III), a 25-μm-thick film can store only a projected charge density of 28 nC/cm^2 over long periods of time. Samples charged to higher levels will therefore initially decay faster. This is actually seen from the results of charge-decay measurements, as depicted in Fig. 19. According to this figure, the decay rate of the charge slows down significantly with decreasing charge density. For charge densities of 20 nC/cm^2 or less, time constants in excess of 100 yr are found.

Electrodes may also affect the charge decay of electrets in certain cases. The customary vacuum-deposited aluminum or gold layers are blocking contacts on Teflon and many other polymers and therefore have no influence on charge decay (Lewis, 1976; Baum *et al.*, 1978). On the other hand, sputtered-gold electrodes on Teflon were found to inject holes at elevated temperatures and thus cause a lowering of TSC peak temperatures (Sessler *et al.*, 1973). The charge transport through irradiated electrodes, which appear to be blocking in the absence of irradiation suggests

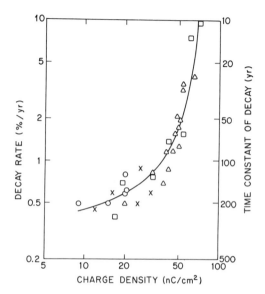

Fig. 19. Decay rate and corresponding time constants of electron-beam-charged Teflon as function of charge density. Energy of injected electrons: O, 10 keV; ×, 15 keV; △, 25 keV; □, 35 keV. [From Sessler and West (1975).]

that the blocking is probably caused by charge trapping in the polymer and not by the electrodes (Gross *et al.*, 1981).

The time constants of the charge decay on Teflon at elevated temperatures can be presented as an Arrhenius plot (Chudleigh *et al.*, 1973). For activated processes the slopes of such plots yield the activation energies of the trapping levels. For Teflon, an activation energy of 1.9 eV has been derived, in agreement with other methods (see above).

While charge decays of electrets are generally well behaved in the sense that the potential–decay curves of samples of the same material charged initially to different potentials do not cross, exceptions to this have been noted. Most prominent is the case of corona-charged polyethylene, which shows marked crossovers (Ieda, 1968). The effect has also been found on liquid-contact-charged polyethylene, on corona-charged PTFE (Gerhard, private communication; see also Sessler, 1978), and on electron-beam-charged polyethylene (Sessler and von Seggern, 1979). A full understanding of the crossovers has not yet been achieved. It appears that in more highly charged samples almost full injection into the bulk occurs, where the charges are rather mobile, while in samples with low charge density the charges remain in deep surface traps. However, the physical reason for this charge-dependent injection process is not clear. Corona light and excited molecules have been held responsible, but the effectiveness of both agents has been subsequently questioned (Baum *et al.*, 1977; Perlman *et al.*, 1979; Sessler and von Seggern, 1979; Kao *et al.*, 1979).

D. Comparison of Experimental and Calculated Potential Decays

The internal decay of real charges in electrets has been analyzed under a variety of assumptions concerning ohmic conductivity, carrier trapping and mobility, injection from surface states, etc. (for references see Sessler, 1980). The simplest model is that of a one-sided metallized dielectric initially carrying a charge on its nonmetallized surface. If the carriers have a field- and space-independent mobility μ, the potential $V(t)$ of the nonmetallized surface decays as

$$V(t)/V(0) = \begin{cases} 1 - \frac{1}{2}t/t_0 & \text{for} \quad t \leq t_0, \qquad (25) \\ \frac{1}{2}t_0/t & \text{for} \quad t \geq t_0, \qquad (26) \end{cases}$$

where $t_0 = s^2/\mu V(0)$ is the transit time and s is the sample thickness. Carrier trapping can be considered in this model by substituting for μ a trap-modulated mobility if the retrapping is fast (*Schubweg* much smaller than sample thickness). This model has been successfully used to describe the charge decay in a number of polymers at low fields (Reiser *et al.*, 1969). Its greatest drawback is the neglect of surface trapping and volume trapping by more than a single trap level.

Another model for the charge transport in a one-sided metallized dielectric, considering deep trapping with a finite capture time, has more recently been discussed and compared with experimental results in Teflon (Chudleigh, 1977). When rigorously applied, this model leads to charge-transport characterized by a constant speed of the free-carrier front.

A more general model considering surface trapping with a finite release time and trap-controlled volume transport, described by a free time between traps and a capture time in a trap, has been solved numerically (von Seggern, 1979b). The charge is assumed to be initially on the nonmetallized surface of the dielectric. The model is particularly tailored to suit the conditions of Teflon electrets. Numerical results of the model are compared in Fig. 20 with experimental data for this material obtained at 145°C. Of the four parameters of the theory, two can be determined independently and two are adjusted for best fit. Over the range of voltage decays measured, the agreement between experiment and theory is excellent. The theoretical model also yields the evolution of the spatial distribution of free and trapped carriers. According to the results, the free-charge density shows a steep front which moves with constant velocity through the material, reaching the electrode at the transit time, while the trapped charge has maximum density on the nonmetallized side.

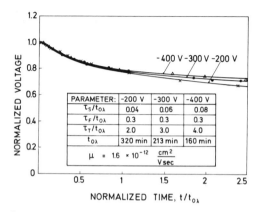

Fig. 20. Decay of normalized voltage of corona-charged Teflon electrets at 145°C. Lines are the best fit obtained with a transport model for the parameters shown in the table. [From von Seggern (1979b).]

E. Radiation Effects

The exposure of polymers to ionizing radiation produces a radiation induced conductivity (RIC) due to the generation of secondary electrons and holes. Since secondaries are lost by recombination, an equilibrium between generation and recombination is eventually reached during irradiation. The secondaries are distributed between various trapping levels and the conduction and valence bands. After termination of the irradiation, a delayed radiation-induced conductivity (DRIC) persists which falls off with time as a result of recombination (Fowler, 1956).

While the RIC and the DRIC, if present in electrets, are detrimental to the temporal stability of the charges, these phenomena are often unavoidable. Examples are electron-beam-charged electrets and electrets exposed to space environments. The RIC and DRIC effects can, however, be used with advantage for the study of electrical properties of polymers and in radiation dosimetry. Radiation effects in polymers have therefore been actively studied [see literature in Gross (1980)].

The RIC in polymers increases rapidly at the onset of irradiation and reaches steady-state values after times on the order of milliseconds to seconds. Most investigations have been devoted to this steady-state conductivity g. It depends on the dose rate ϕ of the irradiation by the power law

$$g = K\phi^\Delta, \tag{27}$$

where K is a material-dependent constant which for polymers lies between 10^{-19} and 10^{-16} and Δ is an exponent between 0.5 and 1 whose value depends on the energy distribution of the trapping levels and also on dose rate. Later, higher recombination probabilities cause a decrease in the conductivity (Gross *et al.*, 1981). For Teflon, which is important for electret applications, the prompt RIC is about 10^{-13} Ω^{-1} cm^{-1} at a dose rate of 10^4 rad/sec (Gross *et al.*, 1974).

Immediately after termination of the irradiation, the RIC is reduced to $g_\mathrm{d} = kg$, with $k \leq 1$. The time dependence of the DRIC is then given by

$$g_d(t) = g_0(1 + bt)^{-1}, \tag{28}$$

where b is a modified recombination coefficient.

The depth dependence of energy and charge-deposition profiles in Teflon was recently computed for 40-keV electrons (Berkley, 1979). In the same investigation and based on these profiles the charge dynamics was also investigated. In particular, charge distributions and centroid locations were evaluated and compared with experimental data. Electrode currents during irradiation can be interpreted to a first approximation by a $1 - \exp(t/\tau)$ time dependence of the RIC (Gross *et al.*, 1980). By considering space- and time-dependent conductivity values, where the time dependence is expressed by two time constants, the observed electrode currents and voltages can be explained quantitatively (Labonte, 1980, 1981). An example is shown in Fig. 21 which depicts the case of a floating rear electrode whose potential is measured.

The DRIC in Teflon shows, a few seconds after irradiation, the $1/t$ dependence expected from Eq. (28). After 10^4 sec, values of about 10^{-17} (ohm cm)$^{-1}$ are reached. Even years after irradiation, the DRIC is still noticeable in TSC experiments by causing peak shifts (Sessler and West, 1979). Results of four different DRIC experiments are shown in Fig. 22. The DRIC can be cured by annealing, since heat accelerates the recombination process (Gross *et al.*, 1975; West, 1979). Detailed knowledge of the

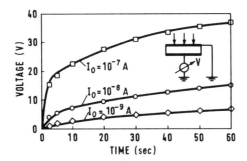

Fig. 21. Open-circuit charging of 25-μm Teflon FEP films with 40-keV electrons. Solid lines are best fit with the proper RIC parameters. $V_\mathrm{d} = 40$ keV. [From Labonte (1981).]

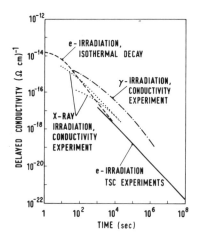

Fig. 22. Comparison of data for DRIC in Teflon obtained with several methods. X-ray irradiation data from Gross *et al.* (1981). For other data see Sessler (1977).

RIC is also available for other electret-forming polymers, such as Mylar (Beckley *et al.*, 1976) and polyethylene (Matsuoka *et al.*, 1976).

VI. APPLICATIONS

Polymer electrets have been utilized in a wide variety of applications. These reach from the technical areas to the biological and medical fields and are in various states of research, development, and production (for a detailed review of electret applications see Sessler, 1980).

A. Electret Transducers

The most widely used electret devices are electret microphones which were first described by Nishikawa and Nukijama in 1928. These transducers, as well as the electret microphones used before and during World War II proved unsatisfactory, since they contained wax electrets which have insufficient electrical stability under normal environmental conditions. With the introduction of the thin-film polymer electret microphone (Sessler and West, 1962, 1966) such transducers gained widespread commercial acceptance.

A cross section of such a microphone, consisting of a metallized acrylonitrile–butadiene–styrene backplate and a thin Teflon electret diaphragm stretched across it, is shown in Fig. 23. An incident sound wave causes vibrations of the diaphragm, which in turn generate an ac signal between the metallizations of the backplate and the diaphragm. The microphone design shown in the figure has an omnidirectional sensitivity

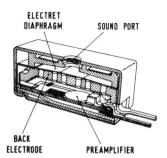

ELECTRET
DIAPHRAGM   SOUND PORT

BACK
ELECTRODE   PREAMPLIFIER

Fig. 23. Cross section of an electret microphone for telephone applications (Baumhauer and Brzezinski, 1979; Khanna and Remke, 1980).

pattern. Systems of this kind are used in telephone applications, particularly speakerphone systems.

Electret microphones have the advantage of being insensitive to mechanical vibrations, shock, and electromagnetic pickup; they have all the favorable properties of condenser microphones but are much simpler in design and thus less expensive. Such microphones are used in cassette recorders, hearing aids, hi-fi setups, sound-level meters, noise dosimeters, movie cameras, and telephony (operator headsets, speakerphones and, very recently, handsets). Worldwide production is about 100 million annually (Zahn, 1981).

The electret transducer principle has also been used for electret earphones. Apart from systems similar to the one in Fig. 23, earphones based on push–pull designs and on the use of nonmetallized electrets carrying only a single-polarity charge, the so-called monocharge electrets are in use (Griese and Kock, 1977). Another interesting application involves earphones that convert digital input signals directly into an acoustic analog output signal (Flanagan, 1980). Other applications of electret transducers are in earphones, loudspeakers, ultrasonic and underwater transducers, and electromechanical transducers such as phonograph pickups and touch, key, and contact sensors.

B. Electrophotography and Electrostatic Recording

Other applications of charge-storage phenomena of great practical importance are in the field of electrophotography. The basic process used in many electrophotographic methods, namely, the production of a charge pattern on an appropriate carrier and its development with powders, was already studied in the early 1930s (Selenyi, 1931). The breakthrough in this field came a few years later when investigations of photoconductive image formation led to the development of xerographic reproduction methods (Carlson, 1940). Apart from nonpolymeric carrier materials

(Schaffert, 1975), the polymer polyvinylcarbazole in the form of a charge-transfer complex with trinitrofluorenone has been used in electrophotography. The photoconductive properties of the PVK-TNF complex have been extensively reviewed (see, e.g., Gill, 1976; Mort, 1980).

Related to electrophotography, but not dependent on photographic methods, are electrostatic recording processes. These are used to record electrical signals, digital information, facsimile, and alphanumeric characters on carrier materials such as polymers and paper (Rothgordt, 1976/77). The recording is performed by electron-beam or electrical discharge from a needle electrode. The patterns may be read by a capacitive-probe arrangement or with a low-current, low-energy scanning electron beam by monitoring secondary emission from the sample (see, e.g.,Feder, 1976).

C. Piezoelectric and Pyroelectric Devices

A number of very recent electret applications are based on the piezoelectric and pyroelectric effects in polarized polymer materials, particularly PVDF (see Table I). Of particular importance are electroacoustic transducers utilizing transverse or longitudinal piezoelectric effects and pyroelectric detectors (Sessler, 1981a).

A microphone based on the transverse piezoelectric effect is shown in Fig. 24 (Lerch and Sessler, 1980). In this transducer, the 25-μm-thick PVDF film is placed over a spherically shaped rigid backplate with ring-shaped supports which provide proper curvature of the membrane. When a sound wave impinges on the membrane, the film changes its length and generates a voltage as a result of the piezoelectric effect. The advantages of this microphone are the mechanical stability and the relatively large capacitance; the latter simplifies amplifying requirements.

A recent application of PVDF transducers is in the multielement imaging array shown in Fig. 25. The individual elements are defined by gold strips deposited on an epoxy substrate. Proper choice of the input signals to the elements yields a line focus which is converted to a spot focus by the cylindrical lens. The spot can be scanned along a line, and two-

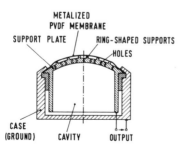

Fig. 24. Schematic cross section of piezopolymer microphone with a PVDF membrane. [From Lerch and Sessler (1980).]

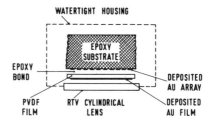

Fig. 25. Schematic cross section of a piezopolymer imaging array with a PVDF transducer. [From Shaw *et al.* (1980).]

dimensional imaging is obtained by linear motion of the system. The array is operated at about 2 MHz. Advantages of using a piezoelectric polymer are large bandwidth and wide-angle response (Shaw *et al.*, 1980).

A number of other transducer applications of piezoelectric polymers have been suggested. Among these are earphones, tweeters, underwater transducers, and electromechanical transducers (Tamura *et al.*, 1975; Lerch, 1978, 1979, 1981; Micheron, 1978; Sullivan and Powers, 1978; Sussner and Dransfeld, 1979; Shotton *et al.*, 1980; Latour and Murphy, 1981; Sessler, 1981a).

The pyroelectric coefficient of PVDF is about an order of magnitude smaller than that of triglycine sulfate, a widely used pyroelectric material. In spite of this, the polymer is attractive because of its excellent mechanical properties, its availability as thin films, and its low thermal conductivity. Currently in use are pyroelectric detectors such as the one shown in Fig. 26. They consist of a PVDF film cemented to a metal heat sink (McFee *et al.*, 1972). Heating of the film by radiation will cause a signal across a terminating resistor. Examples of applications of pyroelectric detectors are in fire detectors and burglar alarm systems. The pyroelectric effect in polymers has also been suggested for use in night-vision targets, photocopying machines, and other applications (see, e.g., Quilliam, 1977).

D. Biological and Medical Applications

Of great future potential are applications of polymer electrets in the biological and medical fields. Of interest in this context are attempts to improve the blood compatibility of polymers by negative charge deposi-

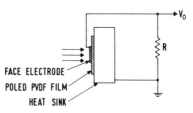

Fig. 26. Schematic view of a pyroelectric detector with a PVDF sensor. [From McFee *et al.* (1972).]

tion (Murphy and Merchant, 1973) and the detection of electret properties of human bones and blood vessel walls (Mascarenhas, 1980). It has also been shown that Teflon electrets placed in contact with bones of animals *in vivo* cause accelerated growth of callus, which is necessary for fracture healing (Fukada and Takamatsu, 1975). Moreover, electret bandages placed on skin incisions considerably improve the tensile strength of the wound over a given period of time and thus hasten the healing process (Konikoff and West, 1978). In an endodontic application, Teflon was shown to enhance the formation of esteodentin when applied to core-filling material (West *et al.*, 1979). Microphone-like transducers were also used as Korotkoff-sound pickups to detect arterial pressure fluctuations (West *et al.*, 1980).

E. Other Applications

An important recent application of polymer electrets has been in gas filters (van Turnhout *et al.*, 1976; 1980). These filters are made of polypropylene films which are corona-charged and fibrillated. Because of repulsive forces, the fibers spread into a broad web. The capture of particles by such electret fibers depends on coulomb and induction forces which act on charged and neutral particles, respectively. Since the operation of electret filters is based on long-range electrical forces, they can have an open structure with fiber spacings much in excess of the particle size. Filtration results of various such filters are compared in Fig. 27. The newly developed electret filters characterized by curled and fine fibers have higher filter efficiencies at about the same resistance as the more conventional ("normal") electret filters.

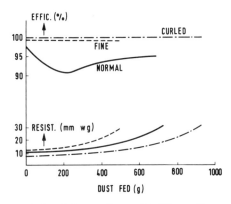

Fig. 27. Filter characteristics of electret filters with different fibers. [From van Turnhout *et al.*, 1979, 1980).]

Other polymer electret devices include relay switches, optical display panels, and radiation dosimeters. While relay-type switches (Perino *et al.,* 1977) utilize the external field of electrets to open or close contacts, optical display devices (Bruneel and Micheron, 1977) depend on the opening or closing of illuminated channels by means of hinged and opaque electrets moved by electrostatic forces. In radiation dosimeters (see, e.g., Bauser and Ronge, 1978), the decay of electret charges in an ionization chamber or the generation of a radiation-induced conductivity in electret materials is employed to measure radiation doses. Very recently, an electret motor based on the slot effect and using thin Teflon electrets has been demonstrated (Gerhard and Kaufhold, to be published).

REFERENCES

Abkovitz, M., and Pfister, G. (1975). *J. Appl. Phys.* **46**, 2559.

Alquie, C., Dreyfus, G., and Lewiner, J. (1981). *Phys. Rev. Lett.* **47**, 1483.

Anderson, E. W., Blyler, L. L., Johnson, G. E., and Link, G. L. (1973). *In* "Electrets, Charge Storage, and Transport in Dielectrics" (M. M. Perlman, ed.), pp. 424–435. Electrochemical Society, Princeton, New Jersey.

Baum, E. A., Lewis, T. J., and Toomer, R. (1977). *J. Phys. D Appl. Phys.* **10**, 487, 2525.

Baum, E. A., Lewis, T. J., and Toomer, R. (1978). *J. Phys. D Appl. Phys.* **11**, 703.

Baumhauer, J. C., and Brzezinski, A. M. (1979). *Bell Syst. Tech. J.* **58**, 1557.

Bauser, H. (1972). *Kunststoffe* **62**, 192.

Bauser, H., and Ronge, W. (1978). *Health Phys.* **34**, 97.

Beckley, L. H., Lewis, T. J., and Taylor, D. M. (1976). *J. Phys. D Appl. Phys* **9**, 1355.

Beers, B. L., and Pine, V. W. (1980). *Proc. Spacecr. Charg. Technol. Conf.* pp. 17–32.

Bergman, J. G., McFee, J. H., and Crane, G. R. (1971). *Appl. Phys. Lett.* **18**, 203.

Berkley, D. A. (1979). *J. Appl. Phys.* **50**, 3447.

Blevin, W. R. (1977). *Appl. Phys. Lett.* **31**, 6.

Blyler, L. L., Johnson, G. E., and Hylton, N. M. (1980). *Ferroelectrics* **28**, 303.

Broadhurst, M. G., and Davis, G. T. (1980). *In* "Electrets" (G. M. Sessler, ed.), pp. 285–319. Springer-Verlag, Berlin and New York.

Broadhurst, M. G., and Davis, G. T. (1981). *Ferroelectrics* **32**, 177.

Broadhurst, M. G., Davis, G. T., McKinney, J. E., and Collins, R. E. (1978). *J. Appl. Phys.* **49**, 4992.

Broemme, D., Gerhard, R., and Sessler, G. M. (1981). *Annu. Rep. Conf. Electr. Insul. Dielectr. Phenomena* pp. 129–135. National Academy of Sciences, Washington, D. C.

Bruneel, J. L., and Micheron, F. (1977). *Appl. Phys. Lett.* **30**, 382.

Carlson, C. F. (1940). U.S. Patent 2,221,776.

Carlson, E. V., and Killion, M. C. (1974). *J. Audio Eng. Soc.* **22**, 92.

Chudleigh, P. W. (1972). *Appl. Phys. Lett.* **21**, 547.

Chudleigh, P. W. (1976). *J. Appl. Phys.* **47**, 4475.

Chudleigh, P. W. (1977). *J. Appl. Phys.* **48**, 4591.

Chudleigh, P. W., Collins, R. E., and Hancock, G. D. (1973). *Appl. Phys. Lett.* **23**, 211.

Collins, R. E. (1973). *Proc. IREE* **34**, 381.

Collins, R. E. (1975). *Appl. Phys. Lett.* **26**, 675.

Collins, R. E. (1976). *J. Appl. Phys.* **47**, 4808.
Collins, R. E. (1977). *Rev. Sci. Instrum.* **48**, 83.
Collins, R. E. (1980). *J. Appl. Phys.* **51**, 2973.
Creswell, R. A., Perlman, M. M., and Kabayama, M. A. (1972). *In* "Dielectric Properties of Polymers" (F. E. Karasz, ed.), pp. 295–312. Plenum, New York.
Das Gupta, D. K., and Doughty, K. (1977). *Appl. Phys. Lett.* **31**, 585.
Das Gupta, D. K., and Doughty, K. (1978). *J. Phys. D Appl. Phys.* **11**, 2415.
Das Gupta, D. K., Doughty, K., and Shier, D. B. (1979). *J. Electrostat.* **7**, 267.
Davies, D. K. (1967a). *In* "Static Electrification" (A. C. Strickland, ed.), pp. 29–36. Institute of Physics, London.
Davies, D. K. (1967b). *J. Sci. Instrum.* **44**, 521.
Davies, G. R. (1981). *In* "Physics of Dielectric Solids," pp. 50–63. Institute of Physics Conf. Ser. No. 58.
Davis, G. T. and Broadhurst, M. G. (1977). *Int. Symp. Electrets Dielectr.* (M. S. de Campos, ed.), pp. 299–318. Academia Brasileira de Ciencas, Rio de Janeiro.
Davis, G. T., De Reggi, A. S., and Broadhurst, M. G. (1977). *Annu. Rep. Conf. Electr. Insul. Dielectr. Phenomena.* National Academy of Sciences, Washington, D. C.
Day, G. W., Hamilton, C. A., Peterson, R. L., Phelan, R. J., and Mullen, L. O. (1974). *Appl. Phys. Lett.* **24**, 456.
De Reggi, A. S., Guttman, C. M., Mopsik, F. I., Davis, G. T., and Broadhurst, M. G. (1978). *Phys. Rev. Lett.* **40**, 413.
Eguchi, M. (1919). *Proc. Phys. Math Soc. Jpn.* **1**, 326.
Eisenmenger, W., and Haardt, M. (1982). *Solid State Commun.* **41**, 917.
Engelbrecht, S. (1974). *J. Appl. Phys.* **45**, 3421.
Euler, K. J. (1976). *J. Electrostat.* **2**, 1.
Faraday, M. (1839). "Experimental Researches in Electricity." Taylor, London.
Feder, J. (1976). *J. Appl. Phys.* **47**, 1741.
Fischer, P. (1977/78). *J. Electrostat.* **4**, 149.
Fischer, P., and Röhl, P. (1976). *J. Polym. Sci.* **14**, 531.
Flanagan, J. L. (1980). *Bell Syst. Tech. J.* **59**, 1693.
Fleming, R. J. (1979). *J. Appl. Phys.* **50**, 8075.
Fowler, J. F. (1956). *Proc. R. Soc. London Ser. A* **236**, 464.
Freeman, J. R., Kallmann, H. P., and Silver, M. (1961). *Rev. Mod. Phys.* **33**, 553.
Fridkin, V. M., and Zheludev, I. S. (1961). "Photoelectrets and the Electrophotographic Process." Consultants Bureau, New York.
Fuhrmann, H. (1978). *J. Electrostat.* **4**, 109.
Fukada, E., Takamatsu, T., and Yasuda, I. (1975). *Jpn. J. Appl. Phys.* **14**, 2079.
Gerhard, R. (1982a). Private communication.
Gerhard, R. (1982b). Phys. Rev. B (submitted).
Gerhard, R., and Kaufhold, J. (to be published).
Gill, W. D. (1976). *In* "Photoconductivity and Related Phenomena," pp. 303–334. Elsevier, Amsterdam.
Gray, S. (1732). *Phil. Trans. R. Soc. London Ser. I* **37**, 285.
Griese, H.-J., and Kock, G. (1977). *Funkschau* **49**, 1251.
Gross, B. (1958). *J. Polym. Sci.* **27**, 135.
Gross, B. (1971). *In* "Static Electrification," pp. 33–43. Institute of Physics, London.
Gross, B. (1980). *In* "Electrets" (G. M. Sessler, ed.), pp. 217–284. Springer-Verlag, Berlin and New York.
Gross, B., Sessler, G. M., and West, J. E. (1974). *J. Appl. Phys.* **45**, 2841.
Gross, B., Sessler, G. M., and West, J. E. (1975). *J. Appl. Phys.* **46**, 4647.
Gross, B., Sessler, G. M., and West, J. E. (1977). *J. Appl. Phys.* **48**, 4303.

Gross, B., Sessler, G. M., von Seggern, H., and West, J. E. (1979). *Appl. Phys. Lett.* **34**, 555.

Gross, B., West, J. E., von Seggern, H., and Berkley, D. A. (1980). *J. Appl. Phys.* **51**, 4875.

Gross, B., Faria, R. M., and Leal Ferreira, G. F. (1981). *J. Appl. Phys.* **52**, 571.

Gubkin, A. N., Yegorova, T. S., Kokorin, L. M., and Zitser, N. Y. (1970). *Vysokomol. Soyed.* **A12**, 602.

Hayakawa, R., and Wada, Y. (1973). *Adv. Polym. Sci.* **11**, 1.

Heaviside, O. (1892). "Electrical Papers," pp. 488–493. Chelsea, New York.

Ieda, M., Sawa, G., and Shinahara, U. (1968). *Electr. Eng. Jpn.* **88**, 67.

Ikezaki, K., Fujita, I., Wada, K., and Nakamura, J. (1974). *J. Electrochem. Soc.* **121**, 591.

Ikezaki, K., Wada, K., and Fujita, I. (1975). *J. Electrochem. Soc.* **122**, 1356.

Ikezaki, K., Hattori, M., and Arimoto, Y. (1977). *Jpn. J. Appl. Phys.* **16**, 863.

Jain, V. K., Gupta, C. L., Jain, R. K., and Tyagi, R. C. (1978). *Thin Solid Films* **48**, 175.

Kao, K. R., Bamji, S. S., and Perlman, M. M. (1979). *J. Appl. Phys.* **50**, 8181.

Kawai, H. (1969). *Jpn. J. Appl. Phys.* **8**, 975.

Kepler, R. G., and Anderson, R. A. (1980). CRC critical review, *Solid State Mat. Sci.* **9**, 399.

Khanna, S. P., and Remke, R. L. (1980). *Bell Syst. Tech. J.* **59**, 745.

Kiess, H. (1975). *RCA Rev.* **36**, 667.

Klaase, P. T. A., and van Turnhout, J. (1979). *Proc. Int. Conf. Dielectr. Mater. Measurements Appl., 3rd* pp. 411–414. IEE Conf. Publ. 177.

Konikoff, J. J., and West, J. E. (1978). *Annu. Rep. Conf. Electr. Insul. Dielectr. Phenomena* pp. 304–310. National Academy of Sciences, Washington, D. C.

Kramer, H., and Messner, D. (1964). *Kunststoffe* **54**, 696.

Labonte, K. (1980). *Annu. Rep. Conf. Electr. Insul. Dielectr. Phenomena* pp. 321–327. National Academy of Sciences, Washington, D. C.

Labonte, K. (1981). *Annu. Rep. Conf. Electr. Insul. Dielectr. Phenomena* pp. 52–57. National Academy of Sciences, Washington, D. C.

Lacabanne, C., and Chatain, D. (1978). *Makromol. Chem.* **179**, 2765.

Latour, M. (1980). *J. Phys. Lett.* **41**, L-35.

Latour, M., and Murphy, P. V. (1981). *Ferroelectrics* **32**, 33.

Laurenceau, P., Dryfus, G., and Lewiner, J. (1977). *Phys. Rev. Lett.* **38**, 46.

Lerch, R. (1978). *In* "Fortschritte der Akustik, DAGA '78," pp. 661–664. VDE Verlag GmbH, Berlin.

Lerch, R. (1979). *J. Acoust. Soc. Am.* **66**, 952.

Lerch, R. (1981). *J. Acoust. Soc. Am.* **69**, 1809; **70**, 1229.

Lerch, R., and Sessler, G. M. (1980). *J. Acoust. Soc. Am.* **67**, 1379.

Lewis, T. J. (1976). *Annu. Rep. Conf. Electr. Insul. Dielectr. Phenomena* pp. 533–561. National Academy of Sciences, Washington, D. C.

Lovinger, A. J. (1981). *In* "Developments in Crystalline Polymers—I" (D. C. Bassett, ed.). Applied Science Publishers, London.

Mascarenhas, S. (1980). *In* "Electrets" (G. M. Sessler, ed.), pp. 321–346. Springer-Verlag, Berlin and New York.

Matsukawa, T., Shimizu, R., Harada, K., and Kato, T. (1974). *J. Appl. Phys.* **45**, 733.

Matsuoka, S., Sunaga, H., Tanaka, R., Hagiwara, M., and Araki, K. (1976). *IEEE Trans. Nucl. Sci.* **NS-23**, 1447.

McFee, J. H., Bergman, J. G., and Crane, G. R. (1972). *Ferroelectrics* **3**, 305.

McKinney, J. E., and Davis, G. T. (1978). *Org. Coatings Plast. Chem.* **38**, 271.

Micheron, F. (1978). *Rev. Tech. Thomson* **10**, 445.

Mizutani, T., Yamada, T., and Ieda, M. (1981). *J. Phys. D Appl. Phys.* **14**, 1139.

Mopsik, F. I., and DeReggi, A. S. (1981). *Annu. Rep. Conf. Electr. Insul. Dielectr. Phenomena* pp. 251–259. National Academy of Sciences, Washington, D. C.

Moreno, R. A., and Gross, B. (1976). *J. Appl. Phys.* **47**, 3397.

Mort, J. (1980). *Adv. Phys.* **29**, 367.

Mott, N. F., and Davis, E. A. (1971). "Electronic Processes in Non-Crystalline Materials." Oxford Univ. Press (Clarendon), London and New York.

Murayama, N., and Hashizume, H. (1976). *J. Polym. Sci. Polym. Phys. Ed.* **14**, 989.

Murayama, N., Oikawa, T., Katto, T., and Nakamura, K. (1975). *J. Polym. Sci. Polym. Phys. Ed.* **13**, 1033.

Murayama, N., Nakamura, K., Ohara, H., and Segawa, M. (1976). *Ultrasonics* **14**, 15.

Murphy, P. V., and Merchant, S. (1973). *In* "Electrets, Charge Storage and Transport in Dielectrics" (M. M. Perlman, ed.), pp. 627–649. Electrochemical Society, Princeton, New Jersey.

Naegele, D., Yoon, D. Y., and Boradhurst, M. G. (1978). *Macromolecules* **11**, 1297.

Nishikawa, S., and Nukijama, D. (1928). *Proc. Imp. Acad. Tokyo* **4**, 290.

Ohigashi, H. (1976). *J. Appl. Phys.* **47**, 949.

Perino, D., Lewiner, J., and Dreyfus, G. (1977). *L'onde Electr.* **57**, 688.

Perlman, M. M., and Unger, S. (1974). *Appl. Phys. Lett.* **24**, 579.

Perlman, M. M., Kao, K. J., and Bamji, S. (1979). *In* "Charge Storage, Charge Transport and Electrostatics with Their Applications," pp. 3–9. Elsevier, Amsterdam.

Pfister, G., Abkovitz, M., and Crystal, R. G. (1973). *J. Appl. Phys.* **44**, 2064.

Phelan, R. J., Peterson, R. L., Hamilton, C. A., and Day, G. W. (1974). *Ferroelectrics* **7**, 375.

Quilliam, M. (1977). *Electron. Ind.* **3**, 23.

Reedyk, C. W., and Perlman, M. M. (1968). *J. Electrochem. Soc.* **115**, 49.

Reiser, A., Lock, M. W. B., and Knight, J. (1969). *Trans. Faraday Soc.* **65**, 2168.

Rozno, A. G., and Gromov, V. V. (1979). *Sov. Phys. Tech.—Phys. Lett.* **5**, 266.

Rothgordt, U. (1976/77). *Philips Tech. Rdsch.* **36**, 98.

Schaffert, R. M. (1975). "Electrophotography." Wiley, New York.

Seanor, D. A. (1972a). *In* "Electrical Properties of Polymers" (K. C. Frisch and A. Patsis, eds.), pp. 37–58. Technomic Publ., Westport Connecticut.

Seanor, D. A. (1972b). In "Polymer Science" (A. D. Jenkins, ed.). North-Holland Publ., Amsterdam.

Seiwatz, H., and Brophy, J. J. (1966). *Annu. Rep. Conf. Electr. Insul. Dielectr. Phenomena* pp. 1–3. National Academy of Sciences, Washington, D. C.

Selenyi, P. (1931). U.S. Patent 1, 818, 760.

Sessler, G. M. (1972). *J. Appl. Phys.* **43**, 408.

Sessler, G. M. (1977). *Int. Symp. Electrets Dielectr.* (M. S. de Campos, ed.), pp. 321–335. Academia Brasileira de Ciencas, Rio de Janeiro.

Sessler, G. M. (1978). *Annu. Rep. Conf. Electr. Insul. Dielectr. Phenomena* pp. 3–10. National Academy of Sciences, Washington, D. C.

Sessler, G. M. (1980). "Electrets." Springer-Verlag, Berlin and New York.

Sessler, G. M. (1981a). *J. Acoust. Soc. Am.* **70**, 1596.

Sessler, G. M. (1981b). *In* "Physics of Dielectric Solids," pp. 133–145. Institute of Physics Conf. Ser. No. 58.

Sessler, G. M., and West, J. E. (1962). *J. Acoust. Soc. Am.* **34**, 1787.

Sessler, G. M., and West, J. E. (1966). *J. Acoust. Soc. Am.* **40**, 1433.

Sessler, G. M., and West, J. E. (1968). *J. Electrochem. Soc.* **115**, 836.

Sessler, G. M., and West, J. E. (1971). *Rev. Sci. Instrum.* **42**, 15.

Sessler, G. M., and West, J. E. (1972). *J. Appl. Phys.* **43**, 922.

Sessler, G. M., and West, J. E. (1975). *J. Electrostat.* **1**, 111.

Sessler, G. M., and West, J. E. (1979). *J. Appl. Phys.* **50**, 3328.

Sessler, G. M., and West, J. E. (1981). *Annu. Rep. Conf. Electr. Insul. Dielectr. Phenomena* pp. 145–150. National Academy of Sciences, Washington, D. C.

Sessler, G. M., and von Seggern, H. (1979). *Annu. Rep. Conf. Electr. Insul. Dielectr. Phenomena* pp. 160–165. National Academy of Sciences, Washington, D. C.

Sessler, G. M., West, J. E., and Gerhard, R. (1982a). *Phys. Rev. Lett.* **48**, 563.

Sessler, G. M., West, J. E., and von Seggern, H. (1982b). *J. Appl. Phys.* (to be published).

Sessler, G. M., West, J. E., Ryan, R. W., and Schonhorn, H. (1973). *J. Appl. Polym. Sci.* **17**, 3199.

Sessler, G. M., West, J. E., Berkley, D. A., and Morgenstern, G. (1977). *Phys. Rev. Lett.* **38**, 368.

Shaw, H. J., Weinstein, D., Zitelle, L. T., Frank, C. W., DeMattei, R. C., and Fesler, K. (1980). *Proc. Ultrason. Symp.* Vol. 2, pp. 927–940. IEEE, Piscataway, New Jersey.

Shotton, K. C., Bacon, D. R., and Quilliam, R. M. (1980). *Ultrasonics* **18**, 123.

Shrivastava, S. K., Ranade, J. D., and Srivastava, A. P. (1980). *Thin Solid Films* **67**, 201.

Southgate, P. D. (1976). *Appl. Phys. Lett.* **28**, 250.

Sullivan, T. D., and Powers, J. M. (1978). *J. Acoust. Soc. Am.* **63**, 1396.

Sussner, H. (1979). *Proc. Ultrason. Symp.* (J. deKlerk and B. R. McAvoy, eds.), pp. 491–498.

Sussner, H., and Dransfeld, K. (1978). *J. Polym. Sci. Polym. Phys. Ed.* **16**, 529.

Sussner, H., and Dransfeld, K. (1979). *Colloid Polym. Sci.* **257**, 591.

Tamura, M., Ogasawara, K., Ono, N., and Hagiwara, S. (1974). *J. Appl. Phys.* **45**, 3768.

Tamura, M., Yamaguchi, T., Oyaba, T., and Yoshimi, T. (1975). *J. Audio Eng. Soc.* **23**, 21.

Tamura, M., Hagiwara, S., Matsumoto, S., and Ono, N. (1977). *J. Appl. Phys.* **48**, 513.

Tashiro, K., Tadokoro, H., and Kobayashi, M. (1981). *Ferroelectrics* **32**, 167.

Tong, D. W. (1980). *Conf. Record 1980 Internat. Symp. Electr. Insul.* pp. 179–183. IEEE Service Center, Piscataway, New Jersey.

Vanderschueren, J. (1974). Thesis, Univ. Liege.

Vanderschueren, J., and Gasiot, J. (1979). *In* "Thermally-Stimulated Relaxation in Solids," (P. Bröunlich, ed.) pp. 135–223. Springer-Verlag, Berlin and New York.

Vanderschueren, J., and Linkens, A. (1978a). *Macromolecules* **11**, 1228.

Vanderschueren, J., and Linkens, A. (1978b). *J. Polym. Sci. Polym Phys. Ed.* **16**, 223.

Vanderschueren, J., Linkens, A., Haas, B., and Dellicour, E. (1978). *J. Macromol. Sci.-Phys.* **B15**, 449.

van Turnhout, J. (1971). *Adv. Static Electr.* **1**, 56–81.

van Turnhout, J. (1975). "Thermally-Stimulated Discharge of Polymer Electrets." Elsevier, Amsterdam.

van Turnhout, J. (1980). *In* "Electrets" (G. M. Sessler, ed.), pp. 81–215. Springer-Verlag, Berlin and New York.

van Turnhout, J., van Bochove, C., and van Veldhuizen, G. J. (1976). *Staub-Reinhalt. Luft* **36**, 36.

van Turnhout, J., Hoeneveld, W. J., Adamse, J. W. C., and Rossen, L. M. (1979). *Conf. Record. IEEE Industry Applications Soc.* pp. 117–125.

van Turnhout, J., Adamse, J. W. C., and Hoeneveld, W. J. (1980). *J. Electrostat.* **8**, 369.

von Seggern, H. (1978). *Appl. Phys. Lett.* **33**, 134.

von Seggern, H. (1979a). *J. Appl. Phys.* **50**, 2817.

von Seggern, H. (1979b). *J. Appl. Phys.* **50**, 7039.

von Seggern, H. (1981a). *J. Appl. Phys.* **52**, 4081.

von Seggern, H. (1981b). *J. Appl. Phys.* **52**, 4086.

Wada, Y., and Hayakawa, R. (1981). *Ferroelectrics* **32**, 115.

Walker, D. K., and Jefimenko, O. (1973). *In* "Electrets, Charge Storage, and Transport in Dielectrics" (M. M. Perlman, ed.), pp. 455–461. Electrochemical Society, Princeton, New Jersey.

Weber, G. (1978). *Angew. Makromol. Chem.* **74**, 187.

West, J. E. (1979). *In* "Charge Storage, Charge Transport, and Electrostatics with their Applications," pp. 392–396. Elsevier, Amsterdam.

West, J. E., von Seggern, H., Nelson, J. R., and Kubli, R. A. (1980). *J. Acoust. Soc. Am.* **68**, S 68.

West, N. M., West, J. E., Revere, J. H., and England, M. C. (1979). *J. Endotont.* **5**, 208.

Williams, R., and Woods, M. H. (1973). *J. Appl. Phys.* **44**, 1026.

Wintle, H. (1970). *J. Phys. E Sci. Instrum.* **3**, 334.

Zahn, R. (1981). *J. Acoust. Soc. Am.* **69**, 1200.

Chapter 7

Contact Electrification of Polymers and Its Elimination

D. Keith Davies
ERA TECHNOLOGY LTD
LEATHERHEAD
SURREY, ENGLAND

I. INTRODUCTION

The electrification of insulating materials, recognized by the Greeks, is now a familiar phenomenon owing to the ubiquity of modern plastics with concomitant "static" problems. The historical developments have been thoroughly described (Chambers Encyclo., 1970) and make interesting reading. It is salutary to appreciate that the early fundamental discoveries

285

in "electricity" were largely in electrostatics. The concept of charge polarity was introduced by Du Fay in 1733, the terms "positive" and "negative" being actually coined by Franklin in 1740. The conservation of charge was discovered independently by Watson and Franklin in 1746 and 1747, respectively, and the law of interaction by Coulomb in 1785. Wilke soon afterward proposed arranging materials into triboelectric series, in which the polarity of charge acquired on contact between any material with another in the series was predictable by its location. The pursuit of arranging such series (largely inconsistently) persists, a recent extensive example by Henniker (1962) being presented in Table I. The particularly fruitful theoretical period of the late eighteenth century culminated in the ideas of Volta, which remain essentially the basis of modern contact charge-transfer theories.

It is still commonly held that a rubbing action is a prerequisite for charge transfer, but it has long been appreciated that only contact is required, rubbing merely improving the intimacy of contact and introducing secondary complications, such as local heating and material transfer.

Ultimately, interfacial charging results from the transfer of electrons or protons, although these can under special circumstances be associated

TABLE I   *Triboelectric Series*[a,b]

Silicone elastomer with silica filler	Styrene–acrylonitrile copolymer
Borosilicate glass, fire polished	Styrene–butadiene copolymer
Window glass	Polystyrene
Aniline–formol resin (acid catalyzed)	Polyisobutylene
Polyformaldehyde	Polyurethane flexible sponge
Poly(methyl methacrylate)	Borosilicate glass, ground surface
Ethyl cellulose	Poly(ethylene glycol terephthalate)
Polyamide II	Polyvinylbutyral
Polyamide 6-6	Formo-phenolique, hardened epoxide resin
Rock salt, NaCl	Polychlorobutadiene
Melamine formol	Butadiene–acrylonitrile copolymer
Wool, knitted	Natural rubber
Silica, fire-polished	Polyacrylonitrile
Silk, woven	Sulfur
Poly(ethylene glycol succinate)	Polyethylene
Cellulose acetate	Poly(diphenylol propane carbonate)
Poly(ethylene glycol adipate)	Chlorinated polyether
Poly(diallyl phthalate)	Poly(vinyl chloride) with 25% D.O.P.
Cellulose (regenerated) sponge	Poly(vinyl chloride) without plasticizer
Cotton, woven	Polytrifluorochlorethylene
Polyurethane elastomer	Polytetrafluoroethylene

[a] From J. Henniker (1962).
[b] The sequence is from positive to negative.

with large molecules, so forming ions. The basic questions of static charging are, therefore, the definition of the transferred species under given circumstances, mechanism of transfer, and environmental factors governing eventual equilibrium. In intermetal or metal–semiconductor contacts, the charge carriers are indubitably electrons; the circumstances at metal–insulator and insulator–insulator interfaces are, however, more ambiguous.

Accepted models of metal–metal and metal–semiconductor contacts are considered briefly here as the basis for a discussion of metal–insulator effects. Theoretical models of insulator charging are reviewed and, following a presentation of methods of charge measurement, a review of experimental data on insulators is presented.

As intimated earlier, "static electrification" is a familiar phenomenon, not the least owing to wide experience with problems ranging from personal discomfort to process interruption and even hazards resulting from the incendivity of sparks generated by charged surfaces in flammable environments. It is largely the case that observed charge levels are defined not by the properties of the materials involved but by the ambient atmosphere, since the potential difference between the separating surfaces generally produces sparks in air at normal pressure. In view of the importance of sparking both in defining the operational charge level and in ignition hazards, and also as a means of charge control through provision of neutralizing species from the atmosphere, the different forms of gaseous electrical breakdown are reviewed briefly.

Charge transfer on contact between dissimilar materials is inevitable; however, resultant electrification may be avoided by increasing the conductivity of the contacting surfaces. This may be achieved by the incorporation of foreign chemical species such as hygroscopic salts, owing to the sensitivity of electrical conductivity of polymers to absorbed water. This, however, is a subject more properly reviewed elsewhere in the present volume, and in the present case charge dissipation by conduction is assumed not to occur and technological means of charge dissipation are reviewed.

II. CHARGE-TRANSFER THEORIES

A. Metal–Metal Contact

There is little controversy with regard to the mechanism of generation of the charge observed on separating dissimilar metals after contact. The conventional electron-band diagrams for metals of differing work functions are shown in Fig. 1a. On contact, the Fermi levels are equalized, electrons near the top of the energy distribution transferring from the metal of lower work function to the other. The resultant rise in potential in the latter (with respect to electrons) counters the electron transfer (Fig. 1b). The charges are distributed throughout the metals, of course, and so a contact potential difference appears between adjacent faces of metals of different work function that are electrically connected as shown in Fig. 1c. Incidentally, variation in the separation of such metals produces an alternating potential on a high-impedance detector, which may be nullified by an applied potential of appropriate polarity and requisite magnitude, thus affording a direct means of determining the contact potential difference.

Harper (1951) investigated charge transfer between metallic spheres both theoretically and experimentally. On separation, a redistribution of charge by tunneling was postulated, thus permitting definition of the retained charge Q in terms of the contact potential V_c and an effective capacitance C defined by the tunneling cutoff distance. Lowell (1975) extended the model for contacting spheres by relating the separation capacitance to the height of surface asperities, having shown that the contribution of "real" asperities to measured separation capacitance to be negligible. Figure 2 presents a plot of measured charge density against the product of measured capacitance and contact potential difference. Back-tunneling on separation seems conceptually difficult. The electrons that are in equilibrium during contact suddenly, on separation, are impelled to

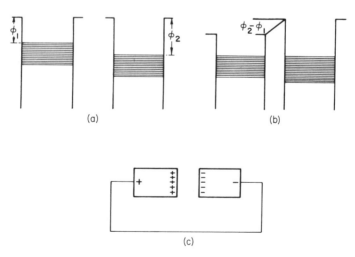

Fig. 1. Conventional band diagrams for metals of differing work functions, values ϕ_1 and ϕ_2 (a) before and (b) after contact with the consequent contact potential difference displayed by (c).

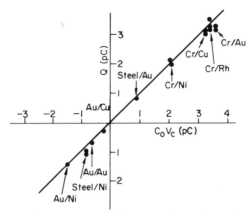

Fig. 2. Contact charge Q plotted against the respective products of measured intermetal capacitance C_0 and contact potential difference V_0 of metal pairs being displayed. [From Lowell (1975).]

return to the other metal across a very small gap by a mechanism shown by the experiment to be unaffected by speed of separation or force, hence area of contact. For these extremely small separations, the field is uniform even for quite irregular surfaces, and really sharp asperities—which might cause local field enhancement—would certainly be flattened anyway. These models evidently need clearer exposition.

B. Metal–Semiconductor Contact

Because of their practical importance, metal–semiconductor interfaces have been thoroughly explored and described. Again, little ambiguity exists. On contact, electrons are transferred across the interface, equilibrating the Fermi levels. The work function of the semiconductor is defined in precisely the same way as that of a metal, being the minimum energy required to remove an electron to a position, at rest, outside the surface. However, an additional parameter is evident in the case of a semiconductor, i.e., the electron affinity χ. This is the difference in energy between an electron that is free to move within the semiconductor, that is, in the conduction band, and one outside the surface. The effect of the electron affinity is to reduce the barrier height for electron transfer. The conventional band diagram before and after contact between a metal and a semiconductor of lower work function is shown in Figs. 3a and 3b, respectively.

Another feature of semiconductors known to play an important role in interfacial phenomena is the surface state that arises owing to interruption

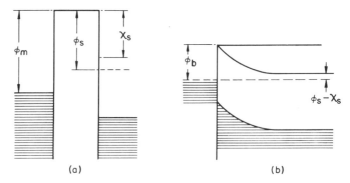

Fig. 3. Conventional diagram showing band bending in metal–semiconductor contact before (a) and after (b) contact, ϕ_m, ϕ_s, and χ being the metal and semiconductor work functions and the semiconductor electron affinity, respectively.

of the periodicity of the lattice. These states may have a range of energies that overlap both valence and conduction bands, although those in the bandgap contribute to contact effects. Surface states are sometimes viewed from the aspect of chemical bonds (Rhoderick, 1978). Each atom at the surface of material with covalent bonds has an unpaired electron in a localized orbital directed away from the surface. This "dangling bond" can either donate or accept an electron; thus there should be twice as many surface states as there are surface atoms, charge neutrality obtaining when they are half-occupied.

Bardeen (1947) originally suggested that the lack of significance of the electrode work function in metal–semiconductor interfacial equilibria resulted from the "pinning" of the barrier by a high density of surface states. A thin insulating layer was postulated to exist between the metal and semiconductor, a charge (density q_m) at the metal surface being balanced by a charge (density q_{ss}) in the surface states and a charge (density q_s) in the semiconductor. Equilibrium results largely from changes in q_m and q_{ss} leaving q_s—and so the barrier height—largely unaffected. This may alternatively be viewed as a screening of the semiconductor from interfacial fields by the surface states.

C. Metal–Insulator Contact

There is considerable current discussion as to the validity of band concepts to polymers. Band theories are evidently valid for wide-bandgap amorphous materials, and numerous theoretical investigations of the valence-band structure of polymers (Wood *et al.*, 1972) provide models consistent with observed electron spectroscopy for chemical analysis

(ESCA) data. Experimental investigations of photoemission from polymers (Less and Wilson, 1973; Davies, 1976) yield photothreshold and bandgap data consistent with the theoretical models. The energy-band structures and interfacial characteristics of insulators are essentially those of wide-bandgap semiconductors. Mott and Gurney (1940) developed a model based on equating diffusion-controlled injection and space-charge-driven injection currents, i.e., $j = 0 = ne\mu E - D(dn/dx)$, where n, μ, D, and e are the carrier density, carrier mobility, diffusion coefficient, and charge, respectively, E being the local electric field. By using the Einstein relationship between mobility and diffusion, replacing the field by the potential gradient dV/dx, and integrating, an expression for the carrier density in terms of local potential is obtained. Substitution into the Poisson equation followed by integration, subject to the boundary condition that the electric field vanishes at infinity, yields an expression for the variation in carrier density with depth x, i.e., $n/n_0 = [x_0/(x_0 + x)]^2$, n_0 being the density at the surface.

With the density of states given by $V^{1/2}$, and using the Fermi distribution function, the transferred charge density σ is given by an expression of the form $\sigma = 10^{19} \exp[(\phi_m - \chi)/kT]$. Van Ostenburg and Montgomery (1958) developed a theory, basing the argument on thermal excitation into the insulator conduction band, and found a similar expression for the charge density. Harper (1967) concluded that the surface barriers were too high for significant transfer from metals to insulator conduction bands and suggested that only transfer to defect states in the bandgap would be feasible.

Many *et al.* (1964) among others postulated a continuous distribution of trap levels in the bandgap, D_v being the trap density per unit energy per unit volume. Hence, if V is the local potential and the local charge density $\rho = -e^2 D_v V$, substitution into and integration of Poisson's equation give $V = V_0 e^{-x/L}$, where $L = (\varepsilon\varepsilon_0/e^2 D_v)^{1/2}$. Here V_0, the potential at the interface, is equivalent to the contact potential difference $(\phi_m - \phi_i)/e$, where ϕ_i is the insulator work function. The surface charge density σ is given by $\sigma = \int_0^\infty \rho\, dx = -e^2 D_v V_0 \int_0^\infty e^{-x/L}\, dx$, hence $\sigma = (eD_v)^{1/2} (\phi_m - \phi_i)$.

This expression evidently indicates a linear dependence of equilibrium contact charge density on metal work function.

Davies (1967a), following Dekker (1958), assumed a constant volume carrier density ρ, integration of the Poisson equation $dE/dx = \rho/\varepsilon\varepsilon_0$ with the boundary conditions of zero field (E_0) at the injecting electrode, and a potential at a maximum charge penetration depth λ equal to the polymer work function–electron affinity difference $\phi_p - \chi$, yielding the expression $\sigma = (2\varepsilon\varepsilon_0/e\lambda)(\phi_m - \phi_i)$ for the charge density.

Chowdry and Westgate (1974) pointed out that complete integration of the Poisson equation gave the expression $\sigma = (2\varepsilon\varepsilon_0/e)\rho(\phi_m - \phi_i)^{1/2}$ and not the earlier linear dependence. They re-presented the distributed trap model (Many et al., 1964) and derived a decidedly nonlinear, complex expression for the charge density. Garton (1974), using essentially thermodynamic arguments and the classical conduction equations, also derived a model exhibiting a nonlinear dependence of charge on work function.

Bauser et al. (1971) and Krupp (1971) have postulated that initial charge transfer involves only surface states. If the surface state density per unit area and energy is D_s, the charge density σ is given by $\sigma = -eD_s \Delta\phi$, where $\Delta\phi$ is the change in local Fermi energy following injection. Equilibrium is given by $\phi_m + eU = \phi_p - \Delta\phi$, the local Fermi level of the insulator coinciding with that of the metal. Since U is the potential difference across the separating surface layer, thickness d (see earlier description of surface states), as shown in Fig. 4, we have $U = -\sigma d/\varepsilon_0$.

Eliminating U and $\Delta\phi$ from these equations gives

$$\sigma = \frac{eD_s}{1 + (e^2dD_s/\varepsilon_0)} (\phi_m - \phi_p),$$

which again relates the surface charge density linearly to the metal work function.

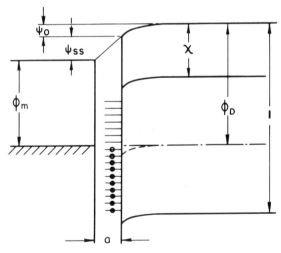

Fig. 4. Energy diagram showing the influence of surface states on metal–insulator contact. Here ϕ_m is the metal work function and ϕ_D, χ, and I are the insulator work function, electron affinity, and ionization energy, respectively. Solid circles are filled surface states, the open circles being states filled on contact causing the surface potential difference ψ_{ss} across layer thickness a.

In summary, therefore, a linear equilibrium correlation is provided by models based on a fixed injection depth and distributed traps within the bandgap or surface states, these of course being indistinguishable experimentally.

D. The Mechanism of Charge Transfer

It is evident that thermionic excitation of electrons from the Fermi level in a metal into the conduction band of an insulator over a potential barrier of 1 eV is an exceedingly slow process. Richardson's equation for current, $j = AT^2 \exp[-(\phi - \chi)/kT]$ gives, for a value of $\phi - \chi$ of 1 eV, a current of about 10^{-7} A m^{-2}. Hence a charge density of 10^{-3} cm^{-2} would require 10^4 sec—which would certainly not be the case for such a low barrier height. Direct transfer into accessible states in the bandgap of the insulator must occur.

Because of the development of field effect semiconductor devices, there is considerable interest in the electrical aspects of metal–insulator structures. The models developed largely restrict the insulator to a passive role, equilibrium being defined by the electron properties of the conductive metal, or semiconductor, electrodes (Walden, 1972). The interpretation of the familiar transient current observed on applying a step voltage to a metal–insulator–metal sandwich is also contentiously described either by "anomalous" polarization effects or space-charge-limited injection currents. Again, modeling concepts are restricted (Wintle, 1975).

The enhanced electron transfer current following field-induced barrier lowering (Schottky effect), tunneling (Fowler and Nordheim), process or preferentially directed motion between localized sites (Poole–Frenkel effect) may all be described by a generic expression of the form $J = a(E) \exp[f(E)]$, where $a(E)$ and $f(E)$ are both relatively slowly varying functions of the interface field E.

If the transferred charge density σ is a function of the field given by $\sigma = A[E(t) - E(0)]$, then $dE/dt = -A^{-1}J(E)$. Following Ferris-Prabhu (1973) and assuming that the functions $a(E)$ and $f(E)$ vary very slowly with respect to the exponential term integration of dE/dt, having used the equation for the current, gives

$$t = A \exp(-\{E(t) - f[E(0)]\} - 1)J[E(0)] \, df(E)/dE.$$

Expansion and manipulation give $\sigma = A[df(E)/dt]_{E_0} \ln(1 + t/t_0)$, hence logarithmic dependence of charge density on time. This, incidentally, also describes the commonly observed inverse time form of the injection current transient (Wintle, 1975).

The Schottky and Fowler and Nordheim equations are, in fact, explicitly integrable in terms of a space-charge-reduced field and yield the above form of expression. These analyses clearly require further development, and certainly the concepts of electron transfer *from* the insulator seem very nebulous.

The quality of theoretical descriptions of interfacial electrification lags seriously behind that of recent experimental data as a result of the rapid development of precise techniques usable in controlled environments.

III. CHARGE MEASUREMENT

An electrode placed near a discrete charge acquires an induced charge and potential dependent on its geometry and location, not only with respect to the initiating charge but also in relation to other local metal surfaces. The electrode also experiences an attractive force toward the originating charge. All these effects present means of determining the primary charge magnitude, the precision of the measurement being predominantly governed by the accuracy of definition of these local geometric factors.

The most direct method is to enclose totally the charge to be measured within the sensing electrode and determine its potential—the Faraday pail. The most complex, but often ambiguous (and commonly used), technique is to sense the electrostatic field by induction electrodes—field mills. Both these methods are of particular relevance to the charge on polymers and are described in detail.

A. Faraday Pail

The apparatus, shown schematically in Fig. 5, comprises two metal containers, one being totally enclosed by—but well insulated from—the other. The outer case is usually earthed. An object with total charge q induces the same charge on the screened container, without necessarily being in contact, and generates a potential V given by $V = q/C$, where C is the combined capacitance of the inner container and the potential measuring system to ground. Clearly, predetermination of the capacitance permits direct determination of the charge q by measurement of the potential by means of an electrometer, electrostatic voltmeter, or other high-input impedance detector.

A practical apparatus would incorporate a switchable range of capacitance, thus providing variable charge sensitivities.

The Faraday pail, while appearing somewhat cumbersome, in fact provides an extremely practical method of measuring charge on particulates.

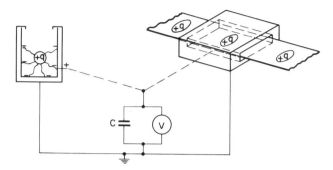

Fig. 5. Schematic diagram showing alternative versions of the Faraday pail, the charge q on the specimen generating a potential V (measured by a high-impedance voltmeter) on the system capacitance C.

A version, about $\frac{1}{2}$ m high and $\frac{1}{2}$ m in diameter, designed to replace the kegs conventionally employed has been used by the author to survey charge generation at a pharmaceutical processing plant. An interesting aspect of this was that the measurements were conducted in a potentially flammable atmosphere, the problem being that the Faraday pail presented an ignition hazard when charged. Dissipation of the charge was safely achieved by means of a sealed-reed relay activated by a bar magnet.

The ideal system, where all field lines from the charged object end on the inner container, can be approximated by using a long, narrow shape, thus obviating the necessity for a lid. This form of device may be used in a dynamic mode, as shown by the variant given in Fig. 5, a continuous input of charged material producing a measurable current rather than a fixed potential. Such systems have been employed to examine the charging of powder (Van Turnhout, 1971) or extruded polymers (Taylor and Lewis, 1974), for example.

B. The Pondermotive Effect

A uniform charge, density σ, generates a field E given by Gauss's equation $E = \sigma/\varepsilon\varepsilon_0$, where ε and ε_0 are the relative permittivity of the medium and the permittivity of free space, respectively. An isolated plane electrode of area A acquires an induced charge density σ and experiences an attractive force $\varepsilon_0 A E^2/2$. The field intensity, hence the charge density, may be determined either by measuring the electrode potential (knowing its capacity to ground) or by measuring the force.

Devices based on the attracted disk principle are rarely employed to determine charge (Van Turnhout, 1971). Essentially, the attracted disk forms an independent central section of a large uniform field electrode. On exposure to a second electrode at a high potential (or a charged surface), the disk is deflected. An equal restoring force is applied electromagnetically, the energy required being measured.

C. The Electrometer Probe

This probe represents a practicable realization of the classical physics concept of measurement of the potential induced on an isolated electrode on the approach of a charge from infinity. The probe comprises an open-ended coaxial capacitor, the outer sheath forming an earthed screen. The potential pulse induced on the center electrode may be readily amplified by means of a FET device, hence measured by a display on an oscilloscope. Clearly, these probes may operate only in a dynamic mode, the charged dielectric being passed rapidly beneath the probe (or vice versa) (Davies, 1967b; Foord, 1969) or an earthed shutter being interposed in between. Both the probe and charged surface in the latter case may, of course, be stationary.

Precise investigations on polymer films have been almost exclusively based on probe techniques. This is because the local geometries, mentioned earlier, can be arranged to be entirely determinable, permitting absolute, accurate, and wholly unambiguous measurement of surface charge density. The charged dielectric is mounted on an earthed plate, thus defining specimen capacitance.

A schematic diagram of a probe system, together with the equivalent circuit, is presented in Fig. 6 (Davies, 1967b). A uniform surface charge, density σ, produces a potential V_s in the presence of the probe. The probe, of effective area A and input capacity C, itself acts as a capacitive divider

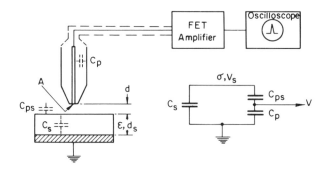

Fig. 6. Diagram showing the principle of electrometer probes. The surface charge, density σ, on the specimen capacitance C_s ($= \varepsilon \varepsilon_0 / ds$) generates a potential V_s which in turn produces a potential V on the probe capacitance C_p via the probe-to-surface capacitance C_{ps}.

and, for a small probe-to-surface separation d, where a uniform field may be assumed, the displayed pulse amplitude V is given by $V = (Ag\sigma/C)(1 + \varepsilon d/d_s)^{-1}$, where g is the amplifier gain and ε and d_s are the specimen permittivity and thickness, respectively. More elaborate analyses that include the contribution of volume distributed charges have been published. However, since generally the precise charge distribution is not known, the assumptions necessary to obtain explicit expressions return to determining an equivalent surface charge.

For a geometrically precise system, absolute charge determination is evidently possible. The unknown quantity in the above equation, having measured the probe input capacitance, is the probe effective area. The capacitance of overlapping parallel electrodes of differing area is defined by that of the smaller. Thus, for a uniformly charged surface of area greater than the geometric area of the probe, the latter defines the probe effective area. However, if the area of charge is smaller than that of the probe, then the effective area is that of the charged surface (Davies, 1973). Usually the charged area is much greater than that of the probe and there is no ambiguity. However, care must be exercised in interpreting the measurements for rapidly varying surface charge distributions. A modified expression for the pulse amplitude has been obtained, assuming a sinusoidal bipolar charge distribution (Murasaki *et al.*, 1970; Elsdon and Mitchell, 1976), but the interpretation of the problem is not eased, particularly since the observed charge distributions rarely conform to the assumed ideal. Calibration experiments, involving examination of the variation of the probe sensitivity with probe–surface distance, are preferable.

Probe systems have proved extremely versatile and have been used to examine both line and area (Foord, 1969; Hughes and Secker, 1971) distributions of surface charge, and novel designs continue to be described (Nordhage and Bäckström, 1976), as shown in Fig. 7. Mapping large areas is achieved by raster scanning, and a recently described experiment (Baum and Lewis, 1975) employed a digital transient recorder. The integrated charge on large areas, however, is more conveniently determined by field mills.

D. Field Mills

Field mills are by far the commonest devices employed for determining electrostatic potential gradients, particularly in practical industrial conditions. The principle of operation of a planar field mill, which is in effect that of a large-scale shuttered probe, is shown in Fig. 8a. Segmented sensor plates are periodically exposed to the field being sampled by a rotating, earthed, segmented shutter. The alternating potential induced on

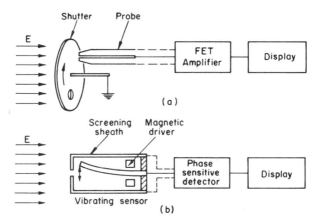

Fig. 7. Variants on shuttered probe systems as described (a) by Davies (1969a) and (b) Nordhage and Bäckström (1976).

the plates is amplified, and after rectification its amplitude is measured. This simple device would be insensitive to the field director (polarity), but in recent versions, in which reference signals are generated in synchronism with the rotating shutter, phase-sensitive detection provides a polarity display.

It has been shown (Mapleson and Whitlock, 1955) that, if the impedance is predominantly capacitive, so that the input time constant CR and shutter rotation speed W satisfy the condition $(WCR)^2 \gg 1$, then the amplitude of the signal V generated on sensor plates of area A is given by $V =$

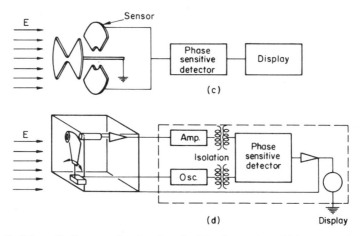

Fig. 8. Schematic diagrams showing the principle of operation of (a) a conventional field mill and (b) a null detecting voltage feedback probe.

$\varepsilon_0 AE/2C$, independently of the shutter speed. The signal is linearly dependent on the field, and sensitivities in the range $10^2 - 10^6$ V m^{-1} have been quoted by Secker (1975). Caution must be exercised in interpreting field mill readings, since the geometry of the field mill in relation to other local metal objects will certainly influence the data. Ideally, the complete capacitive environment of the detector should be defined for accurate and unambiguous measurement but, as described earlier, these conditions usually result in the use of a probe system.

As well as the axial version (Secker, 1975), radial designs (Van de Weerd, 1971; Waters, 1972) have also been described recently, an earlier version (Schuringa and Luttik, 1960) of the latter having been originally designed for use in flowing liquids.

Figure 8b shows a commercial shuttered probe (or field mill) that is distinctive in that the sensed field is nullified by driving the probe head (screen) to the "viewed" potential. This has obvious advantages in incendive atmospheres, since the high potential gradients produced by simply placing a probe adjacent to a charged surface are obviated.

E. Powder Development

Charge images on polymer surfaces may be rendered visible by dusting with opaque powder, as is commonly exhibited by electrographic processing. It is possible to distinguish the distribution of charges of both polarity by the use of appropriately charged powders of different colors. Powder development has achieved a high level of sophistication, including the use of ferrite-covered carrier beads to form an effective magnetic brush, but the technique is, however, necessarily qualitative.

IV. CONTACT CHARGE-TRANSFER EXPERIMENTS

Making and breaking contact is a mechanically complex process involving a large number of uncertainties. Some degree of rubbing or relative motion is probably unavoidable, if only because of the distortion of one or both of the surfaces on a microscopic scale. The area of real contact is invariably indeterminate. In addition to these problems, in the case of surface-charging experiments, complications arise through the effects of the ambient atmosphere, not only owing to modification of the surface characteristics of contamination-sensitive materials but also because electrical discharge between separating surfaces at different electrical potentials is permitted. Discharges (discussed later in detail) have probably been the greatest cause of confusion in electrification experiments conducted in standard atmospheres, since they can not only reduce apparent equilibrium charge levels but may also even cause charge reversal, locally, on nonuniformly charged surfaces. Many early investigations failed to find a correlation between the transferred charge density and recognized surface parameters due these uncertainties. A major improvement in the consistency of experimental data followed from experimentation in controlled environments, in particular, vacuums, and in reproducibility by the use of concurrent surface characterization.

A. Quasi-Static Contact

Arridge (1967), Davies (1967a), and Inculet and Wituschek (1967) established experimentally a correlation between the transferred charge and metal work function for contact between metals and insulators, including both ceramics and synthetic polymers, virtually concurrently despite using radically different techniques. Arridge measured the charge produced on a nylon-66 monofilament in air, using a Faraday cylinder as described earlier, after contact with a number of metals mounted on a rotating wheel under essentially atmospheric conditions. Davies, on the other hand, lowered small metal electrodes into contact with disk samples *in vacuo* and determined the charge density using an electrometer probe. Inculet *et al.* observed the charge injected in sliding contacts between metal styli and plane glass specimens, also *in vacuo*. In all cases the metal work functions were measured by comparison with a reference metal *in situ* and, despite considerable scatter in the data, sensibly linear correlations could be claimed.

Davies (1969a), using an apparatus designed to examine charge transferred on rolling contact between metals and polymer films, confirmed the correlation for a number of common synthetics. In this experiment, shown schematically in Fig. 9, six polymer film specimens were mounted on the periphery of a large earthed metal drum. A small contact wheel comprising five metals of differing work functions, namely, aluminum, gold, zirconium, platinum, and cadmium, was held against the films under light pressure. Rotation of the drum brought all the metals in turn into contact with each of the polymer films sequentially. A probe–shutter system downstream of the contact wheel continuously monitored the injected charge density. Provision was also made to measure continuously the contact metal work functions against a gold reference electrode. The experiment, performed *in vacuo,* was therefore wholly self-contained.

The linear relationship, irrespective of theoretical interpretation, is described by the empirical expression $\sigma = A(\phi_m - \phi_d)$, where σ is the charge density and ϕ_m and ϕ_d are the metal and dielectric work functions, respectively. The work function values obtained for the synthetic poly-

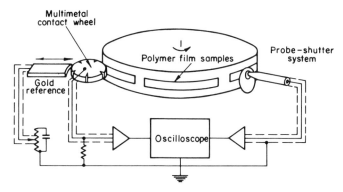

Fig. 9. Diagram of a rolling contact experiment showing both charge and metal contact potential measuring systems. [From Davies (1969a).]

mers examined in the latter experiment, ascribing a value of 4.6 eV for that of the gold reference, are given in Table II.

An excellent example of the linear dependence of charge on metal contact potential difference found experimentally is shown by the data obtained by Cunningham and Hood (1970) presented in Fig. 10. This figure shows the charge transferred to a nickel plate when contacted successively for 2 sec by fresh samples of x-ray film (polyester?) conditioned at 20% RH at 75°F. The surface potential of the plate changed naturally after cleaning but could also be radically altered by contamination with a surfactant (cetyldimethylammonium bromide). The points are numbered in order of measurement, the first six being obtained with a new, cleaned plate allowed to stand between contacts. Point 7 shows the effect of surfactant contamination, while points 8–10 were obtained after successive

TABLE II   *The Work Functions of Dielectric Materials*[a]

Material	Work function (eV)
PVC	4.85 ± 0.2
Polyimide	4.36 ± 0.06
Polycarbonate	4.26 ± 0.13
PTFE	4.26 ± 0.05
Mylar (PET)	4.25 ± 0.10
Polystyrene	4.22 ± 0.07
Nylon 66	4.08 ± 0.06

[a] From Davies (1969a).

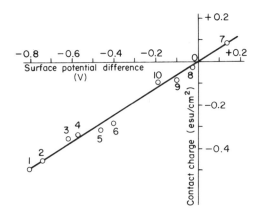

Fig. 10. Charge transferred to a nickel surface when contacted by samples of x-ray film plotted against the respective contact potential difference of the nickel with respect to a reference electrode. The numbers indicate the measurement sequence. [From Cunningham and Hood (1970).]

rinses with distilled water. Similar results were obtained using a chromium plate.

The conclusion that contact charges arise owing to electron transfer seems valid; however, a great number of uncertainties remain. Davies (1973), examining charge transfer between spherical metal electrode and plane specimens of poly(ethylene terephthalate), showed that, while the contact area increased with force according to the Hertz equation for elastic deformation, the charge *density* was independent of the contact pressure. The charge density increased with contact duration, although whether this resulted from changes in microscopic contact area or was due to temporal changes in the injection current could not be distinguished.

Increasing charge transfer is generally observed with repeated short-duration contacts. Lowell (1976), also examining sphere–plane contact *in vacuo,* observed that the charge density was independent of metal work function but not inconsistent in polarity with earlier contact potential difference models, all the polymers usually charging negatively with respect to most metals. A model based on electron tunneling into a surface layer followed by mechanical transfer of the trapped electron into the polymer by molecular motion was proposed, but this did not describe the observed charge saturation. This, it was suggested, resulted eventually because of the electrostatic effect of the injected charge. It may be argued, of course, that if the existing charge exerts a retarding effect on subsequent electron transfer, then an additional mechanism for describing temporal phenomena is unnecessary.

Spherical and conical electrodes were used by Lowell (1977) in an attempt to establish the importance of surface states, the principle being that the latter penetrated the surface, thereby contacting fresh bulk material, and so if the observed charge transfer differed by an amount larger than the predictable area difference, then the surface was evidently important. Only the data obtained for polyethylene were unambiguous, the charge density being significantly different and also highly influenced by prior environmental conditioning of the material. This confirmed the earlier results of Hays (1974), displayed in Fig. 11, who showed a radical increase in the charging of polyethylene on contact with mercury following oxidation of the polymer by exposure to either air or ozone. The density of vinyl groups in the samples of unsaturated polyethylene used was radically reduced on oxidation, as displayed by infrared spectroscopy. The lack of vinyl groups in high-density linear polyethylene resulted in little change in the contact charge characteristics of this material on exposure to ozone. Incidentally, terminal vinyl groups have been shown to act as traps exerting a significant influence on bulk carrier transport in polyethylene (Davies and Lock, 1973; Creswell *et al.*, 1972), as is described in Chapter 1.

Highly deformable silicon rubber has been used as an experimental material in an attempt to obtain intimacy over large areas in a single contact (Cottrell, 1978). Sphere–plane (in both configurations) and plane–plane structures were employed, the charge density transferred to a smooth rubber sphere on contact with steel plates being found to be smaller with increasing surface roughness of the latter. The negative charge was, however, independent of metal work function, contact

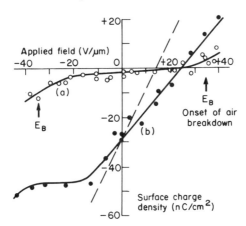

Fig. 11. The surface charge density produced on polyethylene on contact with mercury under applied fields plotted against the respective field, showing the influence of exposure to room air. (a) Data for a freshly prepared film, (b) data for exposed material.

pressure, and duration of contact, with the exception, in the latter case, of the plane–plane configuration, this being attributed (without specific reason) to rubbing at the edges of the plane electrode. A correlation between temporal growth of charge and increase in contact area has also been observed by Homewood and Rose-Innes (1979), although Coste and Pechery (1981) found this to be nonlinear.

Fabish and Duke (1977) interpreted the data obtained for a series of homo- and copolymers of styrene and methacrylate in sequential contact with a number of different metals by intrinsic molecular ion states. However, it could be argued that few of the experimental results exhibit real equilibrium. There is certainly a lack of consistency in multiple short-contact data.

B.   Rubbing Contact

Rubbing was long thought to be a prerequisite for charge transfer; however, it is now considered that it merely modifies the contact phenomena. One undoubted effect is enhancement of the area of actual contact, but the charging observed (Henry, 1953) on rubbing apparently identical surfaces indicates additional effects. The classical experiment, mentioned earlier, in which a rod was "bowed" at a single point by a second rod of the same material, showed the significance of the rubbing asymmetry. The effect was attributed to the temperature rise at the rubbed spot, since reversal of the roles of the two rods reversed the polarity of the transferred charge. A model based on ion transfer caused by differential ionic mobility in the heated layer was proposed. Disruption of an ionic (contaminating) layer compensating for an intrinsic charge in the material has been suggested by Kornfeld (1969). However, in the former the motive for preferred migration out from the heated layer rather than into the bulk was not given, while in the latter some asymmetry—either in thickness or charge density—is necessary to account for the net charge transfer on mixing the layers. Given any form of asymmetry, electron migration is equally possible.

Several investigations have been described (Elsdon and Mitchell, 1976; Lowell, 1976, Wählin and Bäckström, 1974; Nordhage and Bäckström, 1975; O'Neill and Foord, 1975) involving the charges injected into polymers on scribing the surface with metal styli. A typical experiment is shown schematically in Fig. 12. The polymer film is mounted on an earthed plane which can be either rotated or moved linearly, the motion effectively drawing the stylus, held under a known load, across the surface. The stylus is grounded via an electrometer, so that the injected charge is determined from the stylus current. A probe system may also be

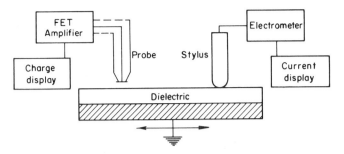

Fig. 12. Representation of the sliding contact experiment showing both the stylus current measurement and scanning probe techniques.

employed to scan the charged track. Data may be presented as charge per unit track length or, if the track width has been determined independently using a microscope, as charge density. A linear dependence of charge on stylus work function has been observed for polytetrafluoroethylene (PTFE) by Wählin and Bäckström (1974) and for polyethylene, for example, by Nordhage and Bäckström (1975), charge being largely independent of the scribing velocity. Contrarily, Elsdon and Mitchell (1976) and O'Neill and Foord (1975) both observed a dependence on velocity and an independence of work function. In all these investigations the observed dependence of charge on the square root of contact force is consistent with charge increasing linearly with contact area. Several analyses of the real area of contact at a spherical electrode–polymer interface have predicted a force dependence by combining both elastic deformation ($F^{0.66}$) and plastic flow ($F^{0.33}$).

Material transfer undoubtedly occurs at high-contact pressure (Briscoe *et al.*, 1975), and this has been advocated as a charging mechanism. It is not clear, however, why the shearing of a surface film should, of itself, continuously produce a net charge of one polarity.

C. Applied Field Enhancement

The charge injected on contact is augmented or diminished by applying an electric field across the interface in accordance with the field direction. This has been confirmed recently for polythene both contacted by a mercury electrode (Hays, 1974) and rubbed by styli of known work function (Nordhage and Bäckström, 1975). No temporal effects were observed, and instantaneous charging was then inferred. This, together with the observed sensitivity to surface modification—particularly by ozone— suggested trapping at surface states.

The latter effect must be considered carefully. The use of mercury as a contact electrode has been critically reviewed by Harper (1967), and its behavior, particularly when contamination or surface oxidation is possible, is far from simple.

In contrast to the apparent instantaneous equilibration observed by Hays (1974) under applied fields using a mercury electrode, charge growth over considerable periods has been observed by the author. With an apparatus designed specifically to minimize rubbing (Davies, 1973) the charge produced by a lowered spherical electrode has been examined *in vacuo* with potentials applied for differing intervals, the electrode being raised with the potential still applied. Uncertainty owing to temporal changes in the contact area was eliminated by maintaining short-circuit contact for at least 15 min before applying the field. Figure 13 shows oscilloscope traces displaying the line scan of charges produced on 10 poly(ethylene terephthalate) specimens for applied potentials of ± 1 kV, respectively, the potential being applied to each specimen, in sequence, for 0, 0.3, 1, 3, 10, 30, 100, 300, 1000, and 3000 sec. The zero field charge ($t = 0$) is negative, reversal for positive potential being evident after 0.3 sec, as is the continued increase over the whole period of the experiment. The increased charges produced by either polarity are plotted in Fig. 14 against the appropriate time interval on a log scale. The increase is approximately linear, but the data are not sufficiently accurate to distinguish among the possible transfer mechanisms.

A similar $\ln(t/t_0)$ dependence of injected space-charge density has been proposed by Hassmyr and Bäckström (1979) to account for the transient currents they observed on applying step voltages to polymer film capacitors.

Considerable charge is injected in a very short time, of course, more than adequate to produce discharges in ordinary atmospheres, and so under these conditions equilibrium appears instantaneous (Davies, 1981). There is evidently considerable scope for further critical experimentation correlating charge, current, and local interfacial distortions in dynamic contact investigations (vide Lowell and Rose-Innes, 1980).

V. PROBLEMS AND CONTROL OF STATIC CHARGES

Industrial problems involving static electrification range from process nuisances, such as adhesion of film to machinery and acquisition of contaminating particles, to serious explosion hazards resulting from charge-initiated incendive sparks.

Electrostatic forces are estimated to contribute only about 10% to the adhesion of materials in intimate contact, the balance being provided by

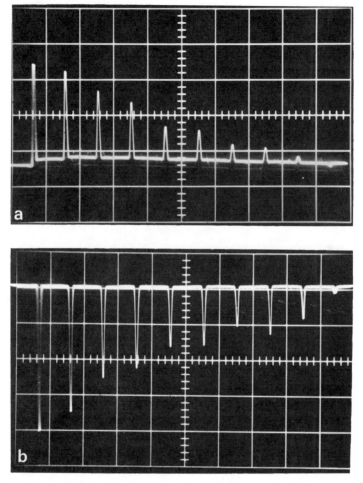

Fig. 13. Photographs of oscilloscope traces displaying the growth of charge on specimens of PET film after contact durations of 0, 0.3, 1, 3, 10, 30, 100, 300, 1000, and 3000 sec for sphere plane contact with +1 kV (a) and −1 kV (b) applied to the electrode.

van der Waals forces or changes in surface energy (Roberts, 1977). However, the significant feature of coulombic forces is that they act over very much larger distances. The force between two surface charge layers of density σ cm^{-2} (or one layer and its image) is $\sigma^2/2\varepsilon\varepsilon_0$, acting over the distance over which the field is uniform. To fix ideas, the adhesive forces calculated (Davies, 1969b) from the charge density measured on films of PTFE and nylon after contact with different metals are presented in Fig. 15 plotted against the respective metal work functions. The contact force

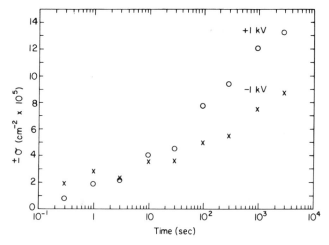

Fig. 14. Charge densities of both polarities calculated from the amplitudes of the pulses displayed in Fig. 13 plotted against the appropriate contact duration.

is approximately equal to atmospheric pressure which is certainly large enough to cause significant problems in film-processing plants.

Polymer films are frequently transported by roller systems. Charging in this circumstance is a dynamic process, the rate of charge acquisition being defined by the polymer–roller metal interface characteristics and the rate of dissipation governed by the carrier migration properties of the polymer. Both these characteristics may be influenced by the speed of

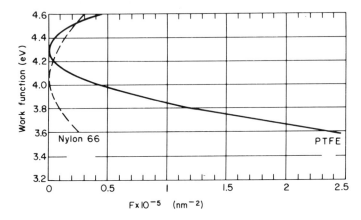

Fig. 15. Attractive forces calculated from charge densities measured for nylon 66 and PTFE in a metal–polymer rolling contact experiment plotted against the respective metal work function ($\phi_{Au} = 4.6$ eV).

film transport and also by ambient atmospheric conditions. The dynamics of charge equilibration on moving webs of material exhibiting significant conductivity is complex, and numerical analyses showing the charging propensity as a function of rolling speed have been described (Horvath and Berta, 1975). However, the simplifying assumption is made here that no charge loss occurs by conduction in the film. Charge control must, therefore, be achieved by the supply of appropriately charged species from the atmosphere. In the absence of conduction, the films are generally charged to the limit imposed by the breakdown strength of the ambient atmosphere.

Electrical discharges in the atmosphere clearly play an extremely important role in surface-charge phenomena, and a review of discharge mechanisms is therefore indicated.

A. Electrical Breakdown of the Atmosphere

1. Uniform Electric Field

Ionized species in the atmosphere will be influenced by the surface-charge field, the energy gained being a function both of field intensity E and the particle-free path length. For sufficiently high fields, an electron may gain sufficient energy to ionize the gas molecules by collision. If α is the number of ionizing collisions per unit path length in the field direction, then

$$\alpha = f(EL)/L,$$

where L is the electron mean-free-path length. Hence, if n electrons cross the unit area per second in traveling a distance dx, the number of new ions dn is

$$dn = n\alpha \, dx.$$

The energy required to ionize a molecule is eV_i, where V_i is an effective ionization potential. An electron can, therefore, ionize if the energy gain exE is greater than eV_i, i.e., if $x = V_i/E$. The probability that the free path length is greater than a given length x is given by

$$n_x = ne^{-x/L},$$

and so substituting for x gives

$$n_x = e^{-V_i/EL}.$$

The probable number of ionizing collisions per unit path length (α) is given by multiplying the average number of free paths per unit length

(L^{-1}) by the probability of the free path being of ionizing length; i.e.,

$$\alpha = L^{-1}e^{-V_i/EL}.$$

The mean free path length is inversely proportional to the gas pressure p, thus $L^{-1} = Bp$, where B is a constant, and so

$$\alpha/p = Be^{-Cp/E},$$

where $C = BV_i$.

Electrons may be produced in the gas by processes other than simple electron collision. Under appropriate conditions of gas mixture, ionization by ion collision or by molecular resonance may occur.

If n_c is the number of electrons per unit cross section at a given plane and n is the number distant d in the field direction, then the number of ions produced is $n - n_c$. Let γ electrons be produced per ion at the original plane as a result of ion collision ionization. If n_0 is the number of electrons at the initial plane prior to any gas ionization, then

$$n_c = n_0 + \gamma(n - n_c).$$

In traversing the distance d, the electron amplification is therefore

$$n = n_0 e^{\alpha d}/[1 - \gamma(e^{\alpha d} - 1)].$$

The current in a given gap for an initial electron current j_0 is thus

$$j = j_0 e^{\alpha d}/[1 - \gamma(e^{\alpha d} - 1)].$$

The transition from this equilibrium current to the catastrophic spark clearly occurs when the denominator of Eq. (2) becomes zero; i.e.,

$$1 - \gamma(e^{\alpha d} - 1) = 0.$$

Since $e^{\alpha d} \gg 1$, this is equivalent to $1/\gamma = e^{\alpha d}$, and so substituting gives

$$V_s = Cpd/\ln[Bpd/\ln(1/\gamma)].$$

Thus the sparking potential is a function of pd, the product of gas pressure and gap distance alone. This is an established experimental fact known as Paschen's law.

It is interesting to compute the equivalent field from recent experimental breakdown data for air (Dakin *et al.*, 1974) and, hence, the respective limiting surface charge density σ_s for which sparking would occur in a uniform field under normal atmospheric pressure. The values of σ_s are plotted in Fig. 16 as a function of the respective gap distances. As may be seen, the sparking charge density decreases monotonically as the gap distance increases.

The stable charge density on an extended polymer film may, therefore,

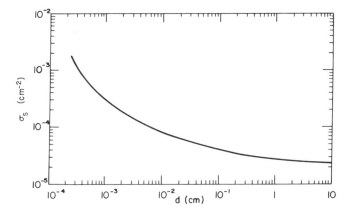

Fig. 16. The charge densities equivalent to the measured uniform sparking electric stress in air at NTP plotted against the electrode separation.

be several orders higher than the frequently accepted limit of about 2×10^{-5} cm^{-2} for normal atmospheres. This is of practical importance in, for example, the utilization of electret transducers where the film–electrode spacing may be very small.

The concepts described apply essentially to uniform electric fields. Under conditions of field nonuniformity such as plane edges or sharply pointed electrodes, the discharge phenomena may be radically different. When the separation between the sharply curved surfaces is greater than the sparking distance, corona discharges may occur in the enhanced electric stress regions adjacent to the electrodes.

2. Corona Discharges

This discharge in air takes the form of a glow appearing on a positive wire as a bluish-white sheath and on a negative wire concentrated in localized reddish tufts. A positive corona occurs at higher voltages than a negative corona in H_2, CO_2, and N_2, while the reverse is true for He, O_2, and air at normal atmospheric pressure.

The minimum field leading to breakdown in normal air is approximately 30 kV/cm. If E_0 is the field at the surface of a wire of radius a, the field E, at any radius r, is given by $E/E_0 = a/r$, hence ionization ceases at radius r, given by $r_1 = E_0 a/30$.

The average field in the ionized sheath is $E_{av} = (E_0 + 30)/2$, but the breakdown stress E_s in a uniform field gap is given empirically by $E_s = 30 + 1.35/d$.

Thus, equating the average field E_{av} to the breakdown stress E_s, and the gap distance d to the sheath width $r_1 - a$ gives $E_0 = 30 + 9/\sqrt{a}$ kV cm^{-1}, hence $r_1 = a + 0.3\sqrt{a}$ is approximately the distance over which the electric stress must exceed the electric strength of air if a corona is to be maintained.

An expression of the form $i = DV(V - V_c)$ may readily be derived to describe the corona current i in terms of the potential difference V between a wire and a concentric cylinder, V_c being the corona onset potential and D a constant. The current from a point to a plane is also described by this expression. For a point of radius a and a plate, the positive discharge starts at the critical stress given by $E_c = 18/\sqrt{a}$ kV cm^{-1}, under normal atmospheric conditions. As intimated earlier, coronas from both points and wires are employed as a means of reducing unwanted static charges.

B. Discharge Eliminators

The normal atmosphere exhibits a conductivity of about 10^{-14} Ω^{-1} m^{-1} (Weast, 1973), owing to natural radioactivity. Thus, a surface charge half-life of about 6 months can be expected for an average polymer film under normal circumstances in an open environment. Clearly, this is inadequate, and a means of augmenting the natural ionization must be obtained. This can be achieved by the corona discharges described earlier or by synthetic nucleonic devices.

It was implicit in the description of corona discharges that the driving potentials were applied to the sharp wire, or point electrode. However, the same discharge conditions evidently exist if the same potential of opposite polarity is applied to the counterelectrode, the sharp electrode being earthed. This principle is employed in so-called passive or inductive eliminators. It is also evident that even further enhanced ionization may be obtained by applying high potentials to the eliminator electrode.

1. Passive Devices

A thorough investigation of these devices has been described by Lövstrand (1975). Measurements were made on a 0.6-m-wide paper band running continuously around four rollers, as shown schematically in Fig. 17. A controlled humidity atmosphere (40% RH) was provided, thus yielding a paper surface resistivity of 6×10^{11} Ω per square. The magnitude of the surface charge, sprayed on the paper by means of a high-voltage corona source, was monitored both upstream and downstream of the eliminator wire. Experimental data are presented in Fig. 18, which show the residual charge density plotted as a function of the original charge density for different wire radii. The onset levels are clearly displayed, and the greater efficiency at higher initial charge levels is also evident.

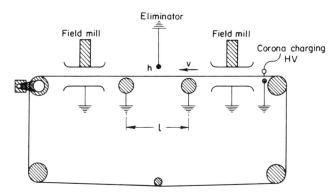

Fig. 17. Schematic diagram of an apparatus for examining the efficiency of passive static eliminators, showing the web driven at velocity v over four rollers and disposition of the charging electrode, field mills, screening rollers, and eliminator set at height h above the web. [From Lövstrand (1975).]

The influence of web velocity has also been examined and comparisons made among the differing eliminator structures—wire, tinsel, and rows of fine points. It was concluded that passive eliminators were not the most efficient, being greatly influenced by adjacent metal surfaces, for instance, but a large reduction in charge density can be achieved particularly for high charge levels.

2. Powered Eliminators

The mechanism of powered eliminators, that is, discharge electrodes to which high potentials are applied, is evident from the earlier account of

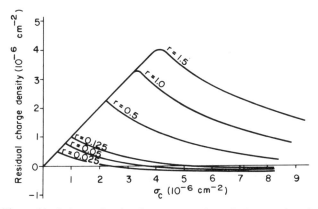

Fig. 18. The residual charge density downstream of an eliminator wire of given radius plotted against the initial positive charge density as measured by Lövstrand (1975).

corona discharges. A form of these devices—the Corotron (Dessauer and Clark, 1965)—has long been used as the means of controlled charging in xerographic machines. A common problem with powered systems is overcompensation; that is, more charge (hopefully of the opposite polarity) than that necessary to neutralize the film charge is provided by the device.

The final charge density downstream of the eliminator is dependent on the power level, charge relaxation rate (and so, cleanliness), speed of winding, and the disposition of extraneous metal surfaces, and so precise compensation is unlikely.

Blythe (1975) has described a novel design with which an attempt was made to control the output current by means of external fields. Essentially the device comprises a partially screened wire to which 4 kV rms at 50 Hz is applied. This structure is located above the charged film as it runs over an insulated metal roller as shown in Fig. 19. The corona regime within the screening cage readily provides a stable current of about 1 mA m^{-1} length of the wire.

Now the charge density on the film is unlikely to be greater than, say, 10^{-4} m^{-2} and so, for film speeds of about 1 cm sec^{-1}, neutralizing current levels of 100 μA m^{-1} width are required, which are an order of magnitude less than the quiescent current within the device. The current extracted from the source is controlled by applying an attractive potential to the insulated roller, thus again providing a stable environment for the device. A further refinement was the provision of a feedback control line between a downstream field mill and the extracting roller, the roller potential being of course that required to provide a minimum charge on the film at the field mill station. Figure 20 presents a plot of the experimentally observed

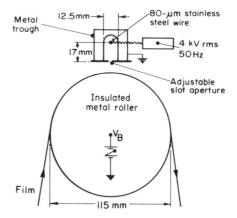

Fig. 19. Diagram of a corona charge control device. [From Blythe (1975).]

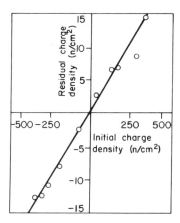

Fig. 20. Residual charge density as a function of initial charge density with automatic operation of a charge controller at a web speed of 10 cm sec⁻¹. [From Blythe (1975).]

residual charge plotted against the initial charge density, for a fixed winding speed, and exhibits a large reduction. This system, while interesting, has evident limitations, particularly with regard to the much higher speeds used in practical film transport systems. Speed is also a limitation for the final type of eliminator—the radioactive source.

3. Radioactive Sources

All radioactive emissions can ionize the atmosphere, but with a varying degree of efficiency depending on the type of radiation and its energy. High-energy, long-range radiation sources are unacceptable owing to health hazards. However, α-particles have a well-defined, limited range of about 4 cm under normal atmospheric conditions, and so devices emitting these particles may be safely employed in controlled industrial environments after proper location.

Commercially available eliminators use the nuclide polonium encapsulated in ceramic beads. Polonium emits particles of 5-MeV energy, and also a very small amount of γ rays, and has a half-life of about 138 days. The typical activity of these devices is about 0.8 mCi cm⁻¹ length. The ionization produced by an α-particle per unit track length in air increases slightly as the particle slows, the total number of ion pairs being about 2×10^5.

The number of particles emitted by a 0.8-mCi source is, by definition, 3×10^7 sec⁻¹. Thus the total number of ion pairs produced per centimeter length of the eliminator bar is about 6×10^{12} sec⁻¹. When these ions are singly ionized, the equivalent current is, therefore, about 1 μA. This, of course, is an absolute maximum, both geometric factors and the self-neutralization of the ion pairs reducing the available effective current.

These devices are often used, however, in conjunction with passive eliminators or in blown-air systems.

VI. POSTSCRIPT

There are many aspects of interfacial charging (which have not been discussed in the present review) such as the influence of temperature or contamination of surfaces on the dynamics and eventual equilibration of the charge-transfer processes. Equally numerous experimental investigations described in the literature have not been mentioned. This arises because of the inconclusive nature of both concepts and data. There is a suggestion that heating reduces the value of the polymer work function, giving rise to a tendency to charge positively. Also, the established reduction of the work function of metals by absorbed species is reflected in the influence of the environment, particularly water content, on the charging of polymers in normal atmospheres. However, these aspects, as well as the description of the charge-transfer dynamics remain to be unequivocally established. It is true that static electrification phenomena now have a sound physical basis and that many of their problematic aspects are well under control.

REFERENCES

Arridge, R. G. C. (1967). *Br. J. Appl. Phys.* **18**, 1311–1316.

Bardeen, J. (1947). *Phys. Rev.* **71**, 717–727.

Baum, E. A., and Lewis, T. J. (1975). "Static Electrification, 1975" (A. R. Blythe, ed.), p. 130. Institute Physics Conf. Ser. No. 27.

Bauser, H., Klöpffer, W., and Rabenhorst, H. (1971). "Advances in Static Electricity," Vol. I, pp. 2–9. Auxilia, Brussels.

Blythe, A. R. (1975). "Static Electrification 1975" (A. R. Blythe, ed.), pp. 238–245. Institute Physics Conf. Ser. No. 27.

Briscoe, B. J., Pooley, C. M., and Tabor, D. (1975). *In* "Advances in Polymer Friction and Wear" (L-H Lee, ed.), p. 191. Plenum Press, New York.

Chambers Encyclopaedia (1970). Vol. 5, p. 52. International Learning Systems Corp.

Chowdry, A., and Westgate, C. R. (1974). *J. Phys. D Appl. Phys.* **7**, 713–725.

Cottrell, G. A. (1978). *J. Phys. D Appl. Phys.* **11**, 681–687.

Coste, J., and Pechery, P. (1981). *J. Electrostat.* **10**, 129–136.

Creswell, R. A., Perlman, M. M., and Kabayama, M. A. (1972). *In* "Dielectric Properties of Polymers" (F. E. Karasz, ed.), pp. 295–312. Plenum Press, New York.

Cunningham, R. G., and Hood, H. P. (1970). *J. Colloid Interface Sci.* **32**, 373–376.

Dakin, T. W. *et al.* (1974). *Electra* (32), 61–83.

Davies, D. K. (1967a). "Static Electrification," pp. 29–36. Institute Physics Conf. Ser. No. 4.

Davies, D. K. (1967b). *J. Sci. Instrum.* **44**, 521–524.
Davies, D. K. (1969a). *Br. J. Appl. Phys. Ser. 2* **2**, 1533–1537.
Davies, D. K. (1969b). *In* "Physics of Adhesion" (*Proc. Conf Karlsrule*), p. 30. Farb. Bayer A. G. Leverkusen.
Davies, D. K. (1973). *J. Phys. D Appl. Phys.* **6**, 1017–1024.
Davies, D. K. (1976). *Nature (London)* **262**, 277.
Davies, D. K. (1981). *Proc.IEEE* **128**, 153–158.
Davies, D. K., and Lock, P. J. (1973). *J. Electrochem. Soc.* **129**, 266–270.
Dekker, A. J. (1958). "Solid State Physics," p. 418. Macmillan, New York.
Dessauer, J. H., and Clark, H. E. (1965). "Xerography and Related Processes." Focal Press, London.
Elsdon, R., and Mitchell, F. R. G. (1976). *J. Phys. D Appl. Phys.* **9**, 1445–1460.
Fabish, T. J., and Duke, C. B. (1977). *J. Appl. Phys.* **48**, 4256–4266.
Ferris-Prabhu, A. V. (1973). *Solid State Electron.* **16**, 1086–1088.
Foord, T. R. (1969). *J. Sci. Instrum. Ser. 2* **2**, 411–413.
Garton, C. G. (1974). *J. Phys. D Appl. Phys.* **7**, 1814–1823.
Harper, W. R. (1951). *Proc. R. Soc. London Ser. A* **205**, 83–103.
Harper, W. R. (1967). "Contact and Frictional Electrification." Oxford Univ. Press, London and New York.
Hassmyr, L., and Bäckström, G. (1979). "Electrostatics 1979" (J. Lowell, ed.), pp. 239–247. Institute Physics Conf. Ser. No. 48.
Hays, D. A. J. (1974). *J. Chem. Phys.* **61**, 1455–1462.
Henniker, J. (1962). *Nature (London)* **196**, 474.
Henry, P. S. H. (1953). *Br. J. Appl. Phys. Suppl. 2* S6-11.
Homewood, K. P., and Rose-Innes, A. C. (1979). "Electrostatics 1979" (J. Lowell, ed.), pp. 233–237. Institute Physics Conf. Ser. No. 48.
Horvath, T., and Berta, I. (1975). "Static Electrification 1975" (A. R. Blythe, ed.), pp. 256–263. Institute Physics Conf. Ser. No. 27.
Hughes, K. A., and Secker, P. E. (1971). *J. Phys. E* **4**, 362.
Inculet, I. I., and Wituschek, E. P. (1967). "Static Electrification," pp. 37–43. Institute Physics Conf. Ser. No. 4.
Kornfeld, M. I. (1969). *Sov. Phys.—Solid State* **11**, 1306–1310.
Krupp, H. (1971). "Static Electrification 1971" (D. K. Davies, ed.), pp. 1–11. Institute Physics Conf. Ser. No. 11.
Less, K. J., and Wilson, E. G. (1973). *J. Phys. C Solid State Phys.* **6**, 1310–1320.
Lövstrand, K. G. (1975). "Static Electrification 1975" (A. R. Blythe, ed.), pp. 246–255. Institute Physics Conf. No. 27.
Lowell, J. (1975). *J. Phys. D Appl. Phys.* **8**, 53–63.
Lowell, J. (1976). *J. Phys. D Appl. Phys.* **9**, 1571–1585.
Lowell, J. (1977). *J. Phys. D Appl. Phys.* **10**, 65–71.
Lowell, J., and Rose-Innes, A. C. (1980). *Adv. Phys.*, **29**, 947–1023.
Many, A., Goldstein, Y., and Grover, N. B. (1964). "Semiconductor Surfaces," Chapter 4. North-Holland Publ., Amsterdam.
Mapleson, W. W., and Whitlock, W. (1955). *J. Atmos. Terr. Phys.* **7**, 61.
Mott, N. F., and Gurney, R. W. (1940). "Electron Processes in Ionic Crystals," Chapter 5. Oxford Univ. Press (Clarendon), London and New York.
Murasaki, K., Kono, M., Matsiu, M., and Ohno, M. (1970). *Electr. Eng. Jpn.* **90**, 187–196.
Nordhage, F., and Bäckström, G. (1975). "Static Electrification 1975" (A. R. Blythe, ed.), pp. 84–94. Institute Physics Conf. Ser No. 27.

Nordhage, F., and Bäckström, G. (1976). *J. Electrostat.* **2**, 91.

O'Neill, B. C., and Foord, T. R. (1975). "Static Electrification 1975" (A. R. Blythe, ed.), pp. 104–114. Institute Physics Conf. Ser. No. 27.

Roberts, A. D. (1977). *J. Phys D Appl. Phys.* **10**, 1801–1819.

Rhoderick, E. H. (1978). "Metal Semiconductor Contacts," p. 20. Oxford University Press, London and New York.

Secker, P. E. (1975). *J. Electrostat.* **1**, 27–36.

Schuringa, A., and Luttick, C. (1960), *J. Sci. Instrum.* **37**, 332–335.

Taylor, D. M., and Lewis, T. J. (1974). Elektrostatische aufladung, *Dechema Monogr.* **72**, 125–136.

Van de Weerd, J. M. (1971). "Static Electrification 1971" (D. K. Davies, ed.), pp. 158–177. Institute Physics Conf. Ser No. 11.

Van Ostenberg, D. O., and Montgomery, D. J. (1958). *J. Textile Res.* **28**, 22–31.

Van Turnhout, J. (1971). *Adv. Static Electr.* **I**, 56–81.

Wahlin, A., and Bäckström, G. (1974). *J. Appl. Phys.* **45**, 2058–2064.

Walden, R. H. (1972). *J. Appl. Phys.* **43**, 1178.

Waters, R. T. (1972). *J. Phys E* **5**, 475–478.

Weast, R. C. (1973). "Handbook of Chemistry and Physics," 53rd ed., p. F175. The Chemical Rubber Co., Cleveland, Ohio.

Wintle, H. J. (1975). Solid-State Electron. **18**, 1039–1042.

Wood, M. H., Barber, M., Hillier, R. H., and Thomas, J. M. (1972). *J. Chem. Phys.* **56**, 1788–1789.

Chapter 8

Dielectric Breakdown Phenomena in Polymers

P. Fischer
SIEMENS AKTIENGESELLSCHAFT
RESEARCH AND DEVELOPMENT CENTER
ERLANGEN, WEST GERMANY

I. INTRODUCTION

The selection of a polymer or other insulating material for a specific application is determined by a number of requirements. The primary electrical function is to prevent current flow between conductors at different potentials. The insulation will furthermore have to perform additional functions and will hence be required to possess certain mechanical, thermal, and chemical properties. These properties must be retained for the anticipated lifetime of the system in which the insulation occurs, limited degradation down to a specific value being perhaps allowed.

The final stage of the electrical failure of a polymeric insulating layer is generally conceded to be thermal in nature. It appears realistic to consider breakdown as the culmination of a series of complex, interacting processes, any one of which may dominate under a given set of circumstances. Even under well-defined laboratory conditions the processes leading up to the high-current, finally destructive phase may be quite varied, and definitive identification and theoretical corroboration are rarely available in polymers. In commercial applications insulation failure is even more complex in that it may be preceded by very gradual chemical, physical, or mechanical degradation. These aging processes will depend upon the design and upon operating and environmental conditions, as well as upon defects and irregularities within or next to the insulation.

II. BREAKDOWN MECHANISMS

A classification of breakdown and prebreakdown phenomena is always to a certain extent personally biased. For this article breakdown mechanisms are defined as depending largely upon the inherent properties of the polymer in question. As such, a distinction is made among purely electronic, thermal, and electromechanical breakdown. As there have been no very recent fundamental advances concerning the respective theoretical aspects of polymer breakdown, only a cursory qualitative discussion is presented. For detailed discussions the reader is referred to the excellent, exhaustive treatises that exist on the subject (Cooper, 1966; Franz, 1953; Klein, 1969; Mason, 1959; O'Dwyer, 1973; Seitz, 1949; Stratton, 1961; Watson *et al.*, 1965; Whitehead, 1951).

Partial discharges occurring in voids, electrical treeing, and the phenomenon of electrochemical treeing represent a degradation of the insulating properties of a polymer or partial breakdown of same. The properties of the respective polymers will, of course, play a role in determining the gravity of such phenomena. Their occurrence is, however, invariably associated with the presence of defects in the insulation, and their effect is, except under very special circumstances, a long-term influence. They have consequently been classified as aging rather than as breakdown phenomena. The vast field of purely chemical and/or mechanical degradation is intentionally excluded except when directly related to discharges and treeing.

A. Electronic Breakdown

Purely electrical or "intrinsic" breakdown depends directly upon the number of mobile electrons in the insulator and upon their energy distribution. This energy in turn depends directly upon the applied electric field strength, whereas thermal breakdown, as commonly defined, depends indirectly upon the electrical field strength via the various dissipative processes occurring in the polymer. The extremely complex nature of the breakdown processes complicates the formulation of a comprehensive and entirely satisfactory theory and necessitates making diverse assumptions and approximations. At present there is indeed no theory of electronic breakdown able to explain satisfactorily the wide variety of experimental observations. Alternatively, measurement of the electrical breakdown strength of a polymer is in the majority of cases so problematic that many of the available data do not really constitute critical tests of existing theories. Direct confirmation, for instance direct observation of the spatial and temporal temperature evolution during thermal breakdown (see Section II.B), and unambiguous identification of physical processes are mostly lacking. The possible applicability of a proposed theory is hence tested by a comparison of theoretically predicted and experimentally observed dependencies (temperature, thickness, time of voltage application, etc.) rather than by a comparison of the absolute magnitudes of the calculated and measured breakdown field strengths.

If mechanical instability caused by the electrostatic attraction of electrodes is ignored, all breakdown must be in a sense thermal, as already mentioned. Just the same, a distinction is made between breakdown as preceded by significant joule and/or dielectric heating and purely electronic breakdown entailing an instability in the free electron distribution and onset of collision ionization. This breakdown strength depends only upon the physical state of the dielectric and is uninfluenced by surrounding conditions. As already mentioned by O'Dwyer (1973), there is no irrefutable evidence that electrical breakdown must be preceded by collision ionization, but its participation appears intuitively correct, and this assumption forms the basis of all present theories of electronic breakdown. The basic question is, in all cases, at what electrical field strength electrons are suddenly produced in sufficient numbers to cause lattice disruption. Two approaches have essentially been followed, namely, that of calculating the field strength for which an instability in electron distribution causes the onset of collision ionization (collective theories) and that which investigates how the carriers produced by chance collision ionization multiply to cause the almost instantaneous build-up of a thermal breakdown current (avalanche theories).

1. Collective Theories

All electronic breakdown mechanisms consist of essentially two basic processes. The first entails the transfer of energy from the electric field to the electron system (mobile and trapped carriers) via mobile electrons; the second process considers the interaction of the electrons with the lattice. The criterion for breakdown is therefore that the rate of energy gain by the electrons will exceed the rate of loss to the lattice. The breakdown field strength is defined as the minimum field strength for which equilibrium cannot be realized. This definition applies irrespective of the particular assumptions made, but the magnitude of the field strength calculated will vary with assumptions and boundary conditions.

The energy balance can be written quite generally as

$$A(F, T, \gamma) = B(T, \gamma), \tag{1}$$

where A is the rate of energy gain and is a function of the electric field strength F, the temperature T, and a variety of parameters describing the electron system that have been lumped into γ; and B is the rate of energy transfer to the lattice and is a function of T and γ. A schematic representation of various situations is shown in Fig. 1. With no electric field applied the electrons are in thermal equilibrium with the lattice ($B = 0$), and they possess thermal energy E_0. With an applied field the average electron energy will increase, as will dissipation to the lattice, the latter reaching a maximum near E_m, the average phonon energy of the lattice. Then there will be a decrease in the rate of energy transfer to the lattice, since times for electron–phonon interaction will be successively reduced with increasing electron energy. The rate of energy gain A will vary with the field strength, and various situations may be envisaged. If the field strength is F_U, then the rate of energy gain will exceed the rate of loss for all energies E; that is, every mobile electron will continually increase its energy. A stable situation is clearly not possible. For a sufficiently low field strength,

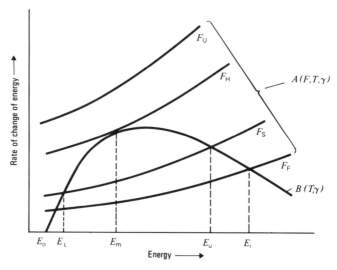

Fig. 1. Schematic diagram of the average rate of energy gain A and energy loss B to the lattice at various applied field strengths F.

F_S say, two stable solutions of Eq. (1) exist at E_1 and E_u. At a field strength $F = F_H$ they merge into one, and F_H is considered by von Hippel (1935) to represent the electronic breakdown field strength. His approach considers that all mobile electrons are accelerated against the retarding influence of the lattice. Since the nonequilibrium situation arises already for relatively low energies of the electrons, this condition for insulator breakdown has been termed the *low-energy criterion*. The breakdown field strengths so calculated show a slight increase with increasing temperature, because the rate of energy loss to the lattice increases as a result of enhanced electron–phonon interaction. Thus F_H is independent of sample thickness down to electrode separations approaching electron mean-free paths, and a dependence upon the rate of voltage application should be negligible. From an experimental viewpoint, these are some of the requirements to be fulfilled by intrinsic breakdown.

Fröhlich (1937) suggests that the criterion imposed by von Hippel is much too stringent. He postulates that the breakdown criterion is not determined by the average behavior of all electrons. He suggests that sufficient fluctuation occurs in the energy of the electrons, the energy range in fact extending up to the ionization potential E_i of the dielectric, and that only electrons with energy near E_i need to be considered. The condition for breakdown ($F = F_F$), where ionization processes due to high-energy electrons can no longer be balanced by recombination processes, hence is referred to as the *high-energy criterion*. Within this concept there is no steady state possible at an electron energy E_i, since a chance fluctuation could always increase the energy to just beyond this value. A stable situation exists only at the lower intersection of $B(T, \gamma)$ and $A(F_F, T, \gamma)$. The two approaches assume that the concentration of mobile electrons is so small that interelectronic interaction and energy exchange can be neglected. The inclusion of electron–electron interaction has been treated by Fröhlich and Paranjape (1956). Since these interelectronic processes represent an additional energy dissipation mechanism, especially for the critical high-energy electrons, the breakdown field strength F_E so calculated will be higher than F_F, indeed, $F_F < F_E < F_H$.

The energy gain from the electric field will, of course, proceed via mobile electrons only. In the physically real case of a polymeric insulator with defects, energy may also be transferred to the lattice via trapped carriers in thermal equilibrium with mobile ones (Fröhlich, 1947). Whereas high concentrations of mobile carriers are difficult to visualize in organic polymers and are certainly not supported by conductivity measurements, the presence of carrier traps distributed not only energetically but also spatially is well accepted. Fröhlich's (1947) theory of amorphous or heavily trap-loaded dielectrics leads to two distinct types of electrical breakdown behavior. Below a transition temperature T_c interelectronic interaction can be neglected. The breakdown field strength will show the expected weak increase with rising temperature. Above T_c the interaction between electrons will dominate, and the breakdown field strength in this temperature range has been derived as

$$F_{am} = A \exp(\Delta V/2kT). \tag{2}$$

In Eq. (2) ΔV is the average energy of shallow trap levels below the bottom of the "conduction" band. An absolute calculation of F_{am} cannot be meaningfully made for polymers, since several parameters lumped into the constant A are simply too uncertain. Equation (2), however, has the distinct advantage of predicting a decrease in breakdown field strength above a certain temperature, an experimental observation made in a number of polymers.

2. Avalanche Theories

Avalanche theories investigate the conditions and requirements for a chance high-energy electron causing collision ionization to develop into an avalanche of critical, thermally destructive size. An estimate of the required number of electrons in such a destructive avalanche is due to Seitz (1949) and yields a value of 10^{12}. If recombination is neglected, this represents 40 generations of collision ionization species, from which the theory takes its name. More elaborate calculations by Stratton (1961) yield a comparable number of electron generations, namely, 38. Avalanche theories that solely consider electron multiplication are uniform field theories. It is easily seen that the positive species formed during avalanching will with high probability have a considerably lower mobility than the electrons. The presence of these positive species represents a positive space charge distorting the field and in effect quenching development of the avalanche. This quenching tends to keep avalanches small and essentially nondestructive, leading only to small local temperature rises in the dielectric. The role of holes must obviously be taken into account (Forlani and Minnaja, 1969), and subsequent theories have thus considered current continuity and not field homogeneity as the condition to be satisfied. Thus O'Dwyer (1967) considers field emission from the cathode and subsequent avalanche formation, the accumulating positive space charge in turn increasing cathode emission so as to lead finally to a negative current instability. Impact ionization is thus mainly restricted to the cathode region, in contrast to the earlier theories where conditions at the anode determined breakdown. Whereas the negative resistance instability thus calculated depends essentially upon electron avalanching, a more recent discussion

(DiStefano and Shatzkes, 1975) predicts a similar negative injection current instability without invoking massive avalanching. Klein (1972) suggests that multiple avalanching causes breakdown and postulates chance local events rather than uniform injection and failure. The possibility of internal field emission or zener breakdown (Zener, 1934) should be mentioned for completeness sake. In this theory breakdown may occur when the concentration of mobile electrons increases catastrophically as a result of valence band to conduction band tunneling. For polymers this approach can in all likelihood be discounted, as the prevailing ionization energies on the order of 5–10 eV will lead to breakdown field strengths too large by at least an order of magnitude, i.e., 10^8 V/cm. Experimentally observed breakdown field strengths of polymers rarely exceed 10^7 V/cm.

3. Experimental Observations

Since it is difficult to eliminate external influences that may completely determine or at least influence the measured breakdown field strength, a most critical assessment of data is required if they are to be used as tests of existing theories. In the low-temperature region the possibility of thermal and electromechanical breakdown is reduced, and results may in general be consistently interpreted in terms of Fröhlich's theory. For several polymers the breakdown field strengths show little dependence upon temperature up to a certain break region. Fröhlich's low-temperature theory suggests that an increase in the breakdown field strength is to be expected with rising temperature, but for the high vibrational frequencies of polymer lattices this increase is very slight. In view of the inevitable scatter and uncertainty of breakdown data, this slight increase may well go undetected. A variety of data are furthermore in agreement with Fröhlich's approach that polar groups and defects will act as additional scattering centers for energetic electrons, thus increasing the breakdown strength. Thus the introduction of a heteroatom such as chlorine or of a polar group such as CO (Oakes, 1949) into polyethylene will increase the breakdown field strength. Indeed the highest breakdown strengths measured to date occur in polar polymers such as poly(methyl methacrylate), poly(vinyl alcohol), and poly(vinyl acetate) (Oakes, 1949). In a recent study attempting to measure the intrinsic breakdown strengths of certain epoxy resins (Hauschildt and Nissen, 1977) using so-called cylindrical specimens, breakdown strengths well in excess of 10^7 V/cm were measured. These values, unfortunately, still did not represent intrinsic breakdown, being invariably defect-initiated.

Observations that represent apparent contradictions to Fröhlich's approach are the increase in breakdown field strength of a series of

polyethylenes with increasing molecular weight (Fischer, 1974) and perhaps also the lack of influence of low-molecular-weight additives, that is,
of so-called voltage stabilizers (Fischer and Nissen, 1976). These additives influence only defect-initiated breakdown, presumably via a field-
grading effect.

Qualitatively the decrease in breakdown strength at elevated temperatures is well documented for a number of polymers. Details of the temperature dependencies, however, differ from author to author, and even
experimental details cannot be ignored (as exemplified by Fig. 5). It is
apparent, therefore, that estimations of the trap depth ΔV and of the influence of structural changes are tentative, particularly in view of the fact
that many polymers undergo marked changes in mechanical properties in
these respective temperature ranges. An attempt has been made by
Artbauer (1965, 1967) to relate the electronic breakdown strength of polymers to the molecular relaxation processes that ultimately also determine
the mechanical properties. He has considered that electrons gain sufficient
energy to cause collision ionization within the "free volume" of the
polymer, the latter being a spatially and temporally fluctuating parameter.

The independence of electrode separation represents a minimum requirement on data considered to represent intrinsic breakdown. Many
measurements indicate, however, a dependence of breakdown strength on
sample thickness. It was shown at an early stage (Morton and Stannett,
1968) that the breakdown strength did in fact appear volume-dependent,
not merely thickness-dependent, particularly with ac stressing when defect influences were especially pronounced. These results may be recon-
ciled with more recent investigations (Fischer and Nissen, 1976) which
clearly show that the volume dependence of electrical strength is only an
apparent one in that a continuous transition takes place from intrinsic
breakdowns to defect-initiated breakdowns as the electrically stressed
volume is increased. Within a certain volume range the effects can be
separated by a suitable analysis of the data (see Section IV.B).

Formative time lags on the order of nanoseconds in polyethylene (Arii
et al., 1974) certainly support the idea of breakdown by a purely electronic
mechanism. The simultaneous application of high-energy irradiation and
of breakdown voltage (Tsutsumi and Kako, 1975) indicates, at least in a
qualitative fashion, that the presence of additional high-energy electrons
and of enhanced electron conduction currents have no marked influence
upon the breakdown field strength of polyethylene. This perhaps supports
Klein's (1972) proposal that chance local events cause breakdown. The
extensive and painstaking work of Bradwell *et al.* (1971) and of Cooper
and collaborators (1977) supports the influence of space charges in
polymer breakdown and also indicates what minute details of prior ther-

mal history and conditioning of the samples can influence density and morphology and the resultant breakdown data.

It appears, therefore, that the vast amount of data available on electronic breakdown in polymers does not allow a decision in favor of any one of the postulated theoretical approaches. The data rather support aspects contained in several of them, and it seems likely that the correct interpretation of purely electronic breakdown in polymers will involve elements from all or several of them. The correct interpretation must await the outcome of further experiments aimed more at identifying processes in well-defined polymers rather than of experiments aimed merely at investigating parameter influences.

B. Thermal Breakdown

As stated before, it is presently accepted that the final stage of dielectric breakdown is thermal in nature, irrespective of the processes leading up to this stage. The term *thermal breakdown* is, however, commonly used for a type of insulation failure that can be adequately described in terms of the macroscopical thermal and electrical properties of a given dielectric–electrode configuration.

In any real insulator loss currents will flow upon the application of an electric field. For dc fields this will lead to joule heating of the dielectric, while for ac fields the dielectric losses due to relaxation phenomena will also contribute. These loss currents will raise the temperature of the dielectric and in turn the losses. Eventually a situation may be reached where the energy gain from the electric field can no longer be balanced by heat losses to the surroundings, and thermal breakdown will ensue. The balance between energy gain and heat dissipation is described by the differential equation

$$c\rho \, dT/dt = \sigma F^2 + \text{div}(\kappa \, \text{grad} \, T), \tag{3}$$

where c and ρ are the specific heat and the density of the dielectric and T is the temperature of the lattice. For an actual solution of Eq. (3) it will suffice to calculate T_{loc} for some localized hottest spot within the dielectric from which breakdown may be considered to start. The first term on the right-hand side represents the energy gain per unit volume. The dielectric conductivity σ sums over all the dissipative processes occurring within the dielectric and will be a function of various parameters such as temperature, electric field strength, frequency, and perhaps even locality. The electrical field strength F may vary locally because of electrode geometries on one hand, but also as a result of unrecognized factors such as space-charge injection in low-conductivity solids on the other. The second term on the right-hand side of Eq. (3) represents the energy dissipation, the thermal conductivity κ being a function of temperature also. It is thus apparent that analytical solutions for F_B, the thermal breakdown field strength, will not be possible except for cases with rather simplified boundary conditions. Figure 2 shows schematically the temperature as a function of time for a thick slab of dielectric sandwiched between parallel plate electrodes to which various field strengths are applied. For low field strength, hence lesser rates of heat generation in conjunction with good thermal contact with the surroundings, a true equilibrium situation may be reached with $T_{final} < T_B$. This is represented by solution a for which thermal breakdown will not occur. If the field strength F_B and other conditions are such that the temperature of the hottest part of the insulation approaches T_B asymptotically, curve b results. It is an almost-equilibrium situation derived by putting the time derivative of Eq. (3) equal to zero. The thermal breakdown strength is commonly expressed as F_B. For other conditions, F only slightly larger than F_B say, the temperature T_B will be reached in a time depending upon the detailed experimental conditions and will then increase rapidly until breakdown occurs. Curve c is a representative example. Curve d shows the case of a very rapidly increasing temperature due to the application of an impulse voltage, say. This solution of Eq. (3) corresponds to the case where the heat dissipation term is

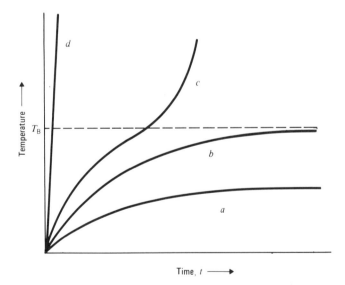

Fig. 2. Schematic representation of stability and instability conditions for thermal breakdown. (a) No breakdown, (b) thermal breakdown at $t \to \infty$, (c) thermal breakdown in finite times, (d) impulse thermal breakdown.

neglected. For these latter two cases thermal breakdown field strengths may be somewhat arbitrarily defined, but the discussion has made it apparent that the thermal electric strength is not only a function of the thermal properties of the dielectric but also of the size, shape, and thermal properties of the contacting electrodes and the surrounding medium, and of the duration and waveform of the applied voltage as well. A thermal breakdown field strength that depends solely on the thermal and dielectric properties of the insulator can in fact not be meaningfully defined, and there are as many solutions to Eq. (3) as there are experimental conditions and arrangements. For detailed general discussions the reader is referred to Whitehead (1951), Franz (1956), and O'Dwyer (1973).

Both under industrial conditions and under the better controlled conditions of laboratory experiments it is not easy to be certain that thermal breakdown has in fact taken place. Certainly a direct observation of the variation in temperature as a function of time and space of the anticipated hottest part of the dielectric and its immediate surroundings is adequate confirmation and highly desirable. The experimental realization is, however, quite difficult to achieve in practice. Hence what is done usually is to calculate the breakdown voltage for a given set of actual or assumed conditions and to compare it with the experimental breakdown voltage. Alternatively, secondary criteria such as the effect of the rate of voltage rise are used. Particularly with respect to a critical assessment of breakdown mechanisms in polymers, a direct measurement of temperature evolution is necessary. In this connection the recent work of Kalkner (1974; Kalkner and Winkelnkemper, 1975) and of Winkelnkemper and Kalkner (1974) merits discussion. These authors not only investigated the conduction current of a series of organic polymers up to breakdown but monitored at the same time the spatial and temporal evolution of temperature profiles of electrically homogeneously stressed regions. The most detailed results were obtained by means of an infrared imaging camera, allowing in fact meaningful quantitative comparison between calculated and measured temperature profiles. As a representative example Fig. 3 shows the temperature reliefs and the respective temperature profiles along a line through the center of a 220-μm-thick poly(vinyl chloride) sample stressed with 50 Hz at the field strengths shown. One clearly recognizes that the temperature distribution is axially symmetrical with a maximum at the sample center, indicating that the sample heating does not progress from some arbitrarily located defect in the polymer. The temporal evolution of temperature along a line through the center of a polyoxymethylene sample is shown in Fig. 4 and correlates nicely indeed with the qualitative description of Fig. 2. It indicates that below a breakdown voltage an appreciable temperature increase occurs, leading, however, to stable equilibria.

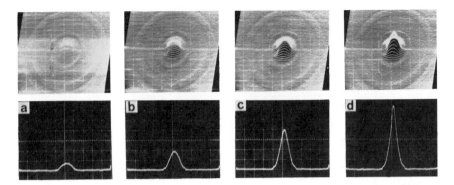

Fig. 3. Temperature reliefs and temperature profiles along lines through the center of a 220-μm-thick poly(vinyl chloride) sample. (a) $\hat{F} = 1.1$ MV/cm, $\Delta T_{max} = 9$ K; (b) $\hat{F} = 1.5$ MV/cm, $\Delta T_{max} = 19$ K; (c) $\hat{F} = \hat{F}_B = 1.6$ MV/cm, $\Delta T_{max} = 28$ K; (d) $\hat{F} = \hat{F}_B = 1.6$ MV/cm, $\Delta T_{max} = 38$ K. [From Kalkner (1974).]

Thus the temperature profiles in region A represent stable equilibria, and they are to be correlated with curve a of Fig. 2. At a voltage level U_B an instability region is reached, and the temperature increases at an accelerating rate until breakdown occurs, even though the stressing voltage is held constant (temperature profiles in this region are shown after equal time intervals). A voltage just below U_B would correspond to curve b of

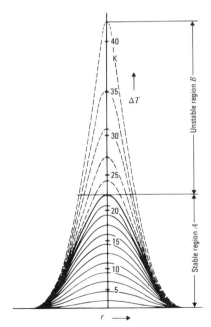

Fig. 4. Temperature profiles along a line through the center of a 190-μm-thick polyoxymethylene sample. Region A shows stationary temperature profiles at various applied voltages $U_i < U_B$; region B shows temperature profiles after equal time intervals after application of the constant breakdown voltage U_B. [From Kalkner (1974).]

Fig. 2, and a voltage of U_B or greater to a series of curves c, depending upon the details of the experiment.

In summing up these measurements, the authors were able to show directly that, for certain high-loss polymers such as poly(vinyl chloride), polyoxymethylene, and cellulose paper breakdown resulted from instability of the combined thermal and electrical equilibrium. It appears as if a prerequisite for this behavior is a loss factor $\varepsilon_r \tan \delta$ increasing with temperature over the temperature interval under investigation. In other polymers, particularly in such low-loss polymers as polystyrene, thermal instability is not the cause of breakdown, negligible or no temperature changes being detected. Processes depending directly upon the electrical field strength appear to be responsible for breakdown in these cases.

C. Electromechanical Breakdown

One of the properties that make polymers attractive for certain manufacturing processes is their ability to become plastically deformed, particularly at elevated temperatures. This property may in fact be responsible for the breakdown behavior of certain polymers in laboratory testing, especially for the often pronounced decrease in electrical strength with increasing temperature. This type of polymer failure, which represents an instability in the balance between mechanical and electrical forces, has been termed electromechanical breakdown (Stark and Garton, 1955).

Consider a laterally unrestrained parallel plate capacitor of initial electrode separation a_0 filled with a dielectric of relative permittivity ε_r. The force due to the attraction of the oppositely charged electrodes will be counteracted by the mechanical rigidity of the polymer in question. With an applied potential U this balance of electrostatic and mechanical forces is given for large strains by

$$\tfrac{1}{2}\varepsilon_r\varepsilon_0(U/a)^2 = Y \ln(a_0/a), \qquad (4)$$

where Y is Young's modulus of the polymer at the respective temperature and a is the momentary reduced electrode separation. The expression $a^2 \ln(a_0/a)$ has a maximum at $\ln(a_0/a) = 0.5$; that is, a mechanically stable situation cannot arise for a voltage larger than a critical voltage U_c for an electrode separation a less than $a_0 \exp(-0.5)$. Referring the critical or electromechanical breakdown voltage U_c to the initial electrode separation a_0, one obtains

$$F_c = U_c/a_0 = (Y/\varepsilon_r\varepsilon_0)^{1/2} \exp(-0.5) \qquad (5)$$

for the breakdown field strength. With this equation satisfactory agreement between calculated and experimentally determined breakdown strengths has been obtained for certain polymers, particularly nonpolar, thermoplastic ones such as polyisobutylene (Stark and Garton, 1955) and polyethylene. Thus Fig. 5 shows the agreement found in the temperature range from about 60–100°C for a low-density polyethylene measured with recessed specimens. Further support for the concept of electromechanical breakdown derives from an increased breakdown strength at elevated temperatures upon improving the mechanical rigidity of polyethylene by cross-linking (Stark and Garton, 1955), from a decreasing breakdown strength of polyethylene with increasing melt flow index, that is, decreasing average molecular weight (Fava, 1965; Oakes, 1948), from a higher breakdown strength upon impulse stressing than upon stressing with a slowly increasing dc voltage (Fava, 1965; Artbauer and Griac, 1965), and somewhat more indirectly from measurements with McKeown (1965) type specimens. In these samples the ball-bearing electrodes are rigidly held by embedding them in a thermosetting resin. The measurement of such samples with polyethylene as the dielectric subjected to breakdown (Lawson, 1966) indicated substantially higher breakdown strength at elevated temperature than for samples in which the electrodes were able to deform the polymer compressively. The increased strength observed in these measurements at 20°C, where deformation of the recessed specimens may be assumed to be negligible (see Fig. 5), was explained in terms of a thin

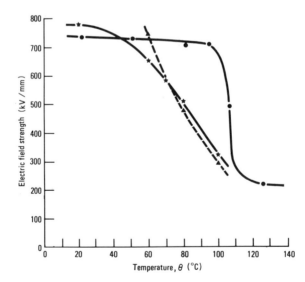

Fig. 5. Electrical breakdown field strength F as a function of temperature θ for recessed and cylindrical low-density polyethylene specimens. ●, cylindrical specimens; ★, recessed specimens; ▲, calculated [Eq. (5)].

layer of epoxy resin that had crept between the electrodes and the dielectric during the molding process.

Direct evidence for the mechanical thinning of thermoplastic polymers prior to breakdown comes from the work of Fava (1965), who used an optical lever system to detect thickness changes, and from Blok and Le-Grand's (1968) measurement of the optical birefringence before breakdown. Recently Sabuni and Nelson (1976) have confirmed field-induced strains in epoxy resins subjected to dc stressing.

Summarizing, there exists sufficient evidence in favor of the possibility of electromechanical breakdown in certain polymers in certain electrode arrangements, but the quest for numerical agreement between calculation and experiment should not be stretched too far. Equation (5) was derived assuming the validity of Hooke's law, whereas it is known that most polymers exhibit nonlinear viscoelastic behavior. Second, a parallel plate, laterally unconstrained sample–electrode arrangement was assumed. This is not given in practice, and the accurate calculation for a real specimen of the recessed type, say, is not feasible. To what extent Young's modulus, the bulk modulus, some intermediary value, or static or dynamic moduli should be used to calculate F_c is a matter of conjecture. Last, the concept can usefully be applied only at elevated temperatures well above the transition range from electronic to electromechanical breakdown, a range in which electronic breakdown of an already compressively thinned sample may occur.

III. AGING OF POLYMERIC INSULATION

The factors involving the deterioration of insulation are most complex, and a full discussion is clearly beyond the scope of this article, particularly when pertaining to chemical and mechanical properties. This section is restricted to three types of aging or prebreakdown phenomena whose occurrence has as a prerequisite the presence of an inhomogeneity and which primarily affects the dielectric properties of the polymer. The first mechanism is internal discharges in gas-filled voids occurring in the insulating system. It causes, in general, a gradual erosion and weakening of the insulation in the vicinity of the defect that may eventually lead to electrical failure. The next two mechanisms occur in the presence of highly divergent fields due to protrusions or inclusions in the insulation. They are referred to as *treeing* because of the treelike structures of quite varied appearance that result. In *electrical treeing* enhanced stress is considered to cause some sort of localized failure involving the production of a gas-filled tubule which then propagates through the insulation as a result of the repetitive action of internal discharges occurring within the tubules. A second type of treeing phenomenon occurs when polymeric insulation is electrically stressed over lengthy periods of time in the presence of a polar liquid, most frequently water. It has hence been termed *water treeing* or *electrochemical treeing*, the usage of the terms being not all that well defined.

A. Discharges in Voids

Voids may occur in insulating materials for various reasons. Apart from contraction voids a large number of voids, albeit in the micrometer range, are produced during the manufacture of cross-linked polyethylene power cables during the vulcanization process with steam. Voids may equally occur in thermosetting resins used widely in the manufacture of medium- and high-voltage equipment as a result of insufficient degassing of the resin during casting or mixing operations. Voids may also occur in impregnated lapped insulation. When such insulation is subjected to an electrical stress, discharges, an electrical breakdown of the gaseous content of these voids, may occur. During this process electrical energy from the applied field will be transferred eventually as thermal energy to the void surfaces and to the neighboring insulation bulk. This may lead, depending upon the duration and severity of the respective processes, to damage of the polymer, which may culminate in breakdown. For obvious reasons considerable effort has been invested in studying such partial discharges. The main aims are ultimately to be able to evaluate the potential danger emanating from the discharges by nondestructive means and to find a correlation between partial discharge activity and the momentary degree of aging of the insulation, i.e., the remaining life of the insulating system. Under ac stressing the destructive effects of partial discharges are particularly pronounced.

The discharge patterns and the temporal dependencies observed from real voids in real insulation show that partial discharges in such systems are very complex indeed. Just the same, the simple capacitive equivalent circuit shown in Fig. 6 for a single idealized void (Gemant and Philipoff, 1932) serves as a useful basis for a qualitative understanding of the phenomenon. Here C_v represents the capacitance of the gas-filled void in series with the capacitance C_s and shunted by the capacitance C_p of sound, discharge-free insulation. If an alternating voltage $U_a = U_0 \sin \omega t$ is applied across the electrodes, the potential across the void without breakdown is given by

$$U_{vo} = U_a C_s / (C_s + C_v) \tag{6}$$

provided the insulator is of sufficiently low conductivity, a prerequisite that is generally fulfilled. If the instantaneous voltage across the gas-filled void reaches the breakdown potential U_{in}, the gap will break down. In Fig. 6 the resistance of the discharging gap is simulated by R. This discharge will in most cases occur before the solid insulation fails, since the field strength within the void will be higher than that in the solid dielectric and because the breakdown strength of the gaseous gap will as a rule be less than that of the adjacent polymer. Provided the breakdown occurs at the constant and polarity-independent potential U_{in}, and assuming the remnant voltage after the breakdown to be zero, Fig. 7 shows the momentary potential U_v across the gap as a function of time. In this approximation the voltage drop U_{in} across the void, occurring in a time on the order of 10^{-7} sec or less, causes a voltage step $U_{in}C_s/(C_p + C_s)$ across the electrodes. These voltage steps at the sample electrodes and the resulting currents that flow in the connected circuit form the basis of partial discharge detection techniques.

The observable quantities may be grossly divided into three categories (Kind and König, 1968): namely, those of single partial discharge pulses, of pulse sequences, and of pulses or pulse sequences in relation to the momentary value of the applied potential. To the first category belong the

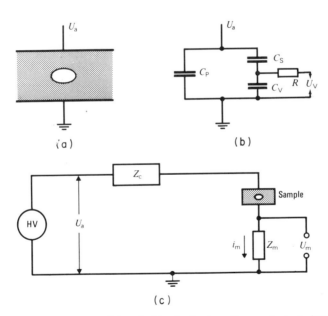

Fig. 6. Single void in a solid dielectric (a), simple capacitive equivalent circuit (b), and basic partial discharge detection circuit (c).

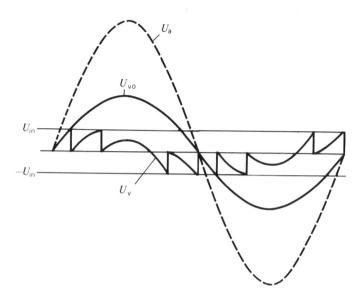

Fig. 7. Time-dependent potential across a void in a solid dielectric (cf. Fig. 6) and across the electrodes. U_a, Applied potential; U_{vo}, Potential across void without gaseous breakdown; U_v, Potential across void with breakdown at U_{in}.

time dependence of the individual partial discharge current pulse $i_m(t)$ and the charge transported by the discharge $\Delta q_a = \int i_m(t)\, dt$. A shape analysis of individual discharge current pulses is rarely useful or for that matter possible, since the shape is a function of sample capacitance and circuit characteristics. Highly stable discharge patterns, which are a prerequisite in order to be able to perform a detailed circuit analysis, are seldom present. The time integral Δq_a is a much more attractive quantity, as it is independent of circuit impedance. It should be realized, however, that the apparent charge so measured externally is not to be equated to the charge actually displaced within the void. To estimate the latter, assumptions about sample capacitances and void geometries need to be made.

To take into account the statistical character of the occurrence of partial discharges, quantities derived from pulse sequences are frequently measured. These include the various partial discharge currents, be they the arithmetic mean of the rectified impulse currents (Widman, 1960), the so-called quadratic rate (Widman, 1960), or weighted quantities such as the radio-interference voltage (Praehauser, 1975). All these quantities have considerable shortcomings, particularly when used in the form of a single measurement. If used in a continuous or repetitive registration mode, even these partial discharge quantities may yield considerable information (König, 1967). One technique of pulse sequence analysis is that of multichannel pulse-height analysis. It constitutes a powerful, modern tool for partial discharge studies in that the temporal changes in, for instance, pulse-amplitude, pulse-interval, phase-angle distribution, and the like may be effectively monitored (Bartnikas, 1973; Kärkkäinen, 1975).

The last group of measurable partial discharge quantities relates the apparent discharge magnitude Δq_a to the instantaneous value of the applied potential, yielding power and energy supplied to the specimen as a result of the discharges. The usefulness of the quantities so determined becomes apparent from the following. With reference to Fig. 7, if a partial discharge occurs at a time t_j, the respective apparent discharge magnitude and the momentary applied voltage being $\Delta q_a(t_j)$ and $U_a(t_j)$, respectively, then

$$W^{\,j}_{\text{ext}} = \Delta q_a(t_j)\, U_a(t_j) \tag{7}$$

represents the energy content of the discharge j. These "external" energies will appear positive or negative depending upon the momentary sign of $U_a(t_j)$, whereas at the discharge site energy is consumed during every discharge pulse. The energy dissipated within the void is not amenable to direct measurement. If a summation of $W^{\,j}_{\text{ext}}$ is, however, performed over a certain time interval, then the average of the external discharge power P_{ext} must be equivalent, by the law of conservation of energy, to the power dissipated within the void P_{int}. The partial discharge power is thus directly related to conditions within the void and is not dependent on either sample or circuit details (Whitehead, 1951).

The apparative requirements, possibilities, and difficulties encountered in partial discharge detection are vast. The interpretation of measurements is particularly difficult when lumped-circuit considerations, which are applicable to most laboratory samples, cannot be applied and when instead distributed circuits must be considered, as in long cables, transformers, and power capacitors. For detailed information the reader is referred to Whitehead (1951), Kreuger (1964), Mason (1965), and Beyer (1978).

Despite the large effort and literature on partial discharges in voids occurring in polymers, a correlation of some particular discharge behavior or of one of the measurable discharge quantities with aging and incipient breakdown has not emerged. This is particularly true when results obtained on laboratory specimens are extrapolated to the insulating polymer operating in commercial equipment under service conditions. Certainly part of the dilemma is seen in the samples commonly employed in laboratory studies. These often consist of stacked films with holes punched in a centrally located film or in one adjacent to an electrode. Polymer-coated electrodes are sometimes used, or samples containing a void that has been machined in some way. If cast, as is often the case for thermosetting resins, the void surface will be endowed with the surface structure of the molding plug. The aging of solid polymeric insulation by discharges in voids will entail physical and chemical changes in the enclosed gas volume as well as changes in the void surface where partial discharge, solid, and gas interact. Changes extending beyond the surface region into the bulk may also occur, for instance, by diffusion of reactive decomposition products. Indeed they must occur at least during the breakdown process. It is consequently not surprising that an artificial void with a geometry and electric field distribution unlikely to occur in a commercial insulation, and with characteristics of the void surface and the gaseous content of the void very unlikely to represent real conditions, cannot simulate aging processes under service conditions correctly.

From the rather crude void model certain basic characteristic discharge parameters may, however, be estimated. Thus it was ascertained at an early stage (Mason, 1953; Hall and Russek, 1954) that the breakdown of a gas bounded on all sides by polymers (or other dielectrics) occurred at potentials very similar to those required for breakdown between metal electrodes and that Paschen's rule was applicable to both cases. Thus the inception voltage and also the apparent discharge magnitude agree with calculated values, at least for voids that are not so large that only part of the void capacity is discharged. Experiments with spherical voids in epoxy resins (Yasui and Yamada, 1967; Weniger and Kübler, 1976) and with ellipsoidal voids in polyethylene samples (Fischer and Nissen, 1978) confirm these results fully for voids more likely to occur in real insulation systems. Quite frequently such agreement is obtainable only with virgin samples and immediately after the initial voltage application or after a conditioning procedure (Densley and Salvage, 1971). As soon as the temporal behavior is taken into account, drastic differences among samples with different void geometries may appear. Consider as an example the partial discharge behavior shown in Fig. 8. It shows the partial discharge activity (for ease of registration the radio-interference voltage was chosen) of two voids completely surrounded by polyethylene. The polyethylene used, the void volumes, and the average stressing field strengths are nearly identical, as is the gross fabrication technique (Fischer and Nissen, 1978). The ellipsoidal void, melt-grown and showing no discontinuities in the curvature of the void wall, shows only intermittent discharge activity, the experimentally observed radio-interference voltage agreeing only initially with the calculated value. It then decreases, and the sample shows only intermittent discharge activity, the interval between measurable discharging ranging from seconds to several tens of hours in extreme cases. Breakdowns do not occur, at least within the time span of 300 h over

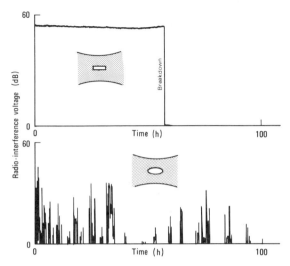

Fig. 8. Time dependence of partial discharge activity in terms of the radio-interference voltage (0.4 μV/60 $\Omega \doteq 0$ dB) of a cylindrical and an ellipsoidal void embedded in polyethylene.

which these experiments were conducted. The cylindrical void shows continuous discharge activity, and breakdowns occur almost with certainty within the 300-h range. It appears, therefore, that the experimental results attempting to relate lifetime to discharge activity in voids in polymers are critically specimen-dependent.

Since the lifetime of polymers even when exposed to the influence of partial discharges may be very long, it is frequently desirable to accelerate the experiments. This is done, for example, by increasing the stressing field strength, but one is always uncertain whether additional or at least altered processes occur. A frequently preferred method is to increase the frequency of the stressing voltage, and it has been shown that such a procedure is justified in some cases (Hogg and Walley, 1970). It should, however, be ascertained that the impulse-amplitude distribution is unchanged by altering the frequency, a fact hardly ever realized even over a limited frequency range and stressing time. For materials that show a self-extinguishing characteristic, such an acceleration is certainly not justified, since the intervals between discharge activity, hence in all likelihood the lifetime, will be governed by real-time effects. Self-extinguishing partial discharge behavior has been observed in many polymers such as polyethylene, poly(vinyl chloride), Perspex, and epoxy resins.

With the initiation of discharges various reactive species will be formed, depending upon the gas composition present. The reactive species will interact with the polymer surface and diffuse into the bulk if stable enough. Thus it is known that discharges in air will produce nascent oxygen, various oxides of nitrogen which will be able to react further with moisture to yield nitrous and nitric acid, carbonate ions, and a variety of hydrated species (Shahin, 1967). The macroscopic degradation and breakdown of polymers by reaction with such reactive species is a possibility but at present does not appear to be a major factor. It would in any case be a very long-term effect, perhaps associated with a loss of mechanical properties. It would also be much more likely in the case of an external corona with an unlimited supply of air and moisture. Indeed, under such conditions the formation and subsequent diffusion of nitric acid into the polymers polyethylene, poly(ethylene terephthalate), poly(vinyl chloride), and acetyl cellulose has recently been measured by Schon (1977). Presumably this may lead to thermal breakdown if sufficient degradation and ionic conductivity are produced. In general, thermal instability due to a cumulative heating effect of the partial discharges is unlikely, except perhaps at the highest frequencies (Whitehead, 1951).

It seems very likely, however, that such chemical processes may be responsible for the pronounced self-extinguishing properties of discharges in enclosed voids in polymers. Whereas in electrical treeing channels pressure has a distinct influence upon the rate of propagation and upon the appearance (Löffelmacher, 1975; McMahon and Perkins, 1964), an increase in pressure in the large voids considered here is unlikely. This can be shown by a simple estimate of the maximum possible pressure increase due to discharges and by direct experimental observation, during which a very narrow channel allows pressure equilibration between an ellipsoidal void in the polyethylene and the ambient (Röhl and Nissen, 1979). Neither a change in the partial discharge parameters, inception voltage, and apparent charge of the individual discharges, nor in the temporal behavior of the discharge activity, was observed for vented and nonvented voids. The explanation of the self-extinction is rather to be seen in an enhanced surface and/or near-surface conductivity decreasing the field strength within the void. This has been postulated for some time (Mason, 1959; Densley and Salvage, 1971) and has been inferred experimentally from the frequency dependence of partial discharge activity by Rogers (1958). A direct confirmation for voids in polyethylene via tan δ measurements has recently appeared (Nissen and Röhl, 1981). It is, however, as yet uncertain what chemical species are responsible, and it is not easy to understand why the extinction at times becomes apparent already seconds after the first discharges occur. On the other hand it is known that hydrated ions of the form $H^+(H_2O)_n$ are formed effectively in a positive corona when only traces of moisture are present. Indeed Moran and Hamill (1963) postulate that a partial pressure of water of merely 10^{-6} mbar in air at atmospheric pressure suffices to form hydrated ions in times on the order of tens of milliseconds.

As far as breakdown initiated by partial discharges is concerned, some sort of complex interaction of erosive processes at the polymer surface is most likely to be responsible. It is known that deposits are formed on the walls of voids exposed to partial discharges (Rogers, 1958; Hiley *et al.*, 1973). Furthermore, erosion of the surface has been observed, varying from small, isolated pits to massive pitting and crater-like formations (Hiley *et al.*, 1973). The complex physical and chemical processes that surely involve the differing dielectric properties of the deposits, lateral charge motion and surface tracking, field distortions, and peculiarities of the void geometries and surface irregularities are at this stage speculative.

It appears that, despite the advances made in the types of instrumentation for partial discharge detection and in their sensitivity, and despite considerable knowledge of the chemistry occurring in the gaseous content of the void, the use of simple laboratory samples with artificial voids often not even reasonably approximating realistic voids will remain of limited use where aspects of aging and lifetime are concerned. These samples may be used for comparative studies. In order possibly to find factors and processes relating quantitatively to the life of the insulation, experiments need to be performed on voids that closely simulate the defect particular to a given application, as well as the environmental and stressing conditions. Similarly a single observation of partial discharge activity is only of limited use in elucidating mechanisms, and the detailed temporal behavior should be investigated.

B. Electrical Treeing

The phenomenon of electrical treeing as distinct from water treeing has long been recognized as one of the main factors involved in the failure of polymeric insulation (Mason, 1955, 1959). An electrical tree consists of hollow, gas-filled channels within the polymer due to the more-or-less massive decomposition of the latter into mainly noncondensable low-molecular-weight fragments. Again, as distinct from nonstained water trees, electrical trees may easily be recognized within many polymers because of the difference in index of refraction. The shapes are most varied, and two representative examples developed under laboratory conditions are shown in Fig. 9. At lower stresses dendritic structures with a more filamentary appearance of the breakdown paths are common; at higher stresses a dense, bushlike appearance prevails. These trees are always observed as arising from some sort of defect, the more common

Fig. 9. Electrical trees in polyethylene grown from needle tips at (a) 10 kV and (b) 30 kV.

Fig. 9 (*Continued*)

type being a conductive inclusion or electrode asperity within the polymer, or the standard defect attempting to simulate these conditions in the laboratory, namely, a needle point inserted into the polymer.

Whereas the field enhancement of naturally occurring defects is in general unknown, laboratory experiments attempt to produce experimental conditions amenable to calculation. Frequently used electrode systems are the needle-to-plane (Mason, 1955; Kitchen and Pratt, 1962) or the needle-to-needle arrangement (McMahon and Perkins, 1964; Ashcraft *et al.*, 1976). Expressions for the electric field distribution or at least the maximum field strength of commonly employed ellipsoidal and hyperboloidal geometries may then be used (Mason, 1955; Ashcraft *et al.*, 1976; Bahder *et al.*, 1974a; Beyer *et al.*, 1972), always assuming the absence of disturbing influences, for instance space charges. For radii of curvature of the needle points on the order of a few micrometers very large field enhancements will result, leading to very high field strengths locally at rather low applied voltages. It has been noticed that a decrease in the radius of curvature below a certain limiting value will, however, not result in a further decrease in the voltage required to cause treeing. This is due to the fact that the stress at the needle points is alleviated by a field-enhanced conductivity and by the creation of a space-charge region around the needle tip. This space charge is also responsible for the observation that for many polymers the voltage required to form a tree at a negative point will often be higher than that required at a positive point. The injection of electrons from contacting metal electrodes into many polymers, particularly polyolefins, is well documented (Davies, 1969; Fischer, 1978; Kryszewski and Szymanski, 1970).

The complex patterns of trees will depend upon various factors, among them the conditions prevailing within the hollow channels with respect to temperature, pressure, and nature of their gaseous content, but also upon chance electronic events, details of the field distribution, and chance local weaknesses of the polymeric structure. The assumption that electrical treeing is basically an electronic phenomenon and the assumption of the involvement of space charges are quite clearly demonstrated by the similar, if not identical, appearances of trees produced by voltage application to needle electrodes embedded in polymers and the discharge patterns observed when a polymer, preferably a transparent one such as Perspex, is exposed to penetrating electron irradiation and the charged sample then grounded via a stress-enhancing point electrode (Dokopoulos and Marx, 1967). Various authors have shown that the structural peculiarities, for instance spherulite formation in certain polyolefins (Wagner, 1975), and mechanical stresses (Billing and Groves, 1974) are of major importance. The gross appearance of trees (bushy or filamentary), as well as their intermittent growth behavior, has long been interpreted in terms of pressure increases due to polymer decomposition and stifling of the partial discharge activity (Mason, 1959). Diffusion of the gaseous channel content causes a pressure decrease and reinception of discharges and growth. Since the amounts of gases produced are very minute, direct pressure measurements are beset with difficulties. Recently Löffelmacher (1975, 1976) used an indirect method of applying external pressure via a hollow needle to simulate the pressure conditions within nonvented voids. He was able to obtain excellent agreement for the discharge patterns observed under nonvented and vented but pressurized conditions. In this study the gaseous products from the treeing decomposition of polyethylene and of an epoxy resin were also analyzed at various applied voltages and after diverse times of treeing progress. The trees are considered to propagate by an extension of the field enhancement region to the channel tip via either conductivity and surface breakdown of the channel walls or via gas breakdown of the channel contents.

The phenomenon of electrical treeing may be divided into two distinct stages, the growth phase being considerably better understood than the initiation or inception phase. The experimental result that electrical trees will continue to grow at a lower voltage than that required to initiate them is persuasive evidence for this point, as is the fact that trees will appear with considerably lower voltages applied for long times than those required with impulse or short-term stressing. With short-term stressing the minimum field strength required to initiate microscopically visible trees is on the order of 10^7 V/cm for many polymers, and it appears that local intrinsic breakdown is the initiating mechanism (Fischer *et al.*, 1979, 1981). For lower stresses, a possibility is that localized overheating and thermal degradation takes place (Mason, 1959), but particularly with extremely low-loss materials such as polyethylene this mechanism is not considered likely. It has been suggested that compressive forces produce mechanical fatigue cracking (Ieda and Nawata, 1973) or that partial discharges, perhaps too weak to be detected, are present initially. The required small voids might be present because of poor adhesion of the polymer to the imperfection, or they might be formed in time as a result of the different thermal expansions of polymer and impurity or metallic needle. The preparation of laboratory samples ensuring perfect contact of needle and polymer is not a trivial problem, particularly when temperature changes occur (Kosaki *et al.*, 1977). Another suggestion relates to chance hot electrons causing ionization and gradual degradation of the polymer (Artbauer, 1965; Patsch, 1975a), a process that may be cumulative in time, thus explaining the pronounced voltage dependence of tree inception times.

Although several possibilities concerning the initiating mechanism have been suggested, there is at present not enough experimental evidence to make a definitive choice. A recent observation is, however, that partial discharges connected with tree inception in polyethylene commence rather suddenly at a discharge level well above the sensitivity limit (Billing and Groves, 1974). At least this suggests that trees as observed microscopically do not grow from an infinitesimally small stage. This observation is corroborated by the study of Shibuya *et al.* (1977) on epoxy resins. These authors observe microscopically, and via scanning electron microscope (SEM) investigation, that samples showing an initially void-free contact of needle and resin develop small voids at the needle tip upon stressing with an ac voltage, the rate of void growth being critically dependent upon the magnitude of the applied voltage. Partial discharges at an extremely low level supposedly accompany this void growth, and discharge activity then increases markedly with the beginning of tree growth out of such a void. The shapes of the voids formed and other experimental evidence such as an unchanged mechanical stress pattern with and without applied voltage suggest that Maxwell stressing causing mechanical failure of the polymer is not the initiating mechanism. In accord with Patsch (1975a) the presence of high-energy electrons is considered more likely.

Particularly with respect to the initiating mechanism further work is required. Likewise, the translation of laboratory results of needle testing should be viewed with caution when application of such results to treeing problems in actual engineering designs is desired. This is evidenced by the numerous factors that can influence the result obtained (McMahon and Perkins, 1973). Nevertheless the practice of needle testing has proven its value in comparative studies on polymer behavior under given sets of circumstances, particularly in laboratory studies on the effectivity of so-called voltage-stabilizing additives.

C. Electrochemical Treeing

Among the various deterioration and possible failure mechanisms of polymeric insulation, the phenomenon of water or electrochemical treeing has recently gained increasing prominence. For lack of a detailed understanding of its origin and nature it is defined as a tree- or bushlike structure of quite variable appearance, which seems to be intrinsically unavoidable in a defect-containing solid polymeric insulation exposed simultaneously to the influence of an electric field and water. Microscopic analyses indicate that water trees consist of a multitude of liquid-filled microvoids (Matsuba and Kawai, 1976; Cherney, 1973; Isshiki *et al.*, 1974). The phenomenon was first observed by Miyashita and Inoue in 1967 and has become a subject of worldwide study since the papers of Vahlstrom (1971, 1972) appeared. It has been shown that the effect will also occur in the presence of polar liquids such as hostapal and ethylene glycol (Bahder *et al.*, 1974b). For this reason the term "electrochemical treeing" appears more applicable than the term "water treeing," which has become widespread, as in the overwhelming majority of cases water or an aqueous solution is the liquid phase involved. Figure 10 shows an example of an electrochemical tree observed within extruded cross-linked polyethylene insulation (termed a "bow tie") and one grown in a labora-

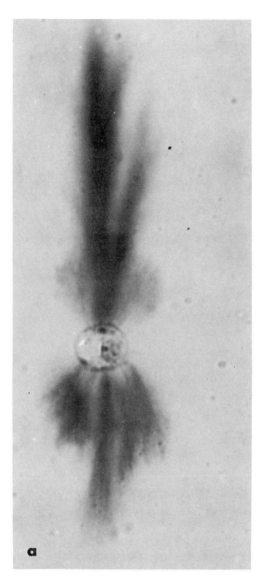

Fig. 10. Electrochemical trees in a polyethylene-insulated power cable (a) and a laboratory specimen (b). [Figure 10b is from Eichhorn (1978).]

Fig. 10 (*Continued*)

tory sample possessing a water electrode as a standard defect (Ashcraft 1977).

Apart from a structure with a rather diffuse appearance, dendritic structures are frequently observed, very much akin to the appearance of electrical trees. Whether these dendritic structures are to be classed in the category of electrochemical trees is not quite certain. What is certain is that electrical trees can grow in the presence of moisture (Eichhorn, 1974).

1. Characteristics of Electrochemical Treeing

Like the processes leading up to the breakdown of a polymeric insulation, the formation and growth stages of the electrochemical treeing phenomenon are very likely also the result of a complex interacting series of electrical, physical, and perhaps chemical processes. The number of variables influencing electrochemical treeing appears to be vast. This is clearly shown by a study of the literature, preferably the Annual Reports of the Conference on Electrical Insulation and Dielectric Phenomena (CEIDP) and the Progress Reports of Working Group 21-11 of CIGRE, to both of which the reader is referred. A further complication arises from the fact that quite generally the chemical and physical nature of the polymers investigated, such as crystallinity, additives, impurities, degree of oxidation, etc., are not sufficiently well defined to allow a meaningful comparison of the data of different authors. The important role that crystallinity and crystalline superstructure may play has recently been discussed by Muccigrosso and Phillips (1978).

Certain characteristics have, however, emerged. Thus electrochemical treeing is always initiated at sites of high and divergent stress, that is, at liquid-filled voids or at impurities within the polymer insulation or at asperities near the polymer–electrode interfaces. Electrochemical treeing appears to be possible in all polymers (Eichhorn, 1976), although the extent and time constants of their appearance will be determined by the properties of the various polymers employed and upon experimental conditions. The majority of investigations have been concerned with polyethylene and cross-linked polyethylene, but laboratory studies have shown the effect to occur in many polymers. These include other polyolefins such as polypropylene (Isshiki *et al.*, 1974) and polybutene-1 (Ashcraft, 1977), silicone rubber, ethylvinyl acetate and ethylene–propylene rubber (Yoshimura *et al.*, 1977) and polystyrene (Ashcraft, 1977). Water trees have recently also been observed in epoxy and polyester resins (Yoshimitsu and Nakakita, 1978). Initially it appeared as if the effect occurred only under ac stressing, but recent experiments indicate that electrochemical trees will also appear under dc stress (Franke *et al.*, 1977; Noto, 1980). The effect is merely much less pronounced in this case. Furthermore it seems certain now that one is justified in viewing electrochemical treeing as a two-stage process, the first step being an incubation phase and the second step the actual growth phase. With respect to the characteristics mentioned so far, electrochemical and electrical treeing behave similarly. Whereas the incubation phase in electrical treeing is considered to be long, the time to electrochemical tree inception appears short under comparable stressing conditions. The growth phases show an opposite behavior. Thus electrochemical trees grow very slowly and appear even to stagnate at times, while electrical tree development after initiation generally proceeds rapidly to failure.

There are also several characteristics that show the electrical and electrochemical treeing phenomena to be quite different, despite the often very similar appearance. The necessity of the presence of a polar liquid in the one case has already been mentioned. In electrical treeing evolution of the tree- or bushlike patterns is accompanied by measurable discharges, whereas with equipment of the highest sensitivity available (detection limit of about 0.01 pC) no partial discharges have thus far been detected during the growth of electrochemical trees of quite considerable size. The treelike patterns of electrical treeing are of a permanent nature, consisting of hollow tubules and channels as a result of polymer decomposition to low-molecular-weight fragments. Electrochemical treeing patterns, on the contrary, are of a temporary nature unless staining techniques (Matsubara and Yamanouchi, 1974; Eichhorn, 1978; Henket *et al.*, 1981) are employed to render them permanent. This indicates that they are perhaps only ruptures of the polymeric structure, with no gross decomposition of the polymer. This is evidenced by the fact that unstained electrochemical treeing structures will eventually disappear by exposure to vacuum or elevated temperatures but will reappear rather rapidly in their original form upon subsequent immersion in water without electrical stress. A final distinguishing feature is that electrochemical trees appear at lower average field strengths than electrical trees. It is uncertain at this time whether this would remain true were the actual local field strength at the respective defects known.

2. Growth Mechanisms

Despite considerable experimental and theoretical effort, the growth mechanism and, as a matter of fact, the detailed identity of electrochemical trees is as yet not completely understood.

The first step of the generation process is obviously the ingress of the required liquid into the polymer. If water is not initially present, and it may well be as the result of a particular manufacturing process such as the vulcanization of polyethylene by exposure to superheated steam, for example, it will enter by diffusion once the polymer is exposed to a moist environment. In polyethylene, to which the following discussion mainly refers, the permeation of water is a relatively fast process because of its very low solubility. The amorphous and interspherulitic regions of a reasonable thickness of the polymer will hence be saturated in relatively short times even at room temperature. The saturation concentration depends upon the temperature and pressure and will be influenced by the detailed chemical structure of the polymer and by the additives invariably present. It is indeed questionable whether the solubility of water in really pure polyethylene has ever been measured. The supermolecular structure (crystallinity, spherulitic structure) will play a decisive role in determining whether moisture remains molecularly disperse or whether condensation to liquid water takes place. The permeation of water into polyethylene has been shown to be an activated process with a thermal activation energy of about 0.3 eV (7 kcal/mole) (Eichhorn, 1970; Simril and Hershberger, 1950). Although the ingress of moisture into the polymer is a necessary requirement for the appearance of electrochemical trees, it is not a sufficient one. Diffusion of moisture alone will not lead to electrochemical treeing, since the presence of a (divergent) electric field is necessary. The first effect of the electrical field will be to increase the saturation content of water. Quite generally the chemical potential μ of water in the polymer in the presence of an electrical field F will be given by

$$\mu = \mu_0 - \tfrac{1}{2}\varepsilon_0 F^2 (\delta\varepsilon_r / \delta\rho)_T, \tag{8}$$

where μ_0 is the chemical potential in the absence of the field, ε_0 is the permittivity of free space, ε_r is the relative permittivity, and ρ is the density of water at temperature T. An estimate shows that, even for high electrical field strengths, here 10^6 V/cm was assumed, the increase in water concentration should be small, maximally on the order of 1–3%. A representative value of the saturation content of water in polyethylene at room temperature is less than 0.01% (Renfrew and Morgan, 1960), so that the increase due to the electrical field is normally below the sensitivity limit of modern analytical methods. Indeed several authors find electrically stressed and unstressed cables to contain the same amounts of water (Chan and Jaczek, 1978; Tabata *et al.*, 1972). Other authors find the water content of electrically stressed polyethylene to be considerably higher than that of unstressed material (Bahder *et al.*, 1974b, Mizukami *et al.*, 1977), an effect attributable to a condensation of moisture on either microvoids already present in the polymer or on wettable impurities (Sletbak, 1977). This condensation of moisture into macroassemblies of water molecules may probably be correlated with the inception stage of electrochemical tree growth. Their presence seems to be a prerequisite for all the growth mechanisms postulated to date.

One suggestion is that local dielectric breakdown occurs at the tips of very fine water channels contacting the electrodes (Tanaka *et al.*, 1974). Even for the theoretically highest resistivity of water a suitable geometry may lead to such field enhancements at the channel tips that the intrinsic breakdown strength, about 800 kV/mm for polyethylene, is exceeded.

This mechanism, however, is not considered very likely. A further suggestion entails the occurrence of partial discharges. Although they have never been found at a measurable level to date, light emission in the visible and ultraviolet regions has been observed (Nitta, 1974; Tanaka *et al.*, 1974). It is suggested that weak discharges occurring in the gas-filled tip of a growing tree decompose water into O_2 and H_2, the resulting large pressure increase leading to local damage in the polyethylene in the form of microcracks. The water then gradually enters these cracks, and the process continues. To support the contention that water is decomposed, the presence of nascent oxygen during tree growth is postulated based on a positive iodine–starch test. Whatever the details of the mechanism, it contains an element that forms the basis of the more recent theoretical treatments, namely, the local loss of mechanical integrity of the polymer as distinct from gross chemical decomposition. Subsequently, movement of liquid into the adjacent regions of the voids or cracks so formed may occur.

A water-filled void will be subject to joule and dielectric heating in an externally applied electric field. This forms the basis of several suggestions as to how water trees grow. Tanaka and co-workers (1974, 1976) suggest that the expansion due to heating of a void elongated in the field direction will tend to make the void more spherical in shape. This shape change would create compressional stresses perpendicular to the field direction and tensile stresses near the void ends pointing in the field direction, with the possibility of creating additional microvoids in the polymer so stressed. These microvoids or an agglomerate thereof would fill with water, and the process would repeat. Eventually a water tree, a stringlike collection of liquid-filled voids (either joined or not yet connected) would result. The direction of growth of such a tree would be determined by the electric lines of force, by the orientation of the initiating nonspherical void in the electric field, and by chance mechanically weak spots in the polymer structure. Model calculations are presented, but it appears that unrealistically high values of the electric field strength within the water-filled voids need to be assumed. Other possibilities suggested, namely, vaporization of water leading to pressure increases and fatigue of the polymer structure, local melting, and also local thermal polymer decomposition suffer from the same shortcoming. An alternative model is also suggested by Tanaka *et al.* (1974, 1976), namely, the influence of electrostrictive forces on the bounding surfaces of water-filled voids. Like the model calculation of the mechanical stresses induced by void heating, these calculations also yield stresses in the polymer which for certain, even if somewhat high, electrical field strengths are in the range of experimentally determined ones and are above the (static) elastic limit of polyethylene. Fedors (1980) suggests osmotic pressure as the driving force for the growth of water trees, while Minnema and co-workers (1980) consider that the joint effect of Coulomb forces and a strongly decreased surface tension at the PE–water interface causes fracture of the polymer; that is, they consider water treeing as a special case of environmental stress-cracking.

A further suggestion for the generation of water trees is that water molecules or ions collide with the polymer matrix. Thus Cherney (1973) suggests that "bow tie" trees propagate by the extension of water-filled voids as a result of field-induced vibration and collision with the polymer matrix of solvated OH^- and H_3O^+ ions, while Tanaka *et al.* (1974) consider the acceleration of neutral but polar or polarizable molecules against the void walls by dielectrophoretic forces (Pohl, 1958).

The most recent considerations are thermodynamical treatments of the behavior of molecularly dispersed water in polyethylene under the influence of an electric field (Heumann *et al.*, 1980; Patsch, 1981). Matsuba and Kawai (1976) show that in a microvoid water will condense, and that the microdrop will grow in size at a rate that depends upon drop size and the electrical field strength to which it is exposed. This mechanism of tree growth presupposes a relatively large microvoid in the vicinity of smaller ones. The field enhancement caused by this larger void (an ellipsoidal shape is assumed for the model calculations) causes another void of requisite size and location to grow preferentially larger than other voids exposed to lesser fields. This growth occurs by attracting water from the surrounding polymer. Eventually a chain of water-filled voids is thus envisaged to be produced. With certain assumptions concerning the void and water content of the insulation, the calculated influence of the electrical field strength, frequency, and temperature correlates with experimental data. The theory, however, also indicates that water tree growth should not take place under dc stressing, a result that is in clear contradiction to experiment (Franke *et al.*, 1977). Mizukami and co-workers (1977) make no assumptions regarding a distribution of larger and smaller microvoids but purport to show that under the influence of an electrical field water will condense in the polymer to form voids. This will obviously occur at random weak spots in the insulation. The polymer around the void will be mechanically stressed, leading eventually to mechanical failure of the void wall and the extension of microfissures into the surrounding material. Again, random weak spots will play a decisive role in this process, but if the polymer matrix around the void may be assumed uniformly mechanically sound, then the fissures will occur at the mechanical stress maximum in the field direction. Water is thought to enter these fissures, forming new voids, and the process repeats. Among several experimental facts, the model predicts some interesting details such as the finding that fewer but larger "bow ties" will be generated at lower field strengths and a larger number of smaller ones at higher field strengths (Mizukami *et al.*, 1977; Sletbak and Botne, 1977).

It is at present not possible to say unequivocally which mechanism or mechanisms are responsible for electrochemical tree growth. All the postulated mechanisms explain some, but not all, of the often apparently contradictory results. What seems to have emerged, however, is that local mechanical failure of the polymeric structure plays a dominant role. To what extent chemical processes may be of importance has hardly been touched upon. Recent papers present some evidence for chemical changes having occured in polyethylene insulation containing water trees (Rye *et al.*, 1975; Garton *et al.*, 1980; Morita *et al.*, 1976; Fournié *et al.*, 1978), but details of the processes taking place and of the changes that have occurred are as yet uncertain.

IV. EXPERIMENTAL ASPECTS

The meaningful measurement and evaluation of data of the dielectric strength of polymers is quite generally, even under the more easily controlled laboratory conditions, a formidable problem beset with numerous difficulties. Results will depend to a large extent upon experimental conditions and upon sample preparation and geometry. Details of the interpretation and treatment of the experimental data also differ from author to author. These data invariably show at times considerable scatter. This may in part be due to the limited measuring accuracy, inability to keep the sample preparation perfectly constant and extraneous, and unrecognized or uncontrollable influences. A scatter of data, however, still remains once these influencing factors have been reduced to a negligible or tolerable level. This is due to defects in the polymer that are impossible or unfeasible to remove or to the inherent indeterminacy of the physical processes involved as a consequence of fluctuations at the point of breakdown in the polymer structure at a microscopic level.

The short-term measurement of electrical breakdown strengths in the laboratory is related to the determination of physical parameters characterizing the polymer, or industrially to the determination in an overall manner of the initial quality of the insulation and the respective manufacturing process. In measurements of the long-term electrical strength additional problems will arise. In such tests the insulation is stressed to breakdown under conditions distinctly below intrinsic breakdown field strengths but nevertheless under enhanced conditions of voltage, frequency, or temperature as compared to service-level stressing. The data thus obtained over a period rarely exceeding 1–2 yr are then extrapolated to the expected lifetime of the insulation, generally on the order of 30–40 yr. The enhanced stressing conditions rarely resemble the situations incurred in service, and the extrapolation via largely empirically derived "life equations" over only limited time spans presupposes breakdown mechanisms unchanged in going from accelerated to service stress. The possible synergistic effects due to several stresses imposed simultaneously are mostly ignored.

In the present section mainly considerations relevant to the physical characterization of short-term breakdown are discussed. For the extensive problems and methods related to lifetime testing the reader is referred to the literature (Augood, 1978; Nelson, 1971, 1972; Dakin and Studniarz, 1978; Fischer *et al.,* 1975, Simoni and Pattini, 1975; Artbauer, 1970; and others).

A. Samples and Methods

Breakdown strengths in the sense of a physical property describing the electrical failure of a polymer should be independent of extraneous influences and shortcomings. In general the measurement of dc strengths is less critical than the determination of ac and dc strengths obtained with samples of small electrically stressed volume, namely, electrode separation, are often considered to represent or closely approach values characterizing the pure material. For such determinations thin-film samples, recessed specimens (Luy and Oswald, 1971), McKeown-type specimens (McKeown, 1965) and more recently so-called cylindrical specimens (Andress *et al.,* 1973) have been used. They are shown in Fig. 11. To ensure dc breakdown of the polymer at the point of highest field strength the condition $(\sigma F_B)_{pol} < (\sigma F_B)_{med}$ must be realized, where σ is the conductivity and F_B the breakdown field strength of the sample investigated and of the required embedding medium. In practice, however, external influences due to two weak spots, the polymer–embedding medium interface and the electrically weaker embedding medium, can rarely be completely eliminated. Thus Fig. 12 shows the cumulative failure probability P in Weibull coordinates as a function of the field strength F for thin-film, recessed, and cylindrical samples of low-density polyethylene. The polymer used, the electrode material and separation, and the embedding medium are identical. As the breakdown area becomes more perfectly embedded in the dielectric to be measured, the average breakdown field strength increases and the scatter of the data decreases. In the McKeown-type samples the weak spots are also relegated to regions of relatively low electrical field

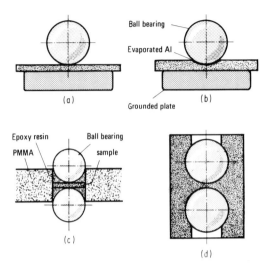

Fig. 11. Common laboratory sample geometries for the determination of the electrical strength of polymers. (a) Film sample, (b) recessed specimen, (c) McKeown sample, (d) cylindrical specimen.

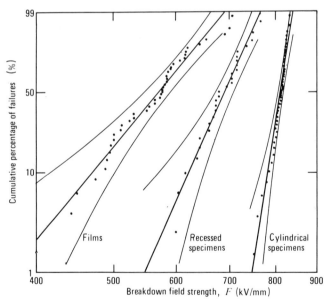

Fig. 12. Dc-failure field strengths of low-density polyethylene at 20°C using various sample geometries. [By permission of etz-a-Schriftleitung (1973).]

strength and, indeed, high values have been reported for the breakdown strengths of a number of polymers. It appears, however, that a casting resin–polymer–casting resin sandwich is being measured in this case, as one cannot prevent creep of the embedding resin into the interelectrode region (Lawson, 1966).

For the case of ac stressing, which is the more interesting from the point of view of industrial application, the requirements on the potential distribution within the sample to be tested and in the surrounding medium are more stringent and difficult to realize. The requirement $(\varepsilon_r F_B)_{pol} <$ $(\varepsilon_r F_B)_{med}$ would necessitate unrealistically high values of the relative permittivity ε_r of the medium for most polymers to be tested, so that the resistivity of the surrounding medium must be reduced or drastically thinned samples be employed in the case of the electrode arrangements of Figs. 11a and 11b. The disadvantage of the embedding medium diffusing into the polymer and affecting the breakdown strength remains. Calculation of the distribution of the electrostatic potential, which for identical or vanishingly small loss tangents of polymer and embedding medium will also represent the distribution of the instantaneous potential of an applied low-frequency ac field, shows that by a proper choice of cylinder dimensions and embedding medium external influences can be completely eliminated. This method of sample preparation will in general also ensure perfect adherence of polymer and electrodes, thus eliminating the undesirable effects of partial discharges. Furthermore, diffusion and creep path lengths for the embedding oil are long, so that the influences, at least in short-term testing, can be safely eliminated. Despite the relatively laborious method of sample preparation, the cylindrical specimen appears to be the sample best suited for studying material properties. It has been employed with success in the investigation of thermoplastic polymers, mainly polyethylene (Fischer and Nissen, 1976).

High electric field strength may be realized at low voltage levels via divergent field arrangements. Such arrangements as the point–plane and point–point, as well as various wedge geometries, are indeed widely used. These arrangements have provided valuable data for the understanding of treeing phenomena (Mason, 1955) and of the effect of additives on the voltage stabilization of polymers (Ashcraft et al., 1976). They should, however, be viewed as standard defects in the polymer, and with sufficient care in sample preparation they are considered of wide applicability for comparative studies on polymer behavior. The use of needle electrodes as a means of studying intrinsic breakdown (Patsch, 1975b) should be viewed with caution. In principle the initial appearance of partial discharges and treeing in rapid-rise experiments should correlate with the intrinsic breakdown strength of the material investigated, but limits on visual observation and partial discharge detection sensitivity allow only gross comparison. More serious difficulties are related to the lack of precise knowledge of electrode geometry, i.e., maximum field strength, to the possible existence of adhesion problems or gaps (Kosaki et al., 1977), and to the influence of time of voltage application until treeing onset (Shibuya et al., 1977). Other details of the experimental arrangement, particularly with a so-called remote ground electrode, also have pronounced effects (Löffelmacher et al., 1977).

B. Statistical Considerations

The electrical strength of an organic polymer cannot be described by a single value, even with due allowance for experimental scatter. It rather requires definition with a distribution of breakdown values. Frequently a Gaussian distribution will be employed. There is, however, no a priori reason why this should be so, and in fact physical and technical reasons may render another distribution function more suitable. Particularly in destructive tests the so-called Weibull distribution function (Weibull, 1949, 1951) has gained prominence. It has the advantage that it is limited to positive values of the random variable, hence the probability of damage to the unstressed specimen is zero, a physically reasonable requirement. Furthermore, the breakdown field strengths or times are frequently asymmetrically distributed, a fact that can be easily accommodated with the Weibull shape parameter. Last, changes in the sizes of the stressed specimens can be dealt with without a change in the form of the distribution function.

1. The Weibull Distribution Function

The probability P with which a random variable X assumes values up to but not exceeding the value x is a nondecreasing, monotone function of x:

$$F(x) = P(X \leq x). \tag{9}$$

$F(x)$ is thus defined and is referred to as the distribution function, cumulative probability density, or failure cumulative distribution function.

The Weibull distribution function is defined as

$$F(x) = \begin{cases} 1 - \exp\{-[(x - x_1)/(x_0 - x_1)]^b\} & \text{for} \quad x \geq x_1, \\ 0 & \text{for} \quad x < x_1, \end{cases} \tag{10}$$

with $x_0 > x_1 \geq 0$ and $b > 0$. The three parameters have the following meaning: x_1 is a lower limit of the random variable, be it field strength, time to breakdown, number of cycles, etc., below which no failures can occur; its value will depend mainly on physical conditions, and it may be used to take into account such factors as incubation periods and other prior history influencing the variable; x_0 is the characteristic value at which $F(x = x_0) = 1 - 1/e$, that is, 63.2% of the failures have occurred; b is the slope or shape parameter describing the distribution, i.e., scatter, of the individual data.

For measurements of electrical strength it is often advisable to put $x_1 = 0$, since for applied field strengths greater than zero there will be a finite probability of failure. Equation (10) then reduces to

$$F(x) = \begin{cases} 1 - \exp[-(x/x_0)^b] & \text{for} \quad x \geq 0, \\ 0 & \text{for} \quad x < 0, \end{cases} \tag{11}$$

which is the commonly employed form. Estimation of the Weibull parameters is generally done graphically on Weibull probability paper on which the distribution function appears as a straight line upon double logarithmic transformation. This is particularly straightforward in the case of $x_1 = 0$, since an independent estimate of either b or x_1 is not necessary. The parameters are obtained directly via least squares or maximum likelihood fitting. All Weibull plots in this article thus show $\ln[-\ln(1 - P_i)]$ as ordinate and $\ln x_i$ as abscissa. Here P_i is approximated by the median rank estimate $(i - 0.3)/(n + 0.4)$ in preference to the mean rank estimate $i/(n + 1)$, n being the total number of specimens in the investigated random sample (Kao, 1960).

2. Mixed Distributions

Normally the experimental data will deviate to some extent from the graphically or otherwise determined distribution function $F(x)$, and the question arises as to whether this deviation is statistically significant or merely random because of the limited number of specimens. This question may be answered via calculation of the respective confidence bands.

Investigations of the electrical strengths of a large number of randomly chosen polymer samples have shown that the simple Weibull distribution function is certainly appropriate in many cases, as shown representatively by Fig. 12. Description of these data by Eq. (11) is certainly justified within the two-sided 90% confidence bands as shown. For many tests such a description is, however, no longer satisfactory. Instead a description in terms of mixed distributions obtained by either additively or multiplicatively mixing simple Weibull distribution functions is possible (Fischer and Röhl, 1974). The relevant considerations that follow refer to destructive tests which can be described in terms of the two complementary events "specimen intact" $\equiv (\bar{A})$ and "specimen failed" $\equiv (A)$.

a. *Additively Mixed Distributions.* The failure event A is described by i mutually incompatible events A_i, that is,

$$A = A_1 + A_2 + \cdots . \tag{12}$$

With $P(A_i)$ the probability for the occurrence of event A_i with weighting factor c_i, one obtains as probability for the breakdown of any one arbitrary sample:

$$P(A) = \sum_i c_i P(A_i); \tag{13}$$

i.e., the distribution function is

$$F(x) = \sum_{i=1}^{m} c_i F_i(x) \quad \text{with} \quad \sum_{i=1}^{m} c_i = 1. \tag{14}$$

The additively mixed Weibull distribution function employing Eq. (11) becomes

$$F(x) = 1 - \sum_{i=1}^{m} c_i \exp[-(x/x_{0i})^{b_i}]. \tag{15}$$

Additively mixed distribution functions are expected when the distinguishable specimens that comprise a random sample may be classified with respect to differences in history, material, sample preparation, etc. The criterion is that an event A leading to breakdown in samples of component lot i cannot be characteristic of the breakdown behavior of component lot j. Consider as an example the experimental data shown in Fig. 13, representing the result of the measurement of the short-term electrical breakdown strength of a lot of 74 low-density polyethylene samples. Of these 33 had been exposed to an ac prestressing well below breakdown voltage for about 4000 h, the remaining 41 specimens, identical in all other aspects, had not been subjected to this "aging." It is obvious that the individual breakdown strengths cannot be meaningfully described within the 90% confidence band of a simple Weibull distribution function (dashed lines). In fact more than 30% of the data lie outside the chosen confidence band, implying that a uniform breakdown behavior does not prevail for the entire group of specimens.

The solid lines show the optimum fit obtained with an additively mixed distribution according to Eq. (15), with $i = 2$. Two-component distributions are indicated with the characteristic values F_{0i} and b_i shown in Table I in the proportions 45.4 and 54.6%. Anyone familiar with the differing electrical history of the samples will associate the lot having the lower electrical strength with the previously stressed specimens and the remainder with the unstressed samples. Now in the particular case discussed, not

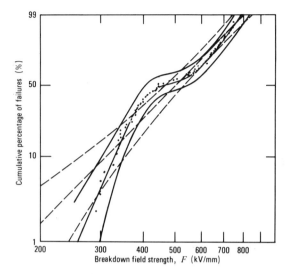

Fig. 13. Short-term electrical breakdown strength of a random sample of low-density polyethylene specimens showing an additively mixed distribution. [By permission of Springer-Verlag. From Fischer and Röhl (1974).]

only were the sizes of the component lots known, but the individual samples were marked to permit identification. As shown in Fig. 14, each component lot is perfectly described by a simple Weibull distribution function, and the agreement between the data so obtained, also given in Table I, and those obtained from an analysis according to Eq. (15) leaves little doubt as to the correctness of the interpretation and analysis.

b. Multiplicatively Mixed Distributions. In contrast to an additively mixed distribution, another type of distribution function, called a multiplicatively mixed one, is physically more reasonable when dealing with an ensemble of specimens considered identical within the accuracy afforded

TABLE I *Comparison of the Weibull Parameters Derived with an Additively Mixed Distribution with the Parameters of the Known Component Distributions*

i	Analysis with mixed distribution			Known component distributions		
	F_{0i} (kV/mm)	b_i	c_i (%)	F_0 (kV/mm)	b	Proportion (%)
1	383	9.9	45.4	388	9.0	44.6
2	647	7.4	54.6	643	6.2	55.4

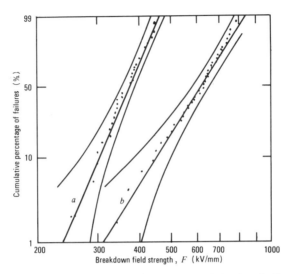

Fig. 14. Weibull plots of the breakdown field strengths of the electrically prestressed (a) and the virgin (b) component lots of Fig. 13. [By permission of Springer-Verlag. From Fischer and Röhl (1974).]

by the overall process of preparation, sample origin, and history. Each sample, however, has the possibility of failure because of different and competing reasons.

The complementary event $\overline{A} = 1 - A$ (i.e., no damage to the specimen) is described by the mutually independent events $\overline{A}_i$ of each individual sample:

$$\overline{A} = \overline{A}_1 \cdot \overline{A}_2 \cdot \overline{A}_3 \cdots . \tag{16}$$

With the transformation

$$1 - P(A) = \prod_i [1 - P(A_i)] \tag{17}$$

this leads to the multiplicatively mixed distribution

$$F(x) = 1 - \prod_{i=1}^{m} [1 - F_i(x)] = 1 - \exp\left[- \sum_{i=1}^{m} (x_0/x_{0i})\right]^{b_i}. \tag{18}$$

Thus Fig. 15 shows the short-term ac breakdown strengths of low-density polyethylene cylindrical samples of small electrically stressed volume. A very pronounced deviation of the data from the description in terms of a simple Weibull distribution (dashed lines) is apparent, with in fact less than 50% of the data lying within the two-sided 90% confidence interval.

With a multiplicatively mixed distribution function entailing two competing breakdown causes, a far superior fit is obtained (solid lines). The multiplicatively mixed distribution always shows a concave curvature toward the ordinate axis and, as such, is in general easily distinguishable from an additive one. It should be realized, just the same, that the characteristic deviation of the mixed distribution may lie outside the visible percentage failure "window" because of the limited number of samples in the ensemble being tested. Plotting of the data will not in all cases provide a reliable means for choosing among single, additively mixed, or multiplicatively mixed functions. It is in any case advisable to determine what technical or physical reasons may lead to breakdown and what type of distribution is hence to be expected.

In the case of Fig. 15 it was possible to assign the flat branch to defect-initiated breakdowns by a visual inspection of several hundred specimens opened after failure. The steep branch was assigned to "intrinsic" breakdown (Andress *et al.*, 1973). This type of analysis was indeed the basis for the understanding of a vast number of seemingly inconsistent breakdown measurements on polyolefins (Fischer and Nissen, 1976).

By varying the number of component distributions $F_i(x)$ and the respective parameters, and by appropriately choosing their combination (i.e., additive and/or multiplicative) any set of data may be described with perfect accuracy. The parameters and relative contributions determined

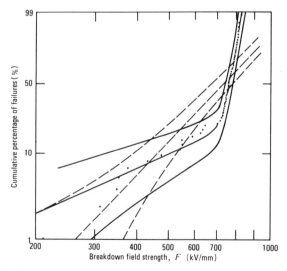

Fig. 15. Ac-breakdown strength of low-density polyethylene specimens showing a multiplicatively mixed distribution. [By permission of Springer-Verlag. From Fischer and Röhl (1974).]

in such a manner would, of course, be utterly meaningless. Because of the finite sample size random deviations from the true distribution function will occur. Therefore the simplest distribution function commensurate with the data and an appropriately chosen confidence interval should be selected. With this chosen statistical reliability more complex form of distribution function cannot be regarded as realistic unless additional criteria for its existence are available. For the examples discussed here, more than two component distributions ($i = 2$) are neither necessary nor meaningful.

ACKNOWLEDGMENTS

I should like to express my gratitude to Springer-Verlag, Berlin, and to etz-a-Schriftleitung, Offenbach, for permission to reproduce and adapt published material. Special thanks are due to W. Kalkner for kindly allowing me to use material from his doctoral thesis as shown in Figs. 3 and 4, and to R. M. Eichhorn for supplying Fig. 10b.

REFERENCES

Andress, B., Fischer, P., and Röhl, P. (1973). *Elektrotech. Z.* **A-94**, 553–556.

Arii, K., Kitanii, I., and Inuishi, Y. (1973). *Tech. Rep. Osaka Univ. Jpn.* **24**, 95–103.

Artbauer, J. (1965). *Kolloid Z. Z. Polym.* **202**, 15–25.

Artbauer, J. (1967). *J. Polymer Sci. Part C* **16**, 477–484.

Artbauer, J. (1970). *Elektrotech. Z.* **A91**, 326–331.

Artbauer, J., and Griac, J. (1965). *Proc. IEEE* **112**, 818.

Ashcraft, A. C. (1977). *World Electrotech. Congr., Moscow, USSR, June.*

Ashcraft, A. C., Eichhorn, R. M., and Shaw, R. G. (1976). *IEEE Symp. Electr. Insul., Montreal, June.*

Augood, D. R. (1978). *Conf. Record, IEEE Int. Symp. Electr. Insul.* pp. 17–21.

Bahder, G., Dakin, T. W., and Lawson, J. H. (1974a). CIGRE Paper 15-05.

Bahder, G., Katz, C., Lawson, J. H., and Vahlstrom, W. (1974b). *IEEE Trans. Power Appar. Syst.* **PAS-93**, 977–990.

Bartnikas, R. (1973). *IEEE Trans. Instrum. Meas.* **IM-22**, 403–407.

Beyer, M. (1978). *Elektrotech. Z.* **A99**, 96–99, 128–131.

Beyer, M., Duarte-Ramos, H., and Meier, N. (1972). *Elektrotech. Z.* **A93**, 475–477.

Billing, J. W., and Groves, D. J. (1974). *Proc. IEEE* **121**, 1451–1456.

Blok, J., and Le Grand, D. G. (1968). General Electric Co. Rep. 68-C-251, pp. 1–5.

Bradwell, A., Cooper, R., and Varlow, B. (1971). *Proc. IEEE* **115**, 247–254.

Chan, J. C., and Jaczek, S. M. (1978). *IEEE Trans. Electr. Insul.* **EI-13**, 194–197.

Cherney, E. A. (1973). Ontario Hydro Research Quarterly, 3rd Quarter, pp. 7–12.

Cooper, R. (1966). *Br. J. Appl. Phys.* **17**, 149–166.

Cooper, R., Varlow, B. R., and White, J. P. (1977). *J. Phys. D Appl. Phys.* **10**, 1521–1529.

Dakin, T. W., and Studniarz, S. A. (1978). *Conf. Record, IEEE Int. Symp. Electr. Insul.* pp. 216–221.

Davies, D. K. (1969). *J. Phys. D Appl. Phys.* **2**, 1533–1537.

Densley, R. J., and Salvage, B. (1971). *IEEE Trans. Electr. Insul.* **EI-6**, 54–62.

DiStefano, T. H., and Shatzkes, M. (1975). *J. Vac. Sci. Technol.* **12**, 37–46.

Dokopoulos, P., and Marx, E. (1967). *Elektrotech. Z.* **A-88**, 617–619.

Eichhorn, R. M. (1970). *Polym. Eng. Sci.* **10**, 32–37.
Eichhorn, R. M. (1974). *Ann. Rep. Conf. Electr. Insul. Dielectr. Phenomena, 1973* pp. 289–298.
Eichhorn, R. M. (1976). *IEEE Trans. Electr. Insul.* **EI-12**, 2–18.
Eichhorn, R. M. (1978). *IEEE Trans. Electr. Insul.* **EI-13**, 198–199.
Fava, R. A. (1965). *Proc. IEEE* **112**, 819–823.
Fedors, R. F. (1980). *Polymer* **21**, 863–865.
Fischer, P. (1974). *Elektrotech. Z.* **A-95**, 316–317.
Fischer, P., and Röhl, P. (1974). *Siemens Forsch. Entwicklungsber.* **3**, 125–129.
Fischer, P., Lukaschewitsch, A., Peschke, E., and Nissen, K. (1975). *Int. High Voltage Symp., Zürich* pp. 686–691.
Fischer, P., and Nissen, K. (1976). *IEEE Trans. Electr. Insul.* **EI-11**, 37–40.
Fischer, P. (1978). *J. Electrostat.* **4**, 149–173.
Fischer, P., and Nissen, K. (1978). *Elektrotech. Z.* **A-99**, 475–480.
Fischer, P., Nissen, K. W., and Röhl, P. (1979). *Annu. Rep. Conf. Electr. Insul. Dielectr. Phenomena, 1979* pp. 539–549.
Fischer, P., Nissen, K. W., and Röhl, P. (1981). *Siemens Forsch. Entwicklungsber.* **10**, 222–227.
Forlani, F., and Minnaja, N. (1969). *J. Vac. Sci. Technol.* **6**, 518–526.
Franke, E. A., Stauffer, J. R., and Czekaj, E. (1977). *IEEE Trans. Electr. Insul.* **EI-12**, 218–223.
Franz, W. (1953). *Ergeb. Exakten. Naturwiss.* **27**, 1–55.
Franz, W. (1956). *In* "Handbuch der Physik" (S. Flügge, ed.), Vol. 17, pp. 155–263. Springer-Verlag, Berlin and New York.
Fröhlich, H. (1937). *Proc. R. Soc. London Ser. A* **160**, 230–241.
Fröhlich, H. (1947). *Proc. R. Soc. London Ser. A* **188**, 521–532.
Fröhlich, H., and Paranjape, B. V. (1956). *Proc. Phys. Soc. B* **69**, 21–32.
Fournié, R., Perret, J., Recoupé, P., and Le Gall, Y. (1978). *IEEE Internat. Symp. Electr. Insul.* pp. 110–113.
Garton, A., Densley, R. J., and Bulinski, A. (1980). *IEEE Trans. Electr. Insul.* **EI-15**, 500–501.
Gemant, H., and Philipoff, W. (1932). *Z. Tech. Phys.* **13**, 425–430.
Hall, H. C., and Russek, R. M. (1954). *Proc. IEEE* **101**, 47–55.
Hauschildt, K. R., and Nissen, K. (1977). Unpublished data.
Henkel, H. J., Kalkner, W., and Müller, N. (1981). *Siemens Forsch. Entwicklungsber.* **10**, 205–213.
Heumann, H., Patsch, R., Saure, M., and Wagner, A. (1980). Cigré-Paper 15-06.
Hiley, J., Nicoll, G. R., Pearmain, A. J., and Salvage, B. (1973). *Annu. Rep. Conf. Electr. Insul. Dielec. Phenomena, 1972* pp. 116–124.
Hogg, W. K., and Walley, C. A. (1970). *Proc. IEEE* **117**, 261–268.
Ieda, M., and Nawata, M. (1973). *Annu. Rep. Conf. Electr. Insul. Dielectr. Phenomena, 1972* pp. 143–150. National Academy of Sciences, Washington, D.C.
Isshiki, S., Yamamoto, M., Chabata, S., Mizoguchi, Y., and Ono, M. (1974). *IEEE Trans. Power Appar. Syst.* **PAS-93**, 1419–1429.
Kalkner, W. (1974). Doctoral Thesis, Technical Univ. Berlin, Germany.
Kalkner, W., and Winkelnkemper, H. (1975). *Int. High Voltage Symp., Zürich* pp. 592–597.
Kao, J. H. K. (1960). *Proc. Nat. Symp. Reliabil. Qual. Contr., 6th, Cornell Univ., Ithaca New York* pp. 190–201.
Kärkkäinen, S. (1975). *Int. High Voltage Symp., Zürich* pp. 244–249.
Kind, D., and König, D. (1968). *IEEE Trans. Electr. Insul.* **EI-3**, 40–46.
Kitchen, D. W., and Pratt, O. S. (1962). *IEEE Trans. Power Appar. Syst.* **PAS-81**, 112–121.

Klein, N. (1969). *Adv. Electron. Electron Phys.* **26**, 309–424.

Klein, N. (1972). *Adv. Phys.* **21**, 605–645.

König, D. (1967). Doctoral Thesis, Technical Univ., Braunschweig, Germany.

Kosaki, M., Shimizu, N., and Horii, K. (1977). *IEEE Trans. Electr. Insul.* **EI-12**, 40–45.

Kreuger, F. H. (1964). "Discharge Detection in High Voltage Equipment." Heywood, London.

Kryszewski, M., and Szymanski, A. (1970). *J. Polym. Sci. Part D* **4**, 245–320.

Lawson, W. G. (1966). *Proc. IEEE* **113**, 197–202.

Löffelmacher, G. (1975). *Elektrotech. Z.* **A96**, 152–154.

Löffelmacher, G. (1976). Doctoral Thesis, Technical. Univ., Hannover, FRG.

Löffelmacher, G., Madry, P., and Ulrich, J. (1977). *In* "ETG Fachberichte," pp. 130–135. VDE-Verlag, Berlin.

Luy, H., and Oswals, F. (1971). *Elektrotech. Z.* **A-92**, 358–363.

Mason, J. H. (1953). *Proc. IEEE* **100**, 149–158.

Mason, J. H. (1955). *IEEE Monograph* **127M**, 254–263.

Mason, J. H. (1959). *Prog. Dielectr.* **1**, 1–58.

Mason, J. H. (1965). *Proc. IEEE* **112**, 1407–1423.

Matsuba, H., and Kawai, E. (1976). *IEEE Trans. Power Appar. Syst.* **PAS-95**, 660–668.

Matsubara, M., and Yamanouchi, S. (1974). *Annu. Rep. Conf. Electr. Insul. Dielectr. Phenomena* pp. 270–278.

McKeown, J. J. (1965). *Proc. IEEE* **112**, 824–828.

McMahon, E. J., and Perkins, J. R. (1964). *IEEE Trans. Power Appar. Syst.* **PAS-83**, 1253–1260.

McMahon, E. J., and Perkins, J. R. (1973). *Annu. Rep. Conf. Electr. Insul. Dielectr. Phenomena, 1972* pp. 133–136.

Minnema, L., Barneveld, H. A., and Rinkel, P. D. (1980). *IEEE Trans. Electr. Insul.* **EI-15**, 461–472.

Miyashita, T., and Inoue, T. (1967). *J. Inst. Electr. Eng. Tokyo* **48**, 161–168.

Mizukami, T., Kuma, S., and Soma, K. (1977). *Conf. Electr. Insul. Dielectr. Phenomena, Colonie, New York* Paper D-11.

Moran, T. F., and Hamill, W. H. (1963). *J. Chem. Phys.* **39**, 1413–1422.

Morita, M., Hanai, M., Shimanuki, H., and Aida, F., *Annu. Rep. Conf. Electr. Insul. Dielectr. Phenomena, 1976* pp. 335–343.

Morton, V. M., and Stannett, H. W. (1968). *Proc. IEEE* **115**, 1857.

Muccigrosso, J., and Phillips, P. J. (1978). *IEEE Trans. Electr. Insul.* **EI-13**, 172–178.

Nelson, W. B. (1971). *IEEE Trans. Electr. Insul.* **EI-6**, 165–181.

Nelson, W. B. (1972). *IEEE Trans. Electr. Insul.* **EI-7**, 36–55, 99–119.

Nissen, K. W., and Röhl, P. (1981). *Siemens Forsch. Entwicklungsber.* **10**, 215–221.

Nitta, Y. (1974). *IEEE Trans. Electr. Insul.* **EI-9**, 109–112.

Noto, F. (1980). *IEEE Trans. Electr. Insul.* **EI-15**, 251–258.

Oakes, W. G. (1948). *Proc. IEEE* **95**, 36–44.

Oakes, W. G. (1949). *Proc. IEEE* **96**, 37–43.

O'Dwyer, J. J. (1967). *J. Phys. Chem. Solids* **28**, 1137–1144.

O'Dwyer, J. J. (1973). "The Theory of Electrical Conduction and Breakdown in Solid Dielectrics." Oxford Univ. Press (Clarendon), London and New York.

Patsch, R. (1975a). *Conf. Electr. Insul. Dielectr. Phenomena, Gaithersburg, Maryland* Paper E-6.

Patsch, R. (1975b). *Int. High Voltage Symp., Zürich* pp. 603–607.

Patsch, R. (1981). *Colloids and Polymer Sci.* **259**, 885–993.

Pohl, H. A. (1958). *J. Appl. Phys.* **29**, 1182–1188.

Praehauser, Th. (1975). *Int. High Voltage Symp., Zürich* pp. 266–270.

Renfrew, A., and Morgan, P. (eds.) (1960). "Polythene." Wiley (Interscience), New York.

Rogers, E. C. (1958). *Proc. IEEE* **105**, 621–630.

Röhl, P., and Nissen, K. W. (1979). *Annu. Rep. Conf. Electr. Insul. Dielectr. Phenomena, 1979* pp. 520–528.

Rye, I., Brown, P., and Feles, W. T. (1975). *J. Phys. D: Appl. Phys.* **8**, 216–217.

Sabuni, H., and Nelson, J. K. (1976). *J. Mat. Sci.* **11**, 1574–1576.

Schon, K. (1977). *Elektrotech. Z.* **A-98**, 504–506.

Seitz, F. (1949). *Phys. Rev.* **76**, 1376–1393.

Shahin, M. M. (1967). *J. Chem. Phys.* **45**, 2600–2605.

Shibuya, Y., Zoledziowski, S., and Calderwood, J. H. (1977). *IEEE Trans. Power Appar. Syst.* **PAS-96**, 198–206.

Simoni, L., and Pattini, G. (1975). *IEEE Trans. Electr. Insul.* **EI-10**, 17–27.

Simril, V. L., and Hershberger, A. (1950). *Mod. Plast.* **27(10)**, 97–102.

Sletbak, J. (1977). *IEEE Trans., Power Appar. Syst.* **PAS-98**, 1358–1365.

Sletbak, J., and Botne, A. (1977). *IEEE Trans. Electr. Insul.* **EI-12**, 383–389.

Stark, K. H., and Garton, G. G. (1955). *Nature (London)* **176**, 1225–1226.

Stratton, R. (1961). *Prog. Dielectr.* **3**, 234–292.

Tabata, T., Fukuda, T., and Iwata, Z. (1972). *IEEE Trans. Power. Appar. Syst.* **PAS-91**, 1361–1370.

Tanaka, T., Fukuda, T., Suzuki, S., Nitta, Y., Goto, H., and Kubota, K. (1974). *IEEE Trans. Power Appar. Syst.* **PAS-93**, 693–702.

Tanaka, T., Fukuda, T., and Suzuki, S. (1976). *IEEE Trans. Power Appar. Syst.* **PAS-95**, 1892–1900.

Tsutsumi, Y., and Kako, Y. (1975). *Proc. IEEE* **122**, 223–224.

Vahlstrom, W. (1971). *IEEE Conf. Underground Distribut., Detroit, Michigan, September.*

Vahlstrom, W. (1973). *Annu. Rep. Conf. Elec. Insulat. Dielectr. Phenomena, 1972* pp. 255–265.

von Hippel, A. (1935). *Ergeb. Exakten. Naturwiss.* **14**, 79–129.

Wagner, H. (1975). *Annu. Rep. Conf. Electr. Insul. Dielectr. Phenomena, 1974* pp. 62–70.

Watson, D. B., Heyes, W., Kao, K. C., and Calderwood, J. H. (1965). *IEEE Trans. Electr. Insul.* **EI-1**, 30–37.

Weniger, M., and Kübler, B. (1976). *Elektrotech. Z.* **A-97**, 477–480.

Weibull, W. (1949). *Trans. R. Inst. Tech. Stockholm* **27**, 1–42.

Weibull, W. (1951). *J. Appl. Mech.* **18**, 293–297.

Whitehead, S. (1951). "Dielectric Breakdown of Solids." Oxford Univ. Press (Clarendon), London and New York.

Widman, W. (1960). *Elektrotech. Z.* **A-81**, 801–807.

Winkelnkemper, H., and Kalkner, W. (1974). *Elektrotech. Z.* **A-95**, 261–265.

Yasui, T., and Yamada, Y. (1967). *Sumitomo Electr. Tech. Rev.* **10**, 60–72.

Yoshimitsu, T., and Nakakata, T. (1978). *Conf. Record, IEEE Int. Symp. Electr. Insul.* pp. 116–121.

Yoshimura, N., Noto, F., and Kikuchi, K. (1977). *IEEE Trans. Electr. Insul.* **EI-12**, 411–416.

Zener, C. (1934). *Proc. R. Soc. London Ser. A* **145**, 523–529.

Index

369